Scientific Data Analysis

Scientific Data Analysis

Graham Currell

Formerly University of the West of England, Bristol

UNIVERSITY PRESS

OXFORD
UNIVERSITY PRESS

Great Clarendon Street, Oxford, OX2 6DP,
United Kingdom

Oxford University Press is a department of the University of Oxford.
It furthers the University's objective of excellence in research, scholarship,
and education by publishing worldwide. Oxford is a registered trade mark of
Oxford University Press in the UK and in certain other countries

© Graham Currell 2015

The moral rights of the author have been asserted

Impression: 1

Published in the United States of America by Oxford University Press
198 Madison Avenue, New York, NY 10016, United States of America

British Library Cataloguing in Publication Data

Data available

Library of Congress Control Number: 2014951237

ISBN 978-0-19-871254-1

Printed in Great Britain by Ashford Colour Press Ltd, Gosport, Hampshire

QR Code images are used throughout this book. QR Code is a registered trademark
of DENSO WAVE INCORPORATED. If your mobile device does not have a QR Code
reader try this website for advice www.mobile-barcodes.com/qr-code-software.

Excel is a registered trade mark of Microsoft Corporation. Microsoft product
screenshots reproduced with permission from Microsoft Corporation.

Portions of information contained in this publication/book are printed
with permission of Minitab Inc. All such material remains the exclusive
property and copyright of Minitab Inc. All rights reserved.

SPSS is a registered trade mark of IBM.

Links to third party websites are provided by Oxford in good faith and
for information only. Oxford disclaims any responsibility for the materials
contained in any third party website referenced in this work.

To my wife Jenny and son Felix for their continued support and encouragement.

About the author

Until his retirement in 2009, Graham Currell was a Principal Lecturer in physics at the University of the West of England, Bristol. During his early career, his particular interest was in the preparation of specialist training programmes to support staff in university science laboratories in Asia, the Middle East, Africa, and Central America but after 2000 he concentrated on the development of data analysis modules and self-study materials for science students, and from 2009 became a part-time research fellow, in which he further explored the development of online learning resources. *Scientific Data Analysis* builds on Graham's previous development of teaching materials for mathematics and statistics, including screen-capture videos in forensic, chemical, biological, and environmental science for the University of the West of England and the Royal Society of Chemistry. The approach reflects his extensive experience of providing tutorial and training support for students and staff carrying out research projects across both the physical and life sciences.

Welcome to *Scientific Data Analysis*

This book was written to satisfy the needs of three main target groups:

- Students following science degrees or masters courses who need support in understanding statistical data analysis, when encountered within taught courses and particularly when applied to their own developing experimental skills.
- University and college science staff wishing to reinforce or extend their own understanding of analytical techniques to help their supervision of student projects, particularly in topics on the border of their own specialisms.
- Teaching staff who are developing courses for science students and wish to structure the curriculum such that it prepares the students for handling the analysis of their own experimental project data.

Students typically learn the techniques of data analysis in two stages:

- Within taught modules during the first and/or second year of undergraduate studies.
- As part of a final year project, when faced with analysing their own experimental data.

This book is divided into two parts to reflect these two different approaches to learning. Part I, 'Understanding the statistics', develops the necessary statistical concepts with the *bottom-up* approach typically used in taught courses, and Part II, 'Analysing experimental data', starts from experimental data, and reviews, *top-down*, the possible techniques that could be used for their analysis.

The content and terminology used in Part I leads into the applications developed in Part II, and, in reverse, the techniques using Minitab and SPSS in Part II are supported through references to the basic statistical concepts in Part I.

For students using this book

If you, as a *first or second year* science student, are in the process of studying the general topic of data analysis, then Part I provides the core statistical concepts, which are developed in this book without complex mathematics but by using statistical models in Excel. The content reflects standard taught courses for science students, but also widens the range of techniques introduced in order to prepare you for the wide variety of different analytical problems encountered in final year projects.

If you, as a *final year* science student, are trying to find an analysis that is applicable to a set of your own experimental data, you should start in Part II by reading Chapter 5, from which you may move to one of Chapters 6 to 9, depending on the type of data involved. The 'Analytical options' feature in each section identifies possible analytical techniques that could be applied to the different types of data, and this approach helps you to select and use possible techniques relevant to your own particular data set. The content in Part II uses references to Part I to reinforce your understanding of the essential statistical concepts.

Video demonstrations

Short video clips are provided throughout the book to demonstrate the analyses using Minitab, SPSS and Excel. Together with the printed instructions, the videos will help you gain confidence in using the software and develop your experience for exploring menu options for other forms of analyses not directly covered in the book. The videos can be accessed directly by scanning the QR code images in the text with your smartphone. Alternatively you can view the videos from the links provided in the Online Resource Centre. Appendix I provides an index of the videos.

Case studies

There are a number of case studies throughout the book, most of which use related examples to develop a specific analytical theme. For example, it can be useful to see how the same or similar data can be analysed in different ways, and the case study links will take you to the relevant sections where the same case study continues. Each case study with more than one appearance starts with an overview which gives an outline of its theme and the locations in which each subsequent step can be found (not necessarily in the linear order of the book). Appendix II provides an index for the case studies.

Analytical software and data files

This book describes the use of **Microsoft Excel 2013**, **Minitab 17**, and **IBM SPSS** Statistics **22** for data analysis, giving the required keystrokes in the text and supported by videos accessible directly via the QR codes. The detailed steps described in Excel and SPSS are also the same as those used in the earlier versions of **Excel 2010** and **IBM SPSS** Statistics **20** respectively. There are some differences between Minitab versions 16 and 17, many of which make little difference to the keystrokes required, except for the use of regression and the general linear models. Minitab 17 was introduced during the final stages of preparation of the book and the keystrokes and videos were updated to reflect the new software, but the legacy keystrokes and videos for **Minitab 16** are still available in the Online Resource Centre. Some examples in the text were particularly relevant to Minitab 16 and these have been retained but clearly identified as such. Data files for the analyses in the book are also available for downloading from the Online Resource Centre.

Excel has widespread use for data handling in addition to its capabilities for statistical data analysis, and Minitab and SPSS have been chosen to demonstrate the 'next level' of analysis beyond Excel because of their easy-to-use, menu-driven operation. The approach developed using examples in Minitab and SPSS will support a greater understanding for solving problems if the student then moves on to using other packages, e.g. 'R', Prism.

Online Resource Centre

The Online Resource Centre can be found at http://www.oxfordtextbooks.co.uk/orc/currell/, and provides:

- Links to every video in the text, plus additional videos for Minitab 16.
- Download links for the modelling files developed in Excel, together with data files for Excel, Minitab, and SPSS.

In addition there is a 'new content 'section which will be updated regularly to accommodate any new content or videos developed.

For the lecturer: about this book

This book evolved from the experience of working with very many students to support them analysing their own data within final year projects across a range of bioscience, forensic, environmental, and chemistry degree courses. It was instructive to discover the mismatch between the standard learning outcomes of first and second year 'statistics' modules and the practical problems faced by the students when they first encounter their own real world data analysis.

The ability to identify and implement the correct analytical technique requires both an *understanding* of the statistics involved together with the *experience* of a range of possible techniques. Unfortunately, for most science students, there is no simple linear path to achieving this combination, and they only develop a confident understanding of the statistics *after* having the experience of using it themselves to analyse their own real, and often somewhat scrappy, data. This book reflects this dichotomy, in that Part I provides a *bottom-up* review of the underlying statistics relevant to data analysis and Part II allows the student to address analytical problems, *top-down*, starting from the need to analyse typical science project data. A key aim of the book is to prepare the student for using his/her first 'solo' research project as an effective learning experience in data analysis, bringing together understanding of the statistics with its practical applications.

Part I of the book, 'Understanding the statistics', is targeted at the 'taught course' or 'fundamental concepts' phase of learning. The first two chapters develop the basics of experimental uncertainty and statistics within a strong scientific context. They also introduce terminology (e.g. sums of squares, confidence intervals, ANOVA tables) that links directly into the analytical techniques developed in Chapters 3 and 4. The aim of these chapters is then to expose the reader to a *wider range* of possible analytical approaches than is provided by most books concentrating on the standard *t*-test, ANOVA, and chi-squared analyses.

As examples, the first application of the *t*-statistic is not to develop the traditional *t*-test for mean values but to demonstrate its wider use by testing for a difference in the slopes of bacterial growth, and the 'repeated measures' ANOVA is not introduced in a questionnaire but used in a forensic test to differentiate between black inks.

Where it is important to understand the underlying statistics of the techniques, these are developed with a modelling approach using Excel supported by videos, which not only gives a more visual clarity to the analysis but also exposes the reader to wider possibilities in using Excel. With its student-centred approach, this book is an effective text/video resource that provides both content and context for learning the fundamental concepts of data analysis.

Part II, 'Analysing experimental data', is intended to be used mainly in a 'top down' approach to analysing experimental data, starting with Chapter 5 which introduces a phase of reflection to avoid rushing for the first analysis that will produce a (possibly irrelevant) result. The subsequent chapters and sections are then defined by the structure of the particular data set, allowing the student to investigate a wider set of analytical techniques than might have been considered initially.

The book prints keystroke instructions for SPSS and Minitab, together with a discussion of the resultant output, both of which are supported with step-by-step videos. Using this approach, the book provides self-study support for the individual reader, either student or lecturer, and would also be useful within the library of a statistics support centre for science students.

Contents

Part I

Understanding the statistics

Part I approaches an understanding of analytical techniques in science from a *statistical* perspective. It develops an understanding of how key analytical techniques work and the scientific interpretation of their results. With this approach, it supports first and second year modules in statistical data analysis, but also acts as a reference resource for students subsequently meeting a technique for the first time. The implementation of many of the analytical techniques, developed in Part I, is then described in Part II from a *scientific* perspective using SPSS and Minitab.

Chapter 1. Statistical concepts provides the key topics and statistics that underpin the analytical techniques developed in subsequent chapters. The content reflects the standard elements of an introductory course in statistics, but the approach and terminology is designed to link into later applications.

Chapter 2. Regression analysis builds on the familiar 'best-fit straight line' analysis as an introduction to important analytical techniques. It provides an understanding of its practical implementation in experimental science using Excel, and develops the approach and terminology used in Minitab and SPSS, leading to the introduction of general linear models of analysis.

Chapter 3. Hypothesis testing provides an understanding of the process and significance of hypothesis testing in science, covering a wide range of underlying parametric and nonparametric techniques, from *t*-tests to Monte Carlo re-sampling. The specific concepts are developed, not through extensive statistical theory, but through the use of modelling in Excel, which provides a more relevant perspective for most science students, coupled with (possibly) new skills in Excel.

Chapter 4. Comparing data considers a range of analytical techniques that are often neglected in teaching but do address important questions in science. These relate to the strengths of agreement, association and interaction between the factors and variables in a scientific system.

Statistical concepts

Introduction

This chapter develops the underlying statistical concepts from the perspective of experimental data, emphasizing the link between experimental variability and the role of statistics in quantifying and managing this variability. The topics are introduced at a level suitable for the first year of an undergraduate science course, but developed with an approach which emphasizes the equations and terminologies that are used later in the book.

Section 1.1 introduces the value of visualizing experimental data through a variety of graphs, including the boxplot for raw data and the interval plot for calculated mean values.

Section 1.2 reviews the key terminology used to describe the factors and variables that influence the scientific system being analysed.

Section 1.3 uses the histogram to describe data variations, and introduces important standard distributions.

Section 1.4 discusses the uncertainty and error in measurement and develops the mathematics for combining experimental uncertainties.

Section 1.5 develops the statistics for analysing data samples and their application in the interpretation of experimental results.

Section 1.6 outlines the generic issues associated with hypothesis testing, leading to the relevance of p-values and Types I and II errors.

The following case studies develop the core statistics in this chapter:

Case study: Blood alcohol / 1. Overview

This case study is based around the measurement of alcohol (in mg) per 100 ml of blood. Some examples assume that the standard deviation experimental uncertainty is given by $\sigma = 2.0$ mg/100 ml.

1.1.2 / 2. Simple boxplot: Describes the *ranking* of data and the meaning of the simple box and whisker plot.

1.1.3 / 3. Boxplots and interval plots: Compares the presentation of *raw data values* using a boxplot and then *calculates best estimates* for the mean using interval plots.

1.3.1 / 4. Data distribution: Demonstrates the use of a *column* (or bar) graph to record the frequency of recorded values.

1.5.1 / 5. Sample statistics: A sample of five values is used to develop the basic statistics of measurement, including the *standard error* and *confidence interval*.

1.5.3 / 6. Samples and populations: Excel is used to randomly generate samples of five values to demonstrate the relevance of *sample* and *population* measurements.

1.6.2 / 7. Hypothesis tests: Illustrates how the *p-value* is calculated from the tail of the frequency distribution.

3.1.2 / 8. One sample *t*-test: Develops the calculations involved in testing whether a sample mean is *significantly* greater than a target value.

8.1.1 / 9. One sample analysis: Example data for the analysis of a single sample of univariate interval data.

Case study: **Experimental uncertainties / 1. Overview**

This case study links together related issues on managing the errors and uncertainties in experimental data.

1.4.4 / 2. Combining uncertainties: Identifies the basic rules for combining uncertainties.

1.4.4 / 3. Propagation of errors: Uses Excel to lay out a complex calculation.

1.4.6 / 4. DIY dice: Considers combining different types of experimental uncertainty.

5.3.4 / 5. Weighting: Demonstrates the use of 'weighting' to combine values with different uncertainties.

1.1 Data visualization

Data visualization is important at all stages of an experimental investigation. The term *data* is a general term for the recorded values that describe the system that we are investigating, and the term *variable* describes any measured quantity that changes within our experimental system.

After taking a set of readings it is often useful to plot the variable values graphically to get both a mental, and a physical, 'picture' of the raw data. This is also valuable in identifying any gross experimental errors and for picking out data regions that would benefit from more experimental study. When finally reporting the research, it is important to remember that the reader will also want to visualize the raw data to help understand the subsequent analysis.

1.1.1 **Graphical information**

In this book we meet a wide range of different graphs produced in Excel, SPSS, and Minitab, where the visualization of the data is a valuable aid in understanding and interpreting the analysis. Some examples are given in Fig 1.1, reproduced from figures later in the book.

Fig 1.1(a): The most familiar graph is the basic *x–y* scatterplot in Excel for showing the interrelations between variables. This graph enables us to decide on possible scientific characteristics for analysis by identifying different *slopes* in growth and *differences* between maximum and minimum values. Do not confuse the *x–y* scatterplot with the 'line' graph in Excel which should only be used if the *x*-axis has *categorical* data.

Fig 1.1(b): The use of the trendline in Excel, together with calculated uncertainty limits, is used in this example as a visual demonstration of the *confidence interval* involved in extrapolating a best-fit line to intercept the *x*-axis within a standard additions calculation.

Fig 1.1(c): The Q–Q normality plot of residuals in SPSS highlights deviations from the diagonal line of *normality* showing the skewness and kurtosis in experimental data.

Fig 1.1(d): A frequency column (bar) graph in SPSS records the numbers of data values that fall into specific categories, showing the relative weighting between the categories.

Fig 1.1(e): The Minitab plot of delta deviance against probability shows graphically the reliability with which measured variables can *predict* (diagnose) the binary state of subjects, with points to the bottom left and right indicating accurate predictions into the two possible states.

Fig 1.1(f): The dendrogram in Minitab gives direct visualization of the *similarity* between multiple input variables in anticipation of reducing their number through principal component analysis.

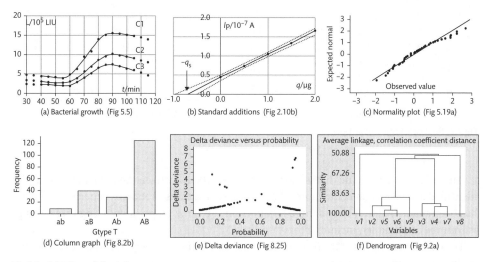

Fig 1.1 Selection of data plots

1.1.2 **Boxplots**

When dealing with repeated measurements, the boxplot shows the *range* of values measured, together with a general indication of their *distribution*. It is a valuable tool, not only in understanding data, but in communicating this understanding in a final report.

Case study: **Blood alcohol / 2. Simple boxplot**

—continued from 1.Introduction, leading to 1.1.3

Column A in Fig 1.2(a) gives a set of nine replicate random measurements of alcohol in blood, *BAlc*, in units of mg of alcohol per 100 ml of blood. The data is *ordered* in value in column B and then *ranked* in Column C.

Fig 1.2(b) gives the boxplot for the data set:

- The *centre line* in the box shows the middle value in the data, called the *median*.
- The *ends of the box* show one quarter and three quarters of the way through the data set, and are called the lower and upper *quartiles*. The distance between these is called the *interquartile range*, IQR.
- The *ends of the whiskers* show the maximum and minimum values of the data, except when the data has outliers.

	A	B	C
1	*BAlc*	Ordered	Ranked
2	83.5	83.5	9
3	80.6	82.4	8
4	79.9	80.6	7
5	78.5	80.2	6
6	78.4	79.9	5
7	80.2	78.7	4
8	78.7	78.5	3
9	82.4	78.4	2
10	77.9	77.9	1

(a) Data values

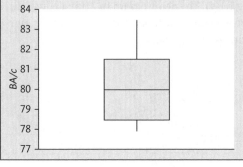

(b) Boxplot using Minitab

Fig 1.2 Replicate data and its boxplot

An *outlier* is a data value that is more than $1.5 \times IQR$ away from the median value, and is identified as a value that could be checked in case a data transcription or experimental error has been made. Fig 6.24 shows three boxplots on the same graph, identifying three groups within the one data set, including three outliers.

The descriptive values of the boxplot are calculated by first *sorting* the raw data (in column A) into ascending (or descending) order (in column B) and then assigning a *rank position* (in column C).

The *location* of a set of, n, data values is then described by:

Median is the *middle* value in a set of ranked data values and gives the *location* of the data. The Median is the value with the rank $0.50 \times (n+1)$.
In the example data, the median has the rank $= 0.5 \times (9+1) = 5$.
Median value with rank '5' $= 79.9$.

Lower quartile, Q_1, is the value *one* quarter of the way from the lowest to the highest value. The lower quartile is the value with the rank $0.25 \times (n+1)$.
In the example data, the lower quartile value has the rank $= 0.25 \times (9+1) = 2.5$.
Lower quartile value with rank '2.5' is halfway between 78.4 and 78.5 $= 78.45$.

Upper quartile, Q_3, is the value *three* quarters of the way from the lowest to the highest value. The upper quartile is the value with the rank $0.75 \times (n+1)$.

In the example data, the upper quartile value has the rank $= 0.75 \times (9 + 1) = 7.5$.
Upper quartile value with a rank '7.5' is halfway between 80.6 and $82.4 = 81.5$.

The *spread* of nonparametric data is described by:

Interquartile range, *IQR*, is the difference in value between the upper quartile and lower quartile.

$$IQR = Q_3 - Q_1 = 3.05$$

Total range is the difference in value between the *lowest value* and the *highest value* = 5.6.

The boxplot in Fig 1.2 shows data that is *symmetrical* around the median value. For skewed data, the median line will appear off-centre in the box.

1.1.3 Raw data and calculated values

It is useful to differentiate between describing raw data values and presenting calculated results.

Case study: Blood alcohol / 3. Boxplots and interval plots

—continued from 1.1.2, leading to 1.3.1

The boxplots in Fig 1.3(a) and interval plots in Fig 1.3(b) describe samples of 10, 40, and 160 values selected randomly from a population of possible measurements, *BAlc*, with a mean value of 80 mg/100 ml and a standard deviation of 2.0 mg/100 ml.

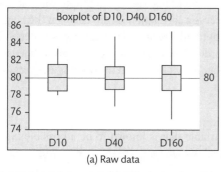
(a) Raw data

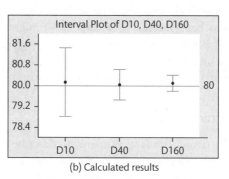
(b) Calculated results

Fig 1.3 Raw data boxplots (a) and calculated confidence intervals (b) (Minitab)

The boxplots in Fig 1.3(a) describe the *raw data* values, and show similar mean and quartile values defined by their common source population. However, as the sample size increases, the probabilities of observing extreme (maximum and minimum) data values increases, giving longer 'whiskers' for the larger samples.

The interval plots in Fig 1.3(b) show the confidence intervals (1.5.2), which are the *calculated* range for each sample within which you could be 95% confident of finding the true value mean of the source population. The shrinking confidence intervals show that the *precision* (1.4.3) of the measurement increases with the square root of the number of measurements according to Eqn 1.21.

1.2 Scientific data

Collection of data in science has developed over many years in many different disciplines, and has resulted in a great variety of styles in which it is classified and described. This section aims to establish some of the categorization and terminology used in this book, and relate these to established descriptors that you will meet elsewhere.

1.2.1 Experimental data

Input and output data

In many of our investigations we are trying to understand the *mechanisms* that might occur within the scientific system. We do this by observing what the system does when it is changed in some way, and consequently the data in a set of experimental results can usually be divided into

- output data that records the response of a system (response or dependent variable) and
- input data that defines specific experimental conditions (independent or factor variable).

For example, in a linear regression calculation (2.1.1) the input (independent) variable is placed on the x-axis and the output (dependent) variable on the y-axis.

However, we need to be careful when categorizing data in this way because we may be measuring two variables that are *both* output data, responding to a separate input variable that is not being recorded. For example, we might observe a correlation (2.1.3) between the height and weight of children plotted on an x–y graph, but both are *output* variables jointly dependent on the age of the child, which is an unrecorded *input* variable. A correlation between observed variables does not necessarily mean that one is a *cause* and the other an *effect*.

Factors and variables

The term *factor* is used mainly to describe a variable that is known to be a possible *input* to the system. A multifactorial analysis is an analysis where more than one *input* variable contributes to the calculation.

We will also meet references to univariate, bivariate, and multivariate types of data, which can broadly be described as follows.

- Univariate data has one *response* variable, with one or more input variables, e.g. the yield of a chemical reaction related to the time and temperature of the reaction.
- Bivariate data has two variables, often without recording any particular input variable, e.g. observing a relationship between the height and weight of children without recording their age.

- Multivariate data occurs in a variety of situations, either as multiple input variables predicting a single output variable, or possibly several variables jointly describing the output state of a system.

1.2.2 **Data types**

Data can be grouped into five main category types:

Interval (or scale) data are values recorded along a numeric scale, e.g. reading the length along a ruler, or noting an output reading on a *pH* meter. The differences (intervals) on the scale represent *true* measures, e.g. the difference between 40°C and 80°C is twice the temperature difference between 0°C and 20°C. Interval data can be **discrete**, where values are recorded to a limited precision (e.g. rounded to two decimal places) or **continuous**, in which values can be recorded to any possible precision (any number of decimal places).

Nominal data records results that fall *directly* into specific categories, e.g. recording the type of fibre (natural / man-made), the fibre shape (round, cylindrical, bilobal, etc.), or the assessment of image quality (poor / satisfactory / good / excellent).

Binary or digital data is categorical data with only two possible values, e.g. presence or otherwise of delustrants in fibres. Digital data could be described by Yes/No, 0/1, True/False, etc.

Ordinal data is nominal data where the different categories show a *sense of progression* that can be represented by numerical values. For example, the assessment of quality defined by poor, satisfactory, good, excellent, could also be defined by 1, 2, 3, 4. However, the values are not interval data because the difference between excellent (4) and satisfactory (2) *cannot* be assumed to be twice the difference between good (3) and satisfactory (2).

Frequency data is simply the result of counting the number of occurrences of some event, e.g. sightings of a particular animal, or numbers of bacteria on a plate. In some cases, larger frequency values can be treated as interval data, e.g. a density of 3,000 cfu (colony-forming units) of bacteria per millilitre of solution.

We also meet several other data descriptors:

Ratio data is *interval* data where the '0' value equates to a *true zero* in a scientific context. For example, the Kelvin and Celsius temperature scales both have the same *interval* of one degree, but the Kelvin scale is a ratio scale because it has 0°K at the absolute thermodynamic zero, whereas the Celsius scale is not ratio data because it only defines 0°C as the triple point of water. 100°K is twice the thermodynamic temperature of 50°K, but 100°C (373°K) is *not* twice the thermodynamic temperature of 50°C (323°K).

Ranking data are *ordinal* data that describe the *order* in which data can be placed, e.g. the ordinal data values of poor, good, very good, excellent can be described by ranking values, 1, 2, 3, 4. We see in 4.1.2 that ordinal data can be analysed by correlation using ranking values.

Categorical data is data that has been put into specific categories. Nominal data is a natural example of categorical data, but ordinal data with a limited number of values can also be treated as categorical data. Interval data can be grouped into categories by *binning* (8.1.8), and the number (frequencies) of values in each category can be 'counted' using *tabulation* (8.1.7). The analysis of categorical data is developed with chi-squared analysis in Section 3.7.

Normal data is scale data whose randomly selected values occur with characteristic bell-shaped probability distributions (1.3.3).

Nonparametric and parametric tests. A nonparametric test uses only ranked data (ordering of values) and can be used for ordinal or scale data, but a parametric test requires interval data because the differences (intervals) between values are used in the calculations.

The **Likert** scale (after Rensis Likert) is a specific example of ordinal data that is frequently used in questionnaires. Responses are given on a balanced scale of options such as:
Disagree strongly, Disagree, Neutral, Agree, Agree strongly,
which can be scored as –2, –1, 0, +1, +2.

1.2.3 Type and value of data

Finally, in this section, it is important to distinguish between the variable *type* and the *values* that can be used to describe it. The main data types–*interval, ordinal, binary, nominal,* and *frequency*–are clearly defined (1.2.2), but there is some flexibility in how they are described and used, which sometimes leads to confusion and problems with some software analyses.

Table 1.1 Variable types and values

Data type:	Value descriptions:
Interval	Described using *scale* values, e.g. 2.56, 0.037.
Nominal	Described using *text (string)* values, e.g. *blue, X*, 2 (using a numeric character as text).
	Can be *coded* using *scale integers* for inclusion in a mathematical model (6.4.1).
Ordinal	Described using *text* and/or *scale integer* values, e.g. *agree*, 2, T3.
	When using text, some analyses may require confirmation of the *order* of values, e.g. agree strongly, agree, neutral, disagree, disagree strongly.
	For some analyses it is necessary to *code* the values using scale integers, e.g. –2, –1, 0, +1, +2 for the Likert scale.
	Ordinal data can be expressed as *ranked* values: 1, 2, 3, 4, etc.
Binary	Described using *text, scale integer*, or *logical* values, e.g. Y/N, 0/1, *True/False*.
Frequency	Described using *scale integer* values, e.g. 3, 56.
	Larger frequencies may be considered as *continuous* data.

1.3 Data distributions

We saw in 1.1.2 that the location and general spread of a distribution can be described by the median plus interquartile range. We now identify parametric descriptors that allow us to describe the *shape* of a distribution in more detail.

1.3.1 **Histogram**

A frequency bar (or column) graph (e.g. Fig 1.1(d)) records the numbers of events that fall into specific *categories*. We now develop this into a **histogram** which identifies the categories as ranges of values along a continuous scale axis.

> ## Case study: **Blood alcohol / 4. Data distribution**
>
> —continued from 1.1.3, leading to 1.5.1
>
> We consider a data set with $N = 500$ replicate blood alcohol, *BAlc*, measurements (in units of mg of alcohol per 100 ml of blood).

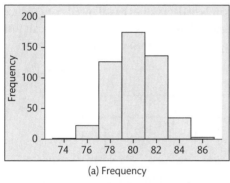
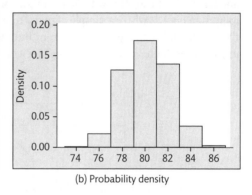

(a) Frequency   (b) Probability density

Fig 1.4 Histograms (Minitab)

Fig 1.4(a) is a **frequency histogram**, where the vertical *height* of the ith column gives the number, or *frequency*, f_i, of data values within the horizontal range given by the width, Δx_i, of that column, e.g. there are 177 values between 79 and 81.

If a specific value, i, is found to occur with a frequency, f_i, out of a total of N values, then the probability, p_i, of this value occurring could be expressed as

$$p_i = \frac{f_i}{N} \tag{1.1}$$

where N is the sum of all frequency values, given by, $N = \Sigma f_i$.

Fig 1.4(b) shows a **probability density histogram**. The *height* of the ith column is called the *probability density*, pd_i, such that the *area* of each column of the histogram gives the probability, p_i, that a randomly selected data value will occur within the width, Δx_i, of that column.

The *area* of each column equals its height times its width, giving the probability of a value occurring in this range:

$$p_i = pd_i \times \Delta x_i \tag{1.2}$$

The *total* area of the histogram is the probability that *any* value will occur, and is therefore the sum of all individual probabilities, giving a total of 1.0, $\Sigma_i p_i = 1.0$.

Taking the central column in Fig 1.4(a), we see that $f_i = 177$, which, using Eqn 1.1, gives a *probability*,

$$p_i = \frac{177}{500} = 0.354$$

Taking the same column in Fig 1.4(b) and using Eqn 1.2, the probability density, $pd_i = 0.177$ and the column width, $\Delta x_i = 2.0$ gives the same probability for category i:

$$p_i = 0.177 \times 2.0 = 0.354$$

1.3.2 Distribution parameters

Starting with a histogram with *broad* columns, as in Fig 1.4(b), we can now develop the histogram in Fig 1.5, where a sample of many thousands of data values has allowed us to reduce the width, and increase the number, of individual columns to the limit when we can just draw a smooth line joining the tops of very many *extremely narrow* columns.

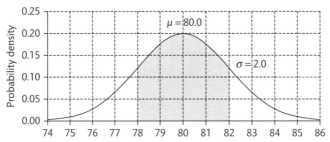

Fig 1.5 Probability density curve

The distribution in Fig 1.5 is an example of a normal distribution (1.3.3), which has an average, or *mean*, value, $\mu = 80$. The spread of a distribution is defined by the *standard deviation* (1.3.2) which, for the distribution in Fig 1.5, has the value, $\sigma = 2.0$.

The total area under the curve equals a total probability of 1.00, and the area of the shaded portion within one standard deviation of the mean (80.0 ± 2.0) is equal to 0.683. From this we can say that the probability that a randomly selected value will fall within one standard deviation of the mean is 0.683 or 68.3%.

In addition to the **median** and **interquartile range** (1.1.2), we can describe a distribution using:

Mean. The mean values, μ for a population and $\bar{x}$ for a sample (1.5.3), are the simple averages of all data values in the distribution.

Mode. The mode is the value which has the greatest frequency of data values, i.e. the position of the *peak* within the distribution. In some cases the distribution may show two or more peaks, in which case it would be referred to as bi- or multi-modal. For example, bimodal distributions can occur in the examination results of a cohort of students which includes two sub-groups with very different abilities.

Standard deviation, σ for a population and s for a sample (1.5.3), measures the spread of the data around the mean value. The calculation of standard deviation is developed in 1.5.1. Skewness is a measure of the extent to which a given distribution has an *unsymmetrical* shape. Skewness describes whether the given distribution has a shape that, compared to the symmetrical bell-shaped normal distribution, is either

- extended to the **left** (negative skewness), Fig 1.6(a), or
- extended to the **right** (positive skewness), Fig 1.6(b).

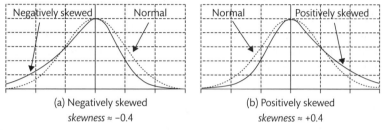

(a) Negatively skewed
skewness ≈ −0.4

(b) Positively skewed
skewness ≈ +0.4

Fig 1.6 Skewness

Kurtosis is a measure of the extent to which the distribution has a central peak that is either

- flatter (platykurtic), Fig 1.7(a), or
- more pointed (leptokurtic), Fig 1.7(b)

than the standard bell-shaped normal distribution (mesokurtic).

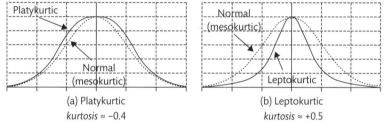

(a) Platykurtic
kurtosis ≈ −0.4

(b) Leptokurtic
kurtosis ≈ +0.5

Fig 1.7 Kurtosis

It is important to note that the *uncertainties* in estimating values of skewness and kurtosis are usually large unless the data sample itself is very large.

It is interesting to note that, from a mathematical perspective:

Mean value is calculated by taking the average of the *data values* themselves.

Standard deviation is calculated by taking an 'average' of the *squares* (power of two) of the *deviations* (1.5.1) from the mean value.

Skewness is calculated by taking an 'average' of the *cubes* (power of three) of the *deviations* from the mean value.

Kurtosis is calculated by taking an 'average' of the *fourth power* (power of four) of the *deviations* from the mean value.

1.3.3 Standard distributions

Normal distribution

This is the classic symmetrical bell-shaped curve, with the probability density described by the equation:

$$pd(x) = \frac{1}{\sigma\sqrt{2\pi}} \exp\left\{-\frac{(x-\mu)^2}{2\sigma^2}\right\} \tag{1.3}$$

Fig 1.5 gives an example of the probability density for a normal distribution with mean, $\mu = 80$, and standard deviation, $\sigma = 2.0$.

The *standard normal* distribution, shown in Fig 1.8, is the specific distribution with mean, $\mu = 0.0$, and standard deviation, $\sigma = 1.0$. The value on the 'x-axis' for this specific distribution is usually referred to as the z-value.

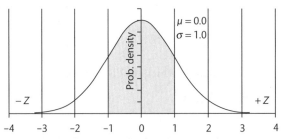

Fig 1.8 Standard normal distribution, with shaded area = 68.3%

The normal distribution is an important element in the analysis of experimental data. The total probability area under the curve = 1.00, but it is also useful to note other specific areas:

Table 1.2 Probability areas under a normal distribution

	Area =	
Within **one** standard deviation, σ, of the mean, from $z = -1.0$ to $z = +1.0$	0.683	68.3%
Within **two** standard deviations, σ, of the mean, from $z = -2.0$ to $z = +2.0$	0.954	95.4%
Within **three** standard deviations, σ, of the mean, from $z = -3.0$ to $z = +3.0$	0.997	99.7%
More than 1.96 standard deviations, σ, from the mean (both sides)	0.05	5.0%
To the right of $z = 1.96 \times \sigma$ (one side)	0.025	2.5%
More than 1.64 standard deviations, σ, from the mean (both sides)	0.10	10.0%
To the right of $z = 1.64 \times \sigma$ (one side)	0.05	5.0%

The mathematical complexity of Eqn 1.3 can be avoided by using functions in Excel for a normal distribution with mean, μ, and standard deviation, σ:

NORM.DIST(x, μ, σ, false) gives the probability density at the point x.

NORM.DIST(x, μ, σ, true) gives the *cumulative* probability from $-\infty$ up to the point x.

NORM.INV(α, μ, σ) gives the value, x, in the distribution such that the cumulative probability up to the value of x is equal to the probability, α.

In many Excel examples developed in this book, we use the function NORM.INV(RAND(), μ, σ) to *select a random value* from a normal distribution with mean, μ, and standard deviation, σ. The RAND() function provides a random value between 0.0 and 1.0, which is used here to provide a random probability for the cumulative probability, α, when selecting the value from the distribution.

Binomial distribution

The binomial distribution gives the probability $p(r)$ of observing r specific outcomes of n trials, where the probability of each r outcome is equal to p.

$$p(r) = {}_nC_r \times p^r \times (1-p)^{(n-r)} \tag{1.4}$$

Fig 1.9 shows three examples, each of which has $n = 20$ trials, but with different individual probabilities, p.

The graph with $p = 0.5$ represents the probability of observing r 'heads' when tossing a balanced coin $n = 20$ times.

The graph with $p = 1/6$ represents the probability of observing r 'sixes' when rolling a six-sided dice.

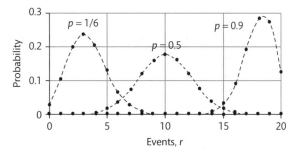

Fig 1.9 Binomial distributions

The distribution of possible values for r will have

Mean value: $$\mu = p \times n \tag{1.5}$$

Standard deviation: $$\sigma = \sqrt{n \times p \times (1-p)} \tag{1.6}$$

The important characteristics of the binomial distribution are that:

- it is defined by the two parameters: n and p.
- the range of values of r is limited between 0 and n.
- it is often skewed, except when $p = 0.5$. When $p < 0.5$ it is positively skewed and negatively skewed when $p > 0.5$.

If $np(1 - p) \geq 5$, the distribution can be represented approximately by a normal distribution.

For a binomial distribution with individual probability, p, and total trials, n, the Excel function:

BINOM.DIST(r,n,p, false) gives the probability value for observing *exactly* r outcomes.

BINOM.DIST(r,n,p, true) gives the cumulative probability *up to* r outcomes.

When considering the data as a *proportion*, $P = r / n$:

Mean value of *proportion*: $$\Pi = \frac{\mu}{n} = \frac{p \times n}{n} = p \tag{1.7}$$

Standard deviation of *proportion*: $$\sigma(P) = \frac{\sqrt{n \times p \times (1-p)}}{n} = \sqrt{\frac{p \times (1-p)}{n}} \tag{1.8}$$

Poisson distribution

The Poisson distribution is a special case of the binomial distribution when the individual probability, p, is very small, $p \ll 1.0$. The probability $p(r)$ of observing r specific outcomes, for a distribution with a mean number of outcomes, μ, is given by:

$$p(r) = \frac{e^{-\mu} \times \mu^r}{r!}. \tag{1.9}$$

For example, if the mean number of occurrences of a rare medical condition in a given population is $\mu = 4$, then the probabilities of observing r occurrences is given by the values in Fig 1.10.

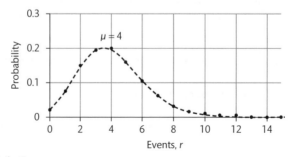

Fig 1.10 Poisson distribution

The important characteristics of the Poisson distribution are as follows.

- There is no theoretical upper limit to r, but the probability for high values becomes vanishingly small.
- The shape of the distribution is defined by the use of a *single* parameter, μ, which is the value of *both* the mean and variance:
  - Mean value $= \mu$.
  - The variance is equal to the mean value. Variance, $\sigma^2 = \mu$.
  - Standard deviation, $\sigma = \sqrt{\mu}$.

For a Poisson distribution with mean, μ, the Excel function:

POISSON.DIST(r, μ, false) gives the probability value for observing *exactly* r outcomes.

POISSON.DIST(r, μ, true) gives the cumulative probability *up to* r outcomes.

It is useful to note that if the mean number of occurrences is N, then the best estimate of the standard deviation uncertainty in that value is $\sqrt{N}$ (1.4.5).

Weibull distribution

The Weibull distribution, given by Eqn 1.10, has a shape that can be adjusted to model different system behaviours for values of $x \geq 0$ (see Fig 1.11). It is often used for time-dependent variations.

$$pd(x) = \frac{k}{\lambda}\left(\frac{x}{\lambda}\right)^{k-1} \times \exp\left\{-\left(\frac{x}{\lambda}\right)^{k}\right\}$$ (1.10)

where

k is the *shape parameter*, and
λ is the *scale parameter* along the x-axis.

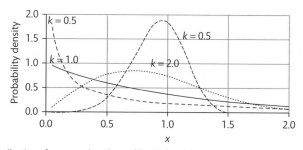

Fig 1.11 Weibull distributions for example values of k with $\lambda = 1.0$

We can interpret the graphs in Fig 1.11 by considering that x represents the time to an event that occurs with a probability, p:

If $k = 1.0$, then the probability, p, is constant in time (e.g. with radioactivity) then the curve is a simple exponential decay which is shown as the solid line.

The curve with $k < 1.0$ represents a situation where the probability, p, decreases with time.

The curves with $k > 1.0$ represent situations where the probability, p, increases with time.

The distribution with $k = 2.0$ produces the Rayleigh distribution which is used to describe the distribution of a value which is the vector combination of two independent normal distributions, e.g. the resultant wind velocity due to two components at right angles to each other.

1.4 Uncertainty and error

When reporting a measured result in science, it is rarely sufficient just to give a value without any indication of the possible uncertainty in that value. Only by combining the best estimates of both the *actual value* and its *uncertainty* can we convey complete information.

We see in 1.5.2 that an effective way of reporting an experimental measurement is given by the *confidence interval*, combining both the best estimate of the unknown value and a calculated value of the possible uncertainty.

1.4.1 Error or uncertainty

In a typical experiment, we aim to arrive at a best estimate of an unknown 'true' value. If we could increase the accuracy of our measurement we would get closer to the unknown 'true' value, but we must accept that there will always be a possible *error* in our result:

Error is the difference between the true value and the best-estimate value.

A key problem is that we never actually know the true value and hence we never truly know the actual magnitude of the error. The best that we can do is to calculate a value for the possible *uncertainty* in our results as the *best-estimate* for that error.

Uncertainty is the best-estimate of possible error.

The term 'error' in statistics normally relates to a statistical 'uncertainty' rather than a *mistake* in the experiment. For example, the standard error of the mean (Eqn 1.21) in a calculated value is a measure of the standard deviation *uncertainty* in that value.

1.4.2 True value

We can identify three main types of 'true' values:

Physical value of a specific characteristic of the system. For example, the *true* concentration of lead in a water sample is a *single* specific value.

Statistical parameter that describes the distribution of values within the system. For example, the weights of all the fish in a tank will have a true *population mean* value, even though no single fish has exactly that weight.

Probability value that describes the probability with which specific events occur. For example, the decay of a radioactive isotope is governed by a true *probability* with which individual nuclei decay. However in this case, we usually prefer to derive a value for the true *half-life* which is the time for half of the nuclei to decay.

1.4.3 **Experimental uncertainty**

The uncertainty in experimental measurements can then be divided into three corresponding categories:

Measurement uncertainty. The *measurement process* itself will vary slightly between repeated measurements, possible due to small differences in experimental procedures, instrumental responses, human observations, etc. For example, repeated measurements of the calcium content of the same sample of mineral water may give different experimental values.

Subject uncertainty. A *subject* is a representative example of the system being measured, but there will often be differences between subjects. For example, similar plants grown under the same conditions may have a variation in their heights.

Probability uncertainty. There is always an inherent *statistical* uncertainty when the occurrence of an event is governed by random probability.

Whatever the source of uncertainty, it is important that any experiment must be designed both to *counteract* the effects of uncertainty, and also to *quantify* the magnitude of that uncertainty. Many real experiments will be subject to a combination of different types of variation (e.g. DIY dice case study), and good experimental design seeks to minimize and manage uncertainty. For example, if it is known that the uncertainty in a system matches a normal distribution, then the uncertainty can be managed more effectively by using a parametric analysis that assumes normally distributed data.

The behaviour of uncertainties can be further classified as causing either:

Random error: Each subsequent measurement has a random error, whose *magnitude* and *direction* is not related to any other measurement. The **precision** of a measurement is the best estimate for the purely *random error* in a measurement. A measurement with a low random error is said to be a *precise* measurement.

Systematic error: Each subsequent measurement has the *same* recurring error. A systematic error causes the measurement to be *biased*, e.g. when setting the liquid level in a burette, a particular student may always set the meniscus of the liquid a little too low. The term **accuracy** is often used in science to describe the amount of *bias* in a measurement.

We can use Fig 1.12 to illustrate these terms, in which samples, A, B, C, and D (each with five replicates) are taken of the *pH* of the same solution. By looking at the closeness of the values, we can say that samples A and C are more *precise* than B and D. However, we can only know the accuracy of the measurements if we know the true value being measured. If the true value were 9.07 then we could say that samples C and D are more *accurate* than A and B.

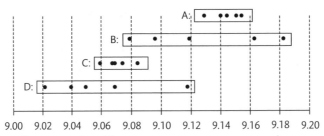

Fig 1.12 Illustration of accuracy and precision

An alternative approach is to use the term **trueness** as the best estimate for the *bias* in a measurement, and then **accuracy** becomes the best estimate for the *overall error* in the final result, and includes both the effects of a lack of precision (due to random errors) and bias (due to systematic errors).

1.4.4 **Combining uncertainties**

It is common to express the uncertainty (or error) either as an

- *absolute* value, which uses the same units as the value itself, e.g. an uncertainty of 0.3mm in a measurement of a length of 200mm, or as a
- *relative* value, frequently given as a *percentage* uncertainty, e.g. 0.3mm in 200mm is equivalent to a percentage of $100 \times 0.3 / 200 = 0.15\%$.

We will express the *absolute* uncertainty in x as u_x. Typically this is given by the *standard deviation* in the possible values of x. The relative *percentage* uncertainty, $u\%_x$, is then obtained by expressing the uncertainty as a percentage of the value itself:

$$u\%_x = 100 \times \frac{u_x}{x} \qquad \text{or} \qquad u_x = \frac{u\% \times x}{100} \qquad (1.11)$$

It is, of course, possible to express a relative uncertainty as a simple ratio without the need to multiply by 100 for the percentage. However, in this book, we will normally use percentage for relative measurement, matching its common use in science.

Case study: **Experimental uncertainties / 2. Combining uncertainties**

—continued from 1.Introduction, leading to 1.4.4

In this case study, we use the Excel worksheet in Fig 1.13 to model the possible combinations of two variables, X and Y, with true values:

X = 9.0 in row 2 with standard deviation uncertainty u_X = 0.7, and

Y = 6.0 in row 3 with standard deviation uncertainty u_Y = 0.5.

The uncertainties in X and Y tell us that, if we were to *randomly repeat the measurements*, we could expect the value of X to be drawn from a normal distribution with a mean of 9.0 and a standard deviation of 0.7, and similarly 6.0 and 0.5 for the value of Y.

We *simulate* these measurements 1,000 times by first generating 1,000 random values of of X and Y in the shaded cells in columns F to ALQ and then we combine X and Y in different ways:

$$X + Y, X - Y, X \times Y, X/Y, X^3$$

We derive the *combined* uncertainties in each of these results by calculating the standard deviations of the 1,000 different values.

Combining uncertainties: Excel analysis for Fig 1.13. Scan here to watch the video or find it via www.oxfordtextbooks.co.uk/orc/currell/

We generate a randomly selected value of X in F2 by using the function:

[F2] = NORM.INV(RAND(), \$B2, \$C2) = 7.77

We then copy this function to all cells F2:ALQ3 to generate 1,000 random values for both X and Y. The '\$' signs lock the columns for the mean and standard deviation values.

	A	B	C	D	E	F	G		ALP	ALQ
1		Value	Uncert, *u*	%Uncert, *u*%		1	2		999	1000
2	*X =*	9.00	0.7	7.78		*7.77*	*8.93*		*9.38*	*9.29*
3	*Y =*	6.00	0.5	8.33		*5.53*	*5.92*		*5.97*	*6.69*
4										
5	*A = X + Y*	15.00	*0.861*	*5.74*		*13.29*	*14.84*		*15.35*	*15.99*
6	*S = X - Y*	3.00	*0.847*	*28.23*		*2.24*	*3.01*		*3.41*	*2.6*
7	*M = X×Y*	54.00	*6.107*	*11.31*		*42.92*	*52.82*		*55.99*	*62.19*
8	*D = X / Y*	1.50	*0.167*	*11.15*		*1.41*	*1.51*		*1.57*	*1.39*
9	*P = X³*	729.00	*177.199*	*24.31*		*468.64*	*711.1*		*824.74*	*802.62*

Fig 1.13 Combinations of random uncertainties (Columns H to ALO are 'hidden' in the worksheet)

We use the 1,000 possible values for X and Y in columns F to ALQ to:

add the variables to give, $A = X + Y$, in row 5
subtract Y from X to give, $S = X - Y$ in row 6
multiply them to give, $M = X \times Y$ in row 7
divide X by Y to give, $D = X / Y$ in row 8
raise X to the power of three, $P = X^3$ in row 9

For each of these calculations in columns F to ALQ, the next step is to calculate the respective standard deviation uncertainties of all 1,000 values and record the *combined uncertainties* in column C. For example, the standard deviation uncertainty in $X + Y$ is calculated:

[C5] = STDEV.S(F5:ALQ5)

Column C now contains the experimentally observed uncertainties, u, (in C5, C6, C7, C8, and C9) for each of the different combinations.

Finally, we use Eqn 1.11 to calculate the percentage uncertainties, $u\%$, in D2 and D3 for X and Y:

$$\%u_X = 100 \times 0.7 / 9.0 = 7.78\%$$
$$\%u_Y = 100 \times 0.5 / 6.0 = 8.33\%$$

and for each of the experimentally observed uncertainties in D5:D9, e.g.:

$$[D5] = 100 * C5 / B5$$

In our model the randomly generated and calculated values, shown in italics, are recalculated every time the F9 button is pressed. Clearly it is unrealistic to use an Excel model like this every time we want to *predict* the uncertainties in calculated values, and we need to know how to *estimate* combined uncertainties from the individual uncertainties, u_X and u_Y.

The **guiding principles** for combining uncertainties are:

- Combine unrelated uncertainties by using the **addition of variances**, which is equivalent to combining standard deviations *squared* in the same way as Pythagoras's equation for the sides of a right-angled triangle.
- For the uncertainty in the **addition** or **subtraction** of values, combine *absolute* uncertainties:

$$u_A = u_s = \sqrt{u_X^2 + u_Y^2} \qquad (1.12)$$

- Note that the uncertainty variances always *add*, even when taking the difference between values.
- For the uncertainty in the **multiplication** or **division** of values, combine *percentage* uncertainties:

$$u\%_M = u\%_D = \sqrt{u\%_X^2 + u\%_Y^2} \qquad (1.13)$$

- For the uncertainty in the **power** of a value, multiply the *percentage* uncertainty by the value, k, of the power:

$$u\%_P = k \times u\%_X \qquad (1.14)$$

- Convert between *absolute* and *relative* uncertainties using Eqn 1.11.

We now compare the use of Eqns 1.12 to 1.14 with the results of the Excel model in Fig 1.13, using the values:

$$X = 9.0 \text{ with } u_x = 0.7 \text{ and } \%u_x = 7.78\%, \text{ and}$$
$$Y = 6.0 \text{ with } u_Y = 0.5 \text{ and } \%u_y = 8.33\%$$

Using Eqn 1.12 for the *absolute* uncertainty in **addition** and **subtraction**:

$$u_A = u_S = \sqrt{0.7^2 + 0.5^2} = 0.860$$

which is consistent with the randomly generated values in cells C5 and C6.

Using Eqn 1.13 for the *percentage* uncertainty in **multiplication** and **division**:

$$u\%_M = u\%_D = \sqrt{7.78^2 + 8.33^2} = 11.40$$

which is consistent with the randomly generated values in cells D7 and D8.

Using Eqn 1.14 for the *relative* uncertainty in taking the **power**:

$$u\%_P = 3 \times 7.78 = 23.33$$

which is consistent with the randomly generated value in cell D9.

We can now use the following example to demonstrate the use of an Excel worksheet to lay out the calculations for a multiple combination of errors.

> ## Case study: **Experimental uncertainties / 3. Propagation of errors**
> —continued from 1.4.4, leading to 1.4.6
>
> In an experiment to measure the specific heat capacity, c, of a material, a body of mass m is heated electrically using a voltage, V, and current, I, for a time, t. The temperature of the body rises from T_1 to T_2 (C).
> The specific heat capacity of the body is given by the equation:
> $$c = \frac{I \times V \times t}{m \times (T_2 - T_1)}.$$
> The experimental values and standard deviation uncertainties of the variables have been entered into rows 2, 3, 6, 7, 8, and 9 of columns B and D in Fig 1.14:

	A	B	C	D	E	F
1		**Value**	**Units**	**Uncert**		**% Uncert**
2	$T_1 =$	40.40	C	0.10		
3	$T_2 =$	45.60	C	0.10		
4				↓		
5	$(T_2 - T_1) =$	5.20	C	0.14	→	2.72
6	$I =$	10.60	A	0.02	→	0.19
7	$V =$	12.00	V	0.10	→	0.83
8	$t =$	120.00	s	1.00	→	0.83
9	$m =$	0.68	kg	0.01	→	1.47
10						↓
11	$c =$	4316.74	J kg^{-1}C^{-1}	143.06	←	3.31

Fig 1.14 Propagation of errors

Propagation of errors: Excel analysis for Fig 1.14. Scan here to watch the video or find it via www.oxfordtextbooks.co.uk/orc/currell/

The best-estimate value for c is calculated in B11 using the given equation:

[B11] = B6*B7*B8 / (B9*(B3−B2)) = 4316.74 Jkg^{-1}C^{-1}

The equation for c has both multiplication and subtraction calculations, which will involve both the *absolute* and *percentage* uncertainty values. We need to split up the calculation into two main stages, following the steps given by the arrows in the worksheet.

First we calculate the value for $(T_2 - T_1)$ in B5, and then the *absolute* uncertainty of $(T_2 - T_1)$ in D5 by combining the absolute uncertainties in D2 and D3 using Eqn 1.12:

[D5] = SQRT(D2^2+D3^2)

The rest of the calculation, with multiplication and division, combines the *relative* uncertainties, which are first calculated using Eqn 1.11, e.g.

[F5] = 100*D5/B5

The combined *relative* uncertainty in c is calculated in F11 using Eqn 1.13:

[F11] = SQRT(F5^2+F6^2+F7^2+F8^2+F9^2)

Finally, the *absolute* uncertainty in c is calculated using Eqn 1.11:

[D11] = F11*B11/100 = 143 $\text{J kg}^{-1}\text{C}^{-1}$

This gives a final result: $c = 4317$ (143 sd) $\text{J kg}^{-1}\text{C}^{-1}$

1.4.5 **Probability uncertainty**

The probability of an event occurring is given by the binomial distribution (1.3.3), such that, if the probability of an individual occurrence is p, then, out of n trials where the event may, or may not, occur:

Mean number of observed events (Eqn 1.5):

$$\mu = p \times n$$

Standard deviation uncertainty (Eqn 1.6):

$$\sigma = \sqrt{n \times p \times (1-p)}.$$

If the probability of observing an individual event is very small, $p \ll 1.0$, then the binomial distribution approximates to the Poisson distribution, and the standard deviation becomes:

$$\sigma \approx \sqrt{n \times p} = \sqrt{\mu}.$$

Thus, for an observed count of N, when the individual probability is *small*, the *best estimate* in the uncertainty of that value is given by

$$u = \sqrt{N}. \tag{1.15}$$

Typical examples include radioactivity, in which the count records the random decay of individual atoms, and the occurrences of a rare, non-communicable disease, observed with a probability given by the Poisson distribution.

1.4.6 **Identifying uncertainties**

An important element of analysis is to separate the different forms of variation and uncertainty that have combined within the experimental data. We see in 3.1.1 how the *t*-statistic compares an observed value with the uncertainty in the measurement, and in 3.2.2 how an ANOVA (ANalysis Of VAriance) separates variances within the data in order to identify significant factor effects.

A statistical analysis needs to be able to calculate the uncertainties in the data if it is able to provide correct estimations of the overall uncertainty. This is demonstrated in the following case study which analyses a combination of *system* and *probability* uncertainties in two different ways, producing different uncertainty estimations in the final result.

DIY dice: Excel analysis for Fig 1.15. Scan here to watch the video or find it via www. oxfordtextbooks. co.uk/orc/ currell/

Case study: **Experimental uncertainties / 4. DIY dice**

—continued from 1.4.4, leading to 5.3.4

We wish to measure experimentally the probability of getting a '6' when we roll a die. Instead of using commercial dice, we imagine that we make our own ten dice by cutting ten cubes from a bar of wood with a square cross-section, and writing a '6' on a randomly chosen face of each die. These DIY (do-it-yourself) dice are unlikely to be perfect cubes, and we can expect that we now have a *subject* uncertainty (the different dice), in addition to the inherent *probability* uncertainty. There is no *measurement* uncertainty as we assume that we can tell accurately whether or not we observe a '6'.

In Fig 1.15 we use Excel to simulate DIY dice by *randomly* allocating (in B3:B12) the *probabilities* with which each die records a '6'. These probabilities have been randomly produced from a distribution with the expected mean for a fair die of 1/6 and with a standard deviation of 20% (in B14) to represent the random differences between the dice. With these individual probabilities Excel then uses the binomial distribution to randomly calculate (in C3:C12) the *frequency* of '6s' that each die might record when it is rolled $n = 200$ times.

	A	B	C	D	E
1	Die	Probability	Frequency	Method	Proportion
2	*i*	*p*	*f*	(2) →	*P*
3	1	0.161	29	/ 200 =	0.145
4	2	0.105	27	/ 200 =	0.135
5	3	0.136	25	/ 200 =	0.125
6	4	0.139	32	/ 200 =	0.160
7	5	0.150	28	/ 200 =	0.140
8	6	0.229	43	/ 200 =	0.215
9	7	0.168	31	/ 200 =	0.155
10	8	0.156	25	/ 200 =	0.125
11	9	0.119	13	/ 200 =	0.065
12	10	0.178	28	/ 200 =	0.140
13			Method		↓
14	Uncert (%) =	20	(1)	Mean =	0.1405
15			↓	Stdev =	0.0371
16		Sum =	281	*SE* =	0.0117
17	Mean proportion =		0.1405	*t* =	2.2622
18	Uncertainties:	*Cd* (1) =	0.0152	*Cd* (2) =	0.0265

Fig 1.15 Rolling ten dice 200 times each

We can now use the 'results' to estimate the probability of getting a specific face (e.g. a '6') when we roll a die, and we can do this in two ways.

1. We can add up the total number of '6s' (in C16) from the 200 rolls of every die, giving a total of 281, and then divide by the overall number of rolls (2,000) to get a final proportion:

   $[C17] = C16 / 2000 = 0.1405$

2. Alternatively, we can calculate the proportion of '6s' for each die, record the results in E3:E12, and then calculate the average of the ten proportions:

[E14]=AVERAGE(E3:E12) = 0.1405

In this particular example, both methods must give the same result because all dice have been rolled the *same number of times*. This is significant because it means that the results of every die (sample) have the same 'importance' when calculating the average of the proportions. See 5.3.4 for the use of *weighting* values when the samples have differing 'importance'.

The important difference between the methods lies in their *estimation of the uncertainty* in the result.

With method 1, the total number of '6s', 281, are counted, and the calculation of the proportion, $P = 0.1405$, is based on $n = 2,000$ trials. However, all information about the *sample* variability has been lost. We can use Eqn 1.8 from the binomial distribution in 1.3.3 to estimate the standard deviation uncertainty just due to the inherent *probability* uncertainty:

$$\sigma(P) = \sqrt{\frac{P \times (1-P)}{n}} \Rightarrow \sqrt{\frac{0.1405 \times 0.8595}{2000}} \Rightarrow 0.00777$$

This then gives an estimated 95% confidence deviation (1.5.2), $Cd(1) = 1.96 \times 0.00777 = 0.0152$.

We could then conclude that, using method 1, we would be 95% confident that the true probability of recording a '6' was within the range 0.140 ± 0.015.

However, if we use method 2 we see a variation in E3:E12 between the different dice (samples), which will be due to both the *sample* and *probability* uncertainty.
We calculate the standard deviation of these values in E15:

[E15] = STDEV.S(E3:E12) = 0.0371

For a sample size of ten values, this gives an estimated 95% confidence deviation (1.5.2):

$$Cd(2) = 2.26 \times \frac{0.0371}{\sqrt{10}} = 0.0265$$

where 2.26 is the *t*-value with $df = 10 - 1 = 9$.

We could then conclude that, using method 2, we would be 95% confident that the true probability of recording a '6' was within the range 0.140 ± 0.027.

We know, both from a theoretical expectation and years of experimentation, that the true probability of recording a specific face on a true die will be $1/6 = 0.167$. The uncertainty calculation in method 1 does not take into account the *system* variations due to the different dice, and the confidence range is too narrow and fails to include the true value. However, method 2 takes into account both variations, and the wider confidence range just includes 0.167 as a possible true value.

It is important to ensure that any initial analysis of the data, e.g. taking totals, means, standard deviations, etc. does not 'hide information' and prevent any subsequent analysis of the data being able to calculate the actual experimental variation in the data (5.1.5).

1.5 Sample data

It is important to start by differentiating the use of the word 'sample', as a *statistical* sample of *n* repeated (replicate) measurements, from the concept of a *chemical* sample, which is a representative amount of material selected for analysis. It is quite common to have a *statistical* sample of several replicate measurements (e.g. calcium content) taken from one *chemical* sample (e.g. of mineral water).

We develop the statistics involved in performing a *statistical* analysis of the repeated measurements in a sample, with the aim of increasing the accuracy of a best-estimate of an unknown true value. This involves calculating the

- *sample mean*, which is the best-estimate for the unknown true value
- *sample standard deviation*, which provides a measure of experimental uncertainty
- *standard error of the mean*, which is a measure of the uncertainty in the sample mean
- *confidence interval* of the mean, which defines a range of possible values within which it is possible to state, with a given confidence, that the true value will lie.

Our initial analysis of the sample statistics assumes that the uncertainty in the measurements is calculated *solely* from the values in the sample. Then, following an outline of the difference between statistical *samples* and *populations*, we then analyse the *scientific* situation where the experimental uncertainty is known using experience from outside of the specific sample of measurements.

1.5.1 Sample statistics

We will develop the key statistics for data samples using a case study of five replicate measurements.

Sample statistics and confidence intervals: Excel analysis for Fig 1.16. Scan here to watch the video or find it via www.oxfordtextbooks.co.uk/orc/currell/

Case study: Blood alcohol / 5. Sample statistics

—continued from 1.3.1, leading to 1.5.3

Column B in Fig 1.16 gives five replicate blood alcohol values, x_i, selected randomly from a normal distribution with a mean of $\mu = 80$ and a population standard deviation of $\sigma = 2.0$. We develop the standard error and confidence interval for this sample.

	A	B	C		D		E	F
1	*i*	x_i	d_i		$d_i{}^2$			
2	1	81.4	1.02		1.04		Sample stdev, *s* =	1.51
3	2	78.8	-1.58		2.50		Standard error of mean , *SE* =	0.68
4	3	79	-1.38		1.90		*t*-value =	2.78
5	4	80.4	0.02		0.00		Confidence deviation, *Cd* =	1.88
6	5	82.3	1.92		3.69		Using CONFIDENCE.T() =	1.88
7								
8	Size =	5	Sum of squares, *SS* =		9.13		Mean square, *MS* =	2.28
9	Mean =	80.38	Deg of freedom, *df* =		4		Sample variance, s^2 =	2.28

Fig 1.16 Statistics of five replicate data values

The separate data values, x_i, are identified by the integer i in column A which runs from 1 to 5.

The sum of values is expressed using capital sigma:

$$\text{Sum of all values} = \sum_i x_i$$

The *mean value*, $\bar{x}$ (x-bar), is calculated as the simple sum divided by the sample size, n:

$$\bar{x} = \frac{\Sigma_i x_i}{n} = \frac{81.4 + 78.8 + 79.0 + 80.4 + 82.3}{5} = \frac{401.9}{5} = 80.38$$

which is calculated directly in cell B9, using the AVERAGE() function in Excel:

$$[B9] = \text{AVERAGE(B2:B6)} = 80.38$$

The sample mean value, $\bar{x}$ is the *best-estimate* for the true blood alcohol value, μ.

In order to quantify the *spread* of data, we first calculate the deviations, $d_{integer}$, of each data value from the mean value:

$$d_i = x_i - \bar{x}$$

which are calculated in column, C, e.g. $[C2] = B2 - B\$9 = 1.02$.

The dollar sign is used to lock the row value for B9 when this formula is copied down to row 6.

If we added the values of all deviations we would find that

$$\sum_i d_i = d_1 + d_2 + d_3 + d_4 + d_5 = 0$$

This will always be the case as the criterion for the mean value is that the deviations will always sum to zero.

To get a positive measure of the data variability we square each deviation in column D,

$$\text{e.g.} [D2] = C2^2 = 1.04$$

and then add all squares to get the total **sum of squares** (SS) of the deviations.

$$SS = \sum_i d_i^2 = \sum_i (x_i - \bar{x})^2 \tag{1.16}$$

calculated in D8 as: $[D8] = \text{SUM(D2:D6)} = 9.13$.

An important factor in many statistical calculations is the **degrees of freedom, df,** in the calculation. This is related to the number of 'bits of information' in the calculation. We started the analysis with $n = 5$ data values, i.e. with five bits of information. We then used one bit of information to calculate the mean value, $\bar{x}$, of this particular sample. Hence the remaining degrees of freedom are:

$$df = n - 1 \tag{1.17}$$

calculated in D9 as: $[D9] = B8 - 1 = 4$.

We can now calculate a **mean square** value, MS, by dividing the sum of squares by the degrees of freedom, df:

$$MS = \frac{SS}{df} \tag{1.18}$$

calculated in F8 as: [F8] = D8 / D9 = 2.28.

For a simple data sample, *MS* is usually called the **sample variance**, s^2:

$$s^2 = MS = \frac{\Sigma_i(x_i - \bar{x})^2}{(n-1)} \tag{1.19}$$

We can also calculate this directly in Excel using the function, VAR.S() in F9:

[F9] = VAR.S(B2:B6) = 2.28

giving the same result as in F8.

The **sample standard deviation,** *s*, is the square root of the variance,

$$s = \sqrt{s^2} = \sqrt{\frac{\Sigma_i(x_i - \bar{x})^2}{(n-1)}} \tag{1.20}$$

Using Eqn 1.20:

$$s = \sqrt{s^2} = \sqrt{2.28} = 1.51$$

We can also calculate this directly in Excel using the function STDEV.S() in F2:

[F2] = STDEV.S(B2:B6) = 1.51

The *sample standard deviation* is the best estimate of the uncertainty in a *single* measurement. However, we will normally take the *mean* of *n* replicate measurements because the average of several values is more likely to be closer to the true value than just a single measurement.

The reduced uncertainty with *n* measurements is described by the **standard error of the mean,** *SE*:

$$SE = \frac{s}{\sqrt{n}} \tag{1.21}$$

We can see that the uncertainty reduces, and the precision increases, in proportion to the *square root* of the number, *n*, of measurements.

We calculate the standard error, *SE*:

[F3] = F2/SQRT(B8) = 0.68

The units for standard deviation and standard error of the mean are the same as the value itself.

1.5.2 **Confidence interval**

The confidence interval, *CI*, is a standard way of presenting the uncertainty in our results by stating that we are 95% (for example) confident that the true value being measured lies within a calculated range of values.

The confidence interval, *CI*, is calculated as the range of values defined by the sample mean value, with a confidence deviation *Cd* on either side.

$$CI = \bar{x} \pm Cd$$

Students are sometimes confused between confidence *interval* and the confidence *deviation*, and it is important to check which value is being presented. For example, Excel 2013 uses the function CONFIDENCE.T() to calculate the confidence *deviation*, Cd.

The confidence deviation (based on the sample standard deviation, s) will depend on:

- the *uncertainty*, which is introduced as the standard error of the mean, SE, and
- the *degree of confidence* that we wish to claim for the calculated range, which is introduced by the *t*-value, t.

$$Cd(\text{using } s) = t \times SE = t \times \frac{s}{\sqrt{n}} \tag{1.22}$$

The ***t*-value** depends on

- the level of confidence required, typically 95% which is often expressed as a *significance* of 0.05 and
- the degrees of freedom, $df = n - 1$.

For a very large sample, and for a confidence of 95%, the *t*-value becomes equal to the limiting *z*-value (1.5.4) = 1.96. For smaller samples (low values for *n* and *df*), the value of *t* increases to make the confidence deviation greater to allow for the greater uncertainty in the calculated standard deviation.

We calculate the *t*-value directly in Excel using the function, T.INV.2T() in F4:

$$[F4] = \text{T.INV.2T}(0.05, D9) = 2.78$$

where D9 is the degrees for freedom and '0.05' is the *significance* equivalent to 95% confidence. We can then calculate the confidence deviation in F5:

$$[F5] = F4 * F3 = 1.88$$

It is also possible to use the function CONFIDENCE.T() to calculate the confidence deviation directly from the sample standard deviation (F2) and the sample size (B8):

$$[F6] = \text{CONFIDENCE.T}(0.05, F2, B8) = 1.88$$

Note that the CONFIDENCE() function does *not* apply to *sample* calculations, as it assumes that the *t*-value is given by the limiting *z*-value for a *population* (1.5.4).

The **confidence interval of the mean** is given by the equation:

$$CI = \bar{x} + Cd = \bar{x} + t \times SE = \bar{x} + \left(t \times \frac{s}{\sqrt{n}} \right) \tag{1.23}$$

The 95% confidence interval, *CI*, for the true blood alcohol in Fig 1.16 is calculated to be:

$$CI = 80.38 \pm 1.88$$

which, rounded to one decimal place, and including units, becomes:

$CI = 80.4 \pm 1.9$ mg per 100 ml.

For a confidence level of 95%, there is a 95% probability that the true value lies *within* the limits of the confidence interval. In other words, there is only a 5% (= 0.05) probability that we would be wrong if we gave the confidence interval as our *best-estimate* of the true value.

It is important to emphasize that the confidence interval expresses the result of a *specific set of sample values*, and it is *not* a prediction of how the mean values could be expected to vary for *different* samples. We see, in 1.5.3, how the calculated confidence interval, *CI*, varies considerably from sample to sample, but, in 95% of samples, the true value is still found within the confidence interval for that particular sample.

1.5.3 Samples and populations

It is important to understand the difference between data **samples** and **populations**. A 'population' describes *all the measurements* that could be made of a particular variable, but a 'sample' represents a *selection of representative measurements* taken from the population.

For example, if we wish to compare the sizes of fish in two fish farm tanks, we could measure the whole *population* of each tank (every fish), but for economy of effort it might be sufficient, within the required accuracy of comparison, just to take a representative *sample* of a *few* fish from each tank.

As another example, we are familiar with making a few replicate measurements of the absorbance of a chemical solution, such that the few measurements are just a *sample* of the almost unlimited *population* of repeated measurements that could be made.

We can test whether a data set is a *sample* or a *population* by considering the effect of *repeating the process by which the data values were identified*. If repeating the process of identification will always give the same values (e.g. selecting every fish), then the data set is a *population*, but if different random values could occur (e.g. selecting a different sub group) then the set is a *sample*.

> A **parameter** is a *variable* that is used to describe some characteristic of a *population*. A parameter is usually given a Greek letter as a symbol, e.g. μ (mu) for population mean and σ (sigma) for population standard deviation.
>
> A **statistic** is a *variable* that is used to describe some characteristic of a *sample*, e.g. $\bar{x}$ for sample mean and s for sample standard deviation.

The value of the sample statistic is a *best-estimate* for the true value of the equivalent population parameter. For example, the sample mean, $\bar{x}$, is the best estimate for the true mean, μ, of the population from which it was randomly drawn. Similarly, the sample standard deviation, s, is the best estimate for the source population standard deviation, σ.

Different samples from the *same* population will typically give different values for the same statistic (e.g. different *sample* means).

The calculations for mean and standard deviation of both samples and populations are summarized in Table 1.3.

Table 1.3 Means and standard deviations

	Sample statistics	Population parameters
Mean	$\bar{x} = \dfrac{\Sigma_i x_i}{n}$	$\mu = \dfrac{\Sigma_i x_i}{n}$
Standard deviation	$s = \sqrt{\dfrac{\Sigma_i (x_i - \bar{x})^2}{(n-1)}}$ (1.24)	$\sigma = \sqrt{\dfrac{\Sigma_i (x_i - \mu)^2}{n}}$ (1.25)

The calculation is the same for both sample mean, $\bar{x}$, and population mean, μ. However, there is a difference in the calculations for the respective standard deviations, s and μ, in that the degrees of freedom divisor is n for the *population* standard deviation but $n - 1$ for the *sample* standard deviation. The sample calculation loses one degree of freedom, or one bit of information, because of the need to calculate the mean value, $\bar{x}$ for each *specific* sample.

We will see in the next case study that the '$n - 1$' is important because the sample standard deviation is used as an *estimate* of the population standard deviation from which it has been derived, and that, without the use of the '-1' in the denominator, it would tend to underestimate the true population standard deviation.

Case study: **Blood alcohol / 6. Samples and populations**

—continued from 1.5.1, leading to 1.6.2

This case study uses Excel to simulate experimental measurements made to establish the blood alcohol level for a sample where the true value, μ, is 80 mg of alcohol per 100 ml of blood, and the measurement uncertainty is equivalent to a standard deviation, σ, of 2.0 mg/100 ml.

Samples and populations: Excel analysis for Fig 1.17. Scan here to watch the video or find it via www.oxfordtextbooks.co.uk/orc/currell/

In principle, there is no statistical restriction to the number of replicate measurements that could be made on the same physical sample of blood. Every time we make a measurement we get a potentially new set of sample results.

In the cells D2:D6 (sample 1) in Fig 1.17, a sample of five values is randomly generated from a normal population with $\mu = 80$ and $\sigma = 2.0$, and then a further 1,999 samples are generated in columns E to BYA. This process of data simulation in Excel uses the values of 80 and 2.0 in B2 and B3 respectively, and then the function NORM.INV(RAND(), \$B\$2, \$B\$3) in every data cell generates values randomly selected on the basis of the defined normal distribution.

	A	B	C	D	E	F	G	BXZ	BYA
1	Population parameters:		Sample:	S1	S2	S3	S4	S1999	S2000
2	Mean =	80	(79.996)	77.9	80.3	80.5	78.6	83	81.7
3	StDev =	2	(2.028)	83.8	81.2	78.7	76.7	78.1	81.9
4				79.4	82.9	76.9	79.4	82.9	76.5
5				81	81.4	78.5	78.1	83.7	83.8
6				80.5	82	80	77.4	82.5	77.6
7	Sample (n = 5) statistics:	Averages:	Calculations:						
8	Mean =	79.996	Mean =	80.52	81.56	78.92	78.04	82.04	80.3
9	Sample stdev =	1.912	Sample stdev =	2.19	0.97	1.41	1.05	2.24	3.1
10	'Population' stdev =	~~1.71~~	'Population' stdev =	~~1.95~~	~~0.86~~	~~1.26~~	~~0.94~~	~~2.01~~	~~2.77~~
11	Test statistics:								
12	Standard error (mean) =	0.855	Standard error =	0.98	0.43	0.63	0.47	1	1.39
13	True value =	80	t-value =	2.78	2.78	2.78	2.78	2.78	2.78
14	Errors:		Cd (with s) =	2.71	1.2	1.75	1.3	2.79	3.85
15	Proportion (using s) =	0.051	Type I error (with s) =	0	1	0	1	0	0
16									
17	Errors:		Cd (with σ) =	1.75	1.75	1.75	1.75	1.75	1.75
18	Proportion (using σ) =	0.053	Type I error (with σ) =	0	0	0	1	1	0

Fig 1.17 Sample statistics (Columns H to BXY are 'hidden' in this worksheet) Numbers in italics show values that change with different randomly generated data sets

The values in cells C2 and C3 record the actual mean (79.996) and standard deviation (2.028) of *all* of the generated values, and can be seen to be consistent with the population values of 80 and 2.0 used to generate them.

The 2,000 columns of data between D and BYA each have five randomly selected values in rows 2 to 6, equivalent to a statistical *sample* of size five. Note that each time the F9 key is pressed, Excel generates completely new sets of sample data. Numbers in italics indicate values that can change every time a new data set is created.

For each of the 2,000 samples, we calculate:

- mean values, $\bar{x}$, of every sample in D8 to BYA8:

  e.g. [D8] = AVERAGE(D2:D6)

- sample standard deviations, s, of every sample in D9 to BYA9 based on Eqn 1.24:

  e.g. [D9] = STDEV.S(D2:D6)

- population standard deviations, σ, of every sample in D10 to BYA10 based on Eqn 1.25:

  e.g. [D10] = STDEV.P(D2:D6)

Note that the *population* standard deviation values are recorded with a *strikethrough*, because it would not be usual to calculate this value for a sample, and is only done here for comparison.

The averages of these three statistics for all 2,000 samples are calculated in B8, B9, and B10 respectively.

As expected, the average of all the sample means, [B8] = 79.996, is very close to the true value, 80, of the population. The mean of the sample, $\bar{x}$, gives the *best estimate* of the unknown true value, μ.

We can also see that the *sample* standard deviations (1.912 in B9), calculated using Eqn 1.24, gives a better estimate of the true standard deviation, 2.00, than the *population* standard deviations (1.710 in B10) calculated using Eqn 1.25. This confirms that the sample standard deviation, *s*, gives the *best estimate* of the unknown true standard deviation, σ, from which the sample has been drawn.

We continue to refer to the Excel worksheet in Fig 1.17 to investigate the use of *standard error* and *confidence interval*.

For each of the 2,000 samples we calculate:

- standard error, *SE* (row 12), based on Eqn 1.21,

  e.g. [D12] = D9 / SQRT(5)

- *t*-value (row 13) for 95% confidence and *df* = 4 (which is the same for all samples),

  e.g. [D13] = T.INV.2T(0.05, 4) = 2.78

- confidence deviation, *Cd* (row 14), based on Eqn 1.22,

  e.g. [D14] = D13 * D12

Note that the confidence deviation, *Cd*(*s*) in row 14, is based on the *sample* standard deviation, *s*.

In B12 we calculate the average value of the standard errors from all 2,000 samples, and record a value of 0.855 which is consistent with Eqn 1.21 using the average sample standard deviation of 1.912 in B9:

$$SE = \frac{1.912}{\sqrt{5}} = 0.855$$

We now wish to *test* whether the calculated confidence interval from Eqn 1.23:

$$CI = \bar{x} \pm Cd$$

provides a correct range for finding the true value for 95% of randomly selected values. According to our theory we would expect that the true value will fall *outside* the confidence interval for only 5% of the samples.

The confidence interval prediction will *fail*, causing a Type I error (1.6.3), if the true value, B13, lies *outside* the confidence interval,

i.e. if $80 > \bar{x} + Cd$ or $80 < \bar{x} - Cd$

We use the IF() function in Excel to test whether this may be true:

e.g. [D15] = IF(OR($B13 > D8 + D14, $B13 < D8 − D14), 1, 0)

where $\bar{x}$ is in D8, *Cd* in D14, and the value 80 in B13.

The IF() function returns a '1' in row 15 if the true value, B13, lies *outside* the confidence interval for each of the samples.

We can see that samples 2 and 4 return a '1':

Sample 2: *CI* = 81.56 ± 1.20, i.e. between 80.36 to 82.76 which does not include 80.

Sample 4: *CI* = 78.04 ± 1.30, i.e. between 76.74 to 79.34 which does not include 80.

We can calculate the *proportion* of failures in B15 by counting the number of '1s' for all 2,000 samples and dividing by 2,000:

[B15] = SUM(D15:BYA15)/2000 = 0.051.

The sampled data reproduced in Fig 1.17, gives a calculated proportion of errors in B15 of 0.051, which is consistent with the expected value of 0.050 for the error probability. Pressing the F9 key to recalculate new samples produces new values, which continue to be consistent with 0.050.

1.5.4 Known experimental uncertainty

In 1.5.3, the experimental standard deviation uncertainty, s, was calculated *only* from the *sample* measurements. For small samples this means that there is a large uncertainty in the value of s, and hence a large uncertainty in the calculated confidence deviation, Cd.

However, there are often practical situations where the true measurement uncertainty, the *population* standard deviation, σ, is already known from previous experience or from a knowledge of the measurement process. In this case, the t-statistic has effectively an infinite number of degrees of freedom and becomes equal to the z-statistic, which for 95% confidence gives $z = 1.96$.

Hence, if we know the true value of the population standard deviation, σ, the confidence deviation becomes:

$$Cd(\text{using } \sigma) = z \times \frac{\sigma}{\sqrt{n}} = z \times SE \tag{1.26}$$

where $z = 1.96$ for 95% confidence.

Returning to Fig 1.17, we now investigate the implications of knowing the experimental uncertainty *separately* from the sample values.

When the population standard deviation, σ, is already known, the t-value for 95% confidence becomes equal to the z-value of 1.96. In this case study, the known population standard deviation, σ, is taken as the value in B3 = 2.0, and the new confidence deviation is the *same* for all 2,000 samples, calculated in row 17:

$Cd(\text{using } \sigma)$: e.g. [D17] = 1.96 * $B3/SQRT(5) = 1.75

In row 18, we again use the IF() function (see also calculation for row 15) to identify which of these confidence intervals $CI(\text{using } \sigma)$ result in an error, and then calculate the overall proportion in B18. The sampled data gives a calculated proportion of errors of 0.053, which is again consistent with the expected value of 0.050.

The *proportion* of errors is actually the same whether we use either *sample* or *population* standard deviations. However, it is not always the same *samples* that record errors. For example, it can be seen that sample 4 gives an error using both methods but not sample 2.

It is now useful to compare the variability of the different samples in Fig 1.17, and in Fig 1.18 we plot the mean values and confidence intervals (using *sample* standard deviations) from 8 of the 2,000 samples.

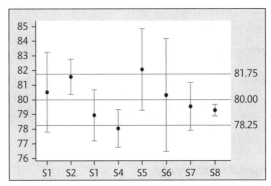

Fig 1.18 Replicate sample confidence intervals (Minitab)

The limits of 78.25 and 81.75, plotted on either side of the true value of 80 in Fig 1.18, are equivalent to the confidence interval, CI(using σ) = 1.75, for a known *population* standard deviation.

Calculations using *sample* standard deviations, *s*, give errors if the value of 80 lies outside the confidence intervals shown with each data point, i.e. for S2, S4, and S8.

Calculations using a known *population* standard deviation, σ, give errors if the mean value of the data point lies outside the lines drawn at 78.25 and 81.75, i.e. for S4 and S5.

We can see from Fig 1.18 that the conditions giving rise to errors can be different for the two types of calculation, although the overall proportions of error are the same for both at 5%.

1.5.5 Presenting results

The Excel model in Fig 1.17 shows that we get a *statistical* error rate of 5% for the confidence interval using either Eqn 1.23 with *s* or Eqn 1.26 with σ. It is then reasonable to ask whether it makes any difference if we present the 'sample' results when we actually know the true experimental uncertainty.

The apparent equivalence of the two types of analyses relates only to the question: 'Does the confidence interval include the true value?' However, they are not equivalent in the information content that is contained within their calculated confidence intervals. A complete result includes *both* the best-estimate value *and* the uncertainty in that value. The values for the uncertainty estimated solely by the sample values can vary widely, particularly for small samples, and, if we fail to use the known experimental uncertainty, then we would be throwing away information that is essential for the quality of the published result.

Further examples of the use of known experimental uncertainty are given in 2.2.1 (Case study: Spectrophotometer calibration / 4. Calibration result) and 2.2.3.

1.6 Hypothesis tests

Many investigations in science seek to answer a simple 'Yes/No' question.
For example:

- Is a new vaccine effective in preventing an infection?
- Do glass fragments at the scene of a crime come from different sources?
- Does a particular training regime improve athletic performance?

However, most experimental results have associated uncertainty and we cannot be absolutely certain of the correct answer. The 'hypothesis test' in statistics is a well-defined and universally accepted procedure for providing an answer based on agreed levels of confidence.

1.6.1 Test procedure

The first step is to define the 'Yes/No' question that we wish to test. The 'Yes' option usually means that we expect to be able to observe some effect or difference in recorded values, whereas the 'No' option shows no effect or difference.

We would like to have a calculation that tells us how confident we can be that 'Yes' is the *correct* answer. Unfortunately, the statistics only tells us how confident we can be that 'No' is *NOT* the correct answer! We often have to infer a *positive scientific* conclusion from a *statistical negative*.

It is then necessary to define the hypotheses of our test:

- Null hypothesis, H_0, as the state where there is no effect or difference, and
- Proposed or alternative hypothesis, H_1, as the state where there is an effect or difference.

Statisticians use the term 'alternative' to differentiate H_1 from the 'null' H_0, but many science students get confused when their 'proposed' *scientific* hypothesis is called a statistical 'alternative'. In this book, we often include the term 'proposed' to identify the scientific objective of the hypothesis test.

H_0 and H_1 should normally be mutually exclusive, such that 'rejecting the null hypothesis' is the same as 'accepting the proposed/alternative hypothesis'.

We must decide on how much confidence we will need to have before deciding that the null hypothesis is not true. This confidence is usually expressed as the *significance level*, α, which is the probability that we would be wrong if we rejected the null hypothesis. For example, the default level of a 95% 'confidence' equates to a significance level (probability of being wrong) equal to 0.05.

After collecting the experimental evidence, we will then either:

- calculate the value of a **test statistic** from the data, and then compare it with a known **critical value** for the statistic to decide whether the evidence suggests that the null hypothesis is unlikely to have produced the observed results. The relevant critical value can be obtained from published tables, or alternatively, and, more usually, with the use of modern software:

- calculate the probability, **p-value**, that the null hypothesis could give the experimental values equal to, or more extreme, than observed. This value is then compared to the **significance level**, α, chosen before the test. The default value for significance in most general analyses is $\alpha = 0.05$.

When using the p-value method:

- If $p \leq \alpha$ we state that, on the basis of the experimental evidence, we would reject the null hypothesis and accept that the proposed (alternative) hypothesis is correct.
- If $p > \alpha$ we state that there is insufficient experimental evidence to reject the null hypothesis and that any observed difference/effect could have occurred by chance.

It should be noted that the test cannot prove that the null hypothesis is true, only that there is not enough evidence to claim that it is not true. As our conclusions are based on a statistical calculation, there is always a chance that we could be wrong, and we examine this possibility in 1.6.3.

1.6.2 **Hypothesis test and *p*-values**

It is important to take care in expressing the scientific objective of your analysis in terms of a hypothesis that can be uniquely tested by experiment. This also requires identifying a measurable value that becomes the *test statistic* to be analysed.

Case study: **Blood alcohol / 7. Hypothesis test**

—continued from 1.5.3, leading to 3.1.2

A scientific objective is to decide if the blood alcohol level in a sample is above the specific value of $\mu_0 = 80$ mg of alcohol per 100 ml of blood. In the experiment we make $n = 5$ replicate measurements which give a sample mean of $\bar{x} = 81.6$, which is the *best estimate* of the true blood alcohol value, μ.

In this investigation, we assume that we know from previous experience that the experimental measurement has a standard deviation uncertainty of $\alpha = 2.0$ mg/100 ml.

We need to decide whether it is possible that the true blood alcohol value, μ, in the sample is actually equal to $\mu_0 = 80.0$, or less, and that the measured higher value only occurred due to the statistical variation in the experimental measurements. (Note that in a legal test for driving in the UK with excess alcohol, the actual level of confidence required for a prosecution is considerably higher).

The test statistic in the case study is the measured blood alcohol level, $\bar{x}$ and the scientific objective is to decide whether the true level is above or below (or equal to) 80 mg /100 ml.

The *scientific* null in this case is that μ is less than, or equal to, 80 mg /100 ml, i.e. $\mu \leq \mu_0$. However, the most likely way in which we might conclude, *incorrectly*, that μ is more than 80 is if μ is actually equal to the limit of 80 and experimental *variation* then records a set of unusually high values. Hence we choose to perform the calculation for the *statistical* null at this limit:

Null hypothesis:

H_0: Blood alcohol level equals 80.0 mg/100 ml, $\mu = \mu_0$.

The proposed *scientific* hypothesis, H_1, is that μ is more than 80, but there are two ways of approaching this *statistically*.

The **two-sided** (often called **two-tailed**) approach first tests whether the observed value, $\bar{x}$, is significantly *different* from 80, and then, if it is also greater than 80, we conclude that $\mu > 80$.

The **one-sided** (often called **one-tailed**) approach looks for a difference in a *predetermined* direction. In this case, we test directly whether the observed value, $\bar{x}$ is significantly *greater* than 80 and conclude that $\mu > 80$.

The one-sided approach should only be used if there is a specific reason for testing for an effect in a particular direction *before* any measurements are made. Some researchers advise against using the one-sided approach at all, because it gives the lower *p*-value and may result in a Type I error if used incorrectly. See the example in 3.1.3.

In general then, the proposed or alternative *statistical* hypothesis has two main options depending on the *symmetry* of the question:

Two sided:

H_1: Blood alcohol level does not equal 80.0 mg/100 ml, $\mu \neq \mu_0$.

One sided, depending on the direction of possible effect:

H_1: Blood alcohol level exceeds 80.0 mg/100 ml, $\mu > \mu_0$.

or

H_1: Blood alcohol level is less than 80.0 mg/100 ml, $\mu < \mu_0$.

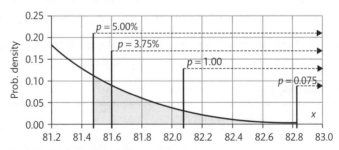

Fig 1.19 One-tailed *p*-values

With a standard deviation, $\sigma = 2.0$, a sample of five values has a standard error of the mean (Eqn 1.21), $SE = 2.0 / \sqrt{5} = 0.894$. We then calculate the probability distribution of measured mean values assuming that the null hypothesis is true, and Fig 1.19 gives the *p*-value probabilities that a randomly selected sample from the distribution (mean = 80, standard deviation = 0.894) could give a measured *mean* value, $\bar{x}$, equal to, or greater than specific values, *x*. The section of the upper tail of this distribution shows that:

Probability of recording a value of 81.47 or greater is 5.0% or 0.05

Probability of recording a value of 81.60 or greater is 3.75% or 0.0375

Probability of recording a value of 82.08 or greater is 1.0% or 0.01

Probability of recording a value of 82.82 or greater is 0.075% or 0.00075

Using the critical value for 5.0%, we could say that we would reject the null hypothesis for any $\bar{x}$ value greater than 81.47 'at a significance level of 0.05'.

Alternatively, for the measured value of $\bar{x} = 81.6$ in the case study example, we see that the p-value = 0.0375, and we reject the null hypothesis at a significance level of $\alpha = 0.05$ because $p < \alpha$.

Fig 1.19 gives the upper tail of the distribution appropriate for the one-tailed hypothesis, but for the two-tailed hypothesis, we must also include the probability area from the *lower* tail of the distribution. Provided that the distribution is *symmetrical*, we are able to calculate:

$$p\text{-value (two-tailed)} = 2 \times p\text{-value (one-tailed)} \quad \text{for symmetric uncertainties.} \qquad (1.27)$$

However, this equation is not true for unsymmetrical distributions (3.8.2).

1.6.3 Errors in hypothesis tests

Any scientific investigation is subject to errors and uncertainties, which means that we might pick the *wrong* option when trying to decide whether or not to accept a hypothesis. This gives two different types of **error** that could be made, each with different consequences:

- **Type I Error**: We 'accept H_1' when in fact the null hypothesis is true. We will be claiming to have identified some effect that is not true.

- **Type II Error**: On the basis of experimental data, we choose to 'NOT accept H_1', when in fact the proposed hypothesis is true. We have failed to identify a real effect. This may be due to poor experimental design or because we did not take enough measurements.

The four possible outcomes from a hypothesis test are illustrated in the Table 1.4. The *columns* give the 'true' situation, and the *rows* give the decision.

Table 1.4 Hypothesis test errors

The decision	True situation:	
	H_1 is true:	H_1 is not true:
Accept H_1:	**Correct**	**Type I Error** We claim to have 'discovered 'an effect that is not true.
Do not accept H_1:	**Type II Error** The experimental data was insufficient to detect the effect.	**Correct**

The probability of recording a Type I Error for a specific data set is given by the calculated *p*-**value**, and in general, we define:

Significance Level, α, as the *largest probability of Type I Error* that is acceptable when choosing the proposed hypothesis, H_1.

The default value for general research is $\alpha = 0.05$ (equal to a 5% probability or '1 in 20'). However, other significance levels can be chosen, depending on the *consequences* of a Type I Error.

The **power** of an experiment is the ability to detect an effect if one exists, i.e. the ability to *avoid* making a Type II Error.

Power $= 1 - \beta$

where β is the probability of a Type II Error.

Although it is possible to calculate the p-value for a given set of data, it is not generally possible to calculate a value for the *power* of the test solely from the data. The power of a particular test can depend on many other factors, and is a main focus for good experiment design.

1.6.4 **Bonferroni correction**

It is quite common to find that our analyses involve *multiple* hypothesis tests, each giving a separate p-value. For example, we may be *independently* testing several different factors to see if they might have a significant effect on a measured response variable. The problem is that, although the probability of a Type I Error is 0.05 for *one* p-value, the probability of seeing at least one Type I Error increases rapidly when we scan more than one p-value. If we record p-values for ten tests for which the null hypothesis is true then the probability that all ten results *correctly* record $p > 0.05$ is equal to $0.95^{10} = 0.599$, which means that the probability that at least one p-value will give a Type I Error and incorrectly 'discover' a significant effect with $p < 0.05$ will be equal to $1.0 - 0.599 = 0.401$ or about 40%.

A common correction for multiple tests, proposed by Bonferroni, is to make the criteria for rejecting the null hypothesis more difficult by reducing the significance level, α, in proportion to the number of tests performed. For n tests, the null hypothesis will be rejected if:

$$p < \frac{\alpha}{n} \quad \text{(Bonferroni correction).}$$

This effectively reduces the number of Type I Errors, but it becomes rather conservative for large values of n giving Type II Errors, failing to detect some real effects.

2 Regression analysis

Introduction

This chapter starts with the familiar process of drawing a *best-fit* straight line through a set of experimental data points on an *x–y* graph, and then derives the statistics in a format that links into the more advanced analyses developed throughout the rest of the book. It concentrates on the practical application of linear regression in science including the derivation of uncertainties and its use for nonlinear data. Finally, the technique of a least squares fit is used as a link into the development of iteration as a method for 'solving' complex mathematical problems. The techniques link into the use of Minitab and SPSS in Part II of the book.

Section 2.1 develops the basic statistics for a least squares regression analysis, deriving the coefficients of the best-fit straight line.

Section 2.2 develops the practical use of linear regression for data analysis including the calculation of derived experimental confidence intervals.

Section 2.3 uses linearization techniques to convert nonlinear data to a straight line relationship for further analysis.

Section 2.4 introduces the process of 'iteration' as a method for finding the best-fit mathematical models using both least squares and maximum likelihood models.

The following case studies develop the core statistics in this chapter:

Case study: **Best-fit straight line / 1. Overview**

This case study develops the statistics and applications related to linear regression, and the use of the best-fit straight line in science.

The basic statistics of 'best-fit' linear regression are developed using Excel in:

2.1.1 / 2. Slope and intercept

2.1.2 / 3. ANOVA table

2.1.3 / 4. Correlation

The use of the straight line as an analytical tool for calculating unknown values, concentrations, etc.:

2.1.4 / 5. Uncertainty in regression

2.2.1 / 6. Confidence interval

2.2.2 / 7. Standard additions

Linear regression is used as an example for the iterative technique using the Excel add-in 'Solver', and demonstrates both the least squares and the maximum likelihood estimation methods of achieving a 'best-fit' result:

2.4.1 / 8. Least squares fit using Solver

2.4.2 / 9. Maximum likelihood using Solver

Case study: **Exponential decay / 1. Overview**

Radioactive decay is used as an example for regression calculations of *exponential* growth and decay.

The technique for the linearization of an exponential curve is developed in:

2.2.4 / 2. Weighted linearization. Uses 'weighting' to reflect the different uncertainties and influences of different data values.

2.3.4 / 3. Linearizing the exponential. Develops the basic technique for linearization and interpreting the results.

Iterative methods for obtaining best-fit lines for nonlinear data are developed in:

2.4.3 / 4. Nonlinear regression using Solver. Uses the iterative analysis of the Excel add-in 'Solver' to solve the problem using different statistical assumptions and to compare their results.

3.4.7 / 5. Generalized linear model. Demonstrates the use of linearization to include the underlying Poisson distribution.

7.2.3 / 6. Nonlinear regression using Minitab and SPSS. Demonstrates the iterative technique in statistical software.

2.1 **Regression statistics**

Linear regression is the statistical process of fitting a *best-fit* straight line through a set of x–y data points. It is a very common analytical procedure in all areas of science.

In this section, we look specifically at the *statistics* of linear regression. We make the assumption that the experimental *uncertainty* in the measurement process is estimated solely from the data being analysed, which, for small sets of data, significantly increases the overall uncertainty in the final result. In 2.2.3, we introduce the wider *experimental context* for an analysis in which an estimate of experimental uncertainty is available from previous experience or from a knowledge of the measurement process itself.

2.1.1 **Slope and intercept**

The equation of a straight line is often written as

$$y = mx + c \quad \text{or} \quad y = b_0 + bx \tag{2.1}$$

which defines the line using the two parameters: the slope, m or b, and the intercept, c or b_0. The variable, x, is the *independent* variable and is considered to *predict* the value of y, the *dependent* variable. The intercept, c or b_0, is the point that the line crosses the y-axis when $x = 0$.

The form of the equation using the 'b' coefficients is useful for multiple regression (9.1.6) when y depends on several 'x' values, and we can extend the number of 'b' coefficients:

$$y = b_0 + b_A x_A + b_B x_B + \cdots \tag{2.2}$$

However, for simple regression we will normally use the more familiar m and c constants.

Statistics of linear regression: Excel analysis for Fig 2.2. Scan here to watch the video or find it via www.oxford textbooks.co.uk/orc/currell/

Case study: **Best-fit straight line / 2. Slope and intercept**

—continued from 2.Introduction, leading to 2.1.2, 2.1.4, and 2.4.1

Fig 2.1(a) presents the x–y data from Fig 2.2 in a scatterplot, together with a best-fit straight line, known as a *trendline* in Excel. The data in the Excel worksheet in Fig 2.2 has five measurements recorded in columns B (x-data) and C (y-data), with each data pair identified by the label, i, in column A.

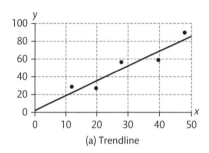

(a) Trendline

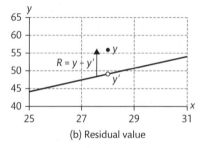

(b) Residual value

Fig 2.1 Best-fit straight line

	A	B	C	D	E	F	G
1	i	x	y	y'	R	R^2	
2	1	12	28	22.65	-5.35	28.58	
3	2	20	27	35.90	8.90	79.25	
4	3	28	56	49.15	-6.85	46.92	
5	4	40	59	69.02	10.02	100.45	
6	5	48	89	82.27	-6.73	45.28	
7		Pairs, n =	5		ΣR^2 =	300.48	
8		Slope, m =	1.656				
9		Intercept, c =	2.782				
10		> ANOVA (Analysis of Variance):					
11			df	SS	MS	F	p
12		Regression	1	2334.32	2334.32	23.31	0.017
13		Residual	3	300.48	100.16		
14		Total	4	2634.80			
15		> Correlation:					
16		Coefficient of determination, r^2 =		0.8860		t =	4.828
17		Correlation coefficient, r =		0.9413		p =	0.017
18		> Uncertainty:					
19		Standard error of regression, SE_{YX} =		10.008			
20		Standard error of slope, SE_{SLOPE} =		0.343			
21		Confidence deviation of slope, Cd_{SLOPE} =		1.092			
22							
23		Standard error of regression, SE_{YX} =		10.008			

Fig 2.2 Statistics of linear regression

The slope and intercept of the best-fit straight line can be calculated directly from the x, y values, and in Fig 2.2 it is convenient to use the specific functions of SLOPE() and INTERCEPT() in Excel to perform these calculations in cells C8 and C9 respectively:

Slope, m: [C8] = SLOPE(C2:C6,B2:B6) = 1.656

Intercept, c: [C9] = INTERCEPT(C2:C6,B2:B6) = 2.782

where C2:C6 describes the range of (y-) values between C2 and C6 and B2:B6 describes the range of (x-) values between B2 and B6.

In column D, we can calculate the y-values, y', *on the best-fit line* for each of the x-values, using the equation for the ith data point:

$$y'_i = mx_i + c_i \tag{2.3}$$

where the 'i' subscript can take values from one to five referring to the data identifier in column A.

For example, the calculation of the first data point, y'_1:

[D2] = C\$8 * B2 + C\$9 = 22.65

The \$ sign in front of the rows in C\$8 and C\$9 *locks* the row values for the slope and intercept when we copy the formula down for the other data points.

The best-fit straight line in Fig 2.1(a) passes through the y' values and can be drawn on the graph as a separate line, or directly in Excel using the *trendline* option.

The extent to which the data does not exactly fit the straight line is shown by the *residual* distances between the points and the line in Fig 2.1(b). The standard process of linear regression assumes that the **only uncertainty exists in the y-values**, with the x-values being known exactly. Hence the uncertainty for the ith data point is given by the *vertical* difference, the *residual*, R_i, between the measured y-value and the value, y'_i, on the best-fit line.

$$R_i = y_i - y'_i = y_i - (mx_i + c_i) \tag{2.4}$$

In cells E2 to E6 we calculate the residual for each point using Eqn 2.4, e.g.

[E2] = D2 − C2 = −5.35

In cells F2 to F6 we square each of the residuals, e.g.

[F2] = E2^2 = 28.58

Finally, in cell F7 we calculate the sum of squares of all the residuals:

$$SS_{RESID} = \sum_i R_i^2 \tag{2.5}$$

by using:

SS_{RESID}:   [F7] = SUM(F2:F6) = 300.48

The sum of squares of residuals is a measure of the error in the best-fit straight line and is sometimes referred to as SS_{RANDOM} or SS_{ERROR}. See also the 'analysis of variance' (3.2.2).

The best-fit straight line is defined as the line that will give the smallest possible value for SS_{RESID} and, for this reason, the process of deriving the best-fit line is often called the 'method of least squares, LS'.

The calculations of SLOPE() and INTERCEPT() in cells C8 and C9 provide the values for the best-fit straight line which has the minimum value of SS_{RESID}.

In 2.4.1 we use this case study to show that it is also possible to use a process of iteration, using the Excel add-in 'Solver', to adjust values of m and c through an *iterative* process until a *minimum* value is reached for SS_{RESID}, producing the same values for m and c as calculated in Fig 2.2. It is also useful to see that the alternative method of 'maximum likelihood estimation' (MLE) (2.4.2) will also derive the same values for m and c.

The calculation of the slope and intercept for the best-fit line has produced a mathematical *regression model* that represents our experimental data. We will see (in Section 3.4) the use of more complex mathematical models to represent other forms of experimental data.

Minitab and SPSS produce results for linear regression in similar formats (e.g. Fig 2.3) to that in Fig 2.2:

Minitab > Stat > Regression > Regression >
Fit Regression Model ... Response: y
Continuous predictors: x
→ Output: Fig 2.3

SPSS > Analyze > Regression > Linear...
Dependent: y
Independent(s): x
→ Output: Gives the same values as in Fig 2.3

In Fig 2.3, the slope, $m = 1.656$, is given by the *Coef* of 'x' and the intercept, $c = 2.78$, by the value of 'constant'. The confidence deviations, Cd, in the slope, m, and intercept, c, can be calculated as the respective values for *SE Coef* multiplied by the appropriates t-value for the degrees of freedom, $n - 2$ (see Eqn 2.16).

More advanced use of Minitab and SPSS for regression can be found in 3.4.3 and 9.1.6.

```
The regression equation is
y = 2.8 + 1.66 x

Predictor    Coef  SE Coef     T      P
Constant     2.78    11.10  0.25  0.818
x          1.6560   0.3430  4.83  0.017

S = 10.0080   R-Sq = 88.6%   R-Sq(adj) = 84.8%

Analysis of Variance

Source          DF      SS      MS      F      P
Regression       1  2334.3  2334.3  23.31  0.017
Residual Error   3   300.5   100.2
Total            4  2634.8
```

Fig 2.3 Extract from linear regression output using Minitab

2.1.2 ANOVA table

We can now extend the basic regression calculations to derive other relevant statistics that will relate to questions and topics elsewhere. These statistics are based on how much the observed data deviates from the regression model. We start with a set of statistics which is often referred to as an analysis of variance (or ANOVA) table (3.2.3). In this, the variability in the data is described by the sums of squares terms that we first met in Section 1.5.

Using Eqn 1.16, the total variability in the y-values can be described by the total *sum of the squares* of the deviations from the mean value, $\bar{y}$:

$$SS_{TOT} = \sum (y_i - \bar{y})^2 \tag{2.6}$$

We have also seen (Eqn 1.19) that the sample variance in the y-values is given by

$$s_y^2 = \frac{\sum (y_i - \bar{y})^2}{df} \tag{2.7}$$

where the degrees of freedom, $df = n - 1$.

Case study: **Best-fit straight line / 3. ANOVA table**

—continued from 2.1.1, leading to 2.1.3

Referring to the data in Fig 2.2, we develop the ANOVA results table in B10:G14.

Combining Eqns 2.6 and 2.7 we can derive

$$SS_{TOT} = s_y^2 \times df \tag{2.8}$$

which is calculated in D14 using:

SS_{TOT}: [D14] = VAR(C2:C6) * (C7 − 1) = 2634.80

The total variability, SS_{TOT}, in the data equals the variability described by the regression model, SS_{REG}, plus the remaining variability described by the residuals, SS_{RESID}:

$$SS_{TOT} = SS_{REG} + SS_{RESID} \tag{2.9}$$

We copy the value of SS_{RESID} from F7 to D13, and as we already know SS_{TOT} and SS_{RESID} we can calculate SS_{REG} in cell D12 by simple subtraction,

SS_{REG}: [D12] = D14 − D13 = 2334.32

Now that we have the three sums of squares we can calculate the *mean square values*, MS_{REG}, and MS_{RESID} in E12 and E13 using the general Eqn 1.18:

$$MS = \frac{SS}{df}$$

The degrees of freedom, df, for the y-value calculations can be understood as follows:

- SS_{TOT} has $df = n - 1$ because one degree of freedom has been used in the prior calculation of $\bar{y}$.

- SS_{RESID} has df $= n - 2$ because two degrees of freedom have been used in the prior calculation of slope and intercept.
- SS_{REG} then has the remaining degree of freedom, $df = 1$.

We can now use the F-statistic (3.2.1) to calculate the ratio of the variance explained by the regression model divided by the variance due to residual experimental variations.

$$F = \frac{MS_{REG}}{MS_{RESID}} \tag{2.10}$$

This value is calculated in F12:

F_{STAT}:   [F12] = E12 / E13 = 23.31

A large value for the F-statistic implies that the variation described by the straight line is more than just random variations, and that the best-fit straight line has a slope that is significantly different from zero. It is possible to calculate the p-value in cell G12 for the F-test using the Excel function F.DIST.RT() with the relevant degrees of freedom C12 and C13 for the numerator and denominator

p-value: [G12] = F.DIST.RT(F12, C12, C13) = 0.017

The p-value is the probability that you would be wrong if you stated that the slope of the best-fit line was not zero.

2.1.3 Correlation

In this section we derive the statistics of correlation, which is a measure of the extent to which the variation in the y-values can be *predicted* by the variation in the x-values (and vice versa).

The coefficient of determination, r^2, is a measure of how much the variation in the y-values is explained by the regression model, and is given by the fraction, from 0.0 to 1.0, of the sum of squares variation explained by the regression:

$$r^2 = \frac{SS_{REG}}{SS_{TOT}}$$

which can be rearranged to give:

$$r^2 = \frac{SS_{REG}}{SS_{TOT}} \Rightarrow \frac{SS_{TOT} - SS_{RESID}}{SS_{TOT}} \Rightarrow 1 - \frac{SS_{RESID}}{SS_{TOT}} \tag{2.11}$$

The square root of the coefficient of determination is, for the simple straight line, called the Pearson's product moment correlation coefficient, r:

$$r = \sqrt{1 - \frac{SS_{RESID}}{SS_{TOT}}} \tag{2.12}$$

We also calculate the correlation coefficient in a different way in 4.1.1, which provides a different approach to the same statistics.

Case study: Best-fit straight line / 4. Correlation

—continued from 2.1.1

Referring to the data in Fig 2.2, we develop the correlation results in B15:G17.

The coefficient of determination is calculated in E16 using Eqn 2.11:

Coefficient of determination, r^2: [E16] = 1 − D13 / D14 = 0.8860

Values of r^2 are often given when a statistical analysis fits a mathematical model to a set of data points, and are used as a measure of 'goodness of fit' of the model for the given data (4.4.1). For example, good calibration lines might expect to have values for r^2 of at least 0.999. It is also important to note the use of an 'adjusted r^2' which takes into account the degrees of freedom of the data in the calculation (2.1.5).

The magnitude of the correlation coefficient is calculated in E17 using Eqn 2.12:

Correlation coefficient, r: [E17] = SQRT(1 − D13 / D14) = 0.9413

The sign (+ or −) of the correlation coefficient is the same as the *sign* of the slope, but the magnitude is **not related** to the *magnitude* of the slope.

It is also possible to get this result directly in Excel by simply using the function CORREL() or PEARSON().

We can use the correlation coefficient, r, as the relevant statistic in a hypothesis test for *linear correlation*, i.e. testing that the best-fit straight line is not zero. We can either compare the value with tables of critical values or calculate the p-value. In calculating the p-value for the significance test for correlation we first calculate a value for an effective t-statistic for the correlation with $n − 2$ degrees of freedom:

$$t_S = r \times \sqrt{\frac{n-2}{1-r^2}} \qquad (2.13)$$

which we have calculated in G16:

t-statistic, t_S: [G16] = E17 * SQRT((C7 − 2) / (1 − E17^2)) = 4.828

The equivalent p-value is then calculated in G17 using the T.DIST.2T() function with $n − 2$ degrees of freedom:

p-value: [G17] = T.DIST.2T(G16, C7 − 2) = 0.017

We can see that the two p-value calculations in G12 and G17 give exactly the same value, which was to be expected as they both relate to a test for a non-zero best-fit slope.

As with the ANOVA calculations, the p-value is only of relevance when *testing* whether or not a non-zero linear relation might exist for the given data points. However, when we are using linear regression to produce a *calibration* line, we would be primarily concerned with *how close* the coefficient of determination, r^2, is to 1.00 (4.4.1).

Minitab and SPSS produce the same results in Fig 2.4 as above for the correlation coefficient and p-value:

Minitab > Stat > Basic Statistics >
Correlation...
Variables: yx
☑ **Display p-values**
→ Output: Gives the same values as in Fig 2.4

SPSS > Analyze > Correlate > Bivariate...
Variables: yx
☑ **Pearson**
→ Output: Fig 2.4

Correlations

		x	y
x	Pearson Correlation	1	.941*
	Sig. (2-tailed)		.017
	N	5	5
y	Pearson Correlation	.941*	1
	Sig. (2-tailed)	.017	
	N	5	5

*. Correlation is significant at the 0.05 level (2-tailed).

Fig 2.4 Pearson's correlation in SPSS

2.1.4 **Regression uncertainties**

When using statistics for deriving scientific results, it is also essential to derive the uncertainties in those results. The key statistic for the uncertainty in regression is given by the *standard error of regression*:

$$SE_{REG} = \sqrt{\frac{\sum R_i^2}{n-2}} = \sqrt{\frac{SS_{RESID}}{n-2}} \qquad (2.14)$$

This value is effectively a best estimate for the standard deviation of the *individual* data values around their 'true' values in the linear model.

> ### Case study: **Best-fit straight line / 5. Uncertainty in regression**
> —continued from 2.1.1, leading to 2.2.1
>
> Referring to the data in Fig 2.2, we develop the uncertainty results in B18:E21.

The standard error of regression can be calculated directly in E19 from the *x–y* data in columns B and C using the Excel function, STEYX():

Standard error of regression: [E19] = STEYX(C2:C6, B2:B6) = 10.008.

The standard error of the *slope* can then be calculated as:

$$SE_{SLOPE} = \frac{SE_{REG}}{\sqrt{\sum (x_i - \bar{x})^2}} \Rightarrow \frac{SE_{REG}}{\sqrt{s_x^2 \times (n-1)}} \qquad (2.15)$$

which is calculated in E20:

Standard error of slope: [E20] = E19 / (SQRT(VAR(B2:B6) * (C7 − 1))) = 0.343

The 95% confidence deviation (1.5.2) in slope can then be calculated by multiplying by the relevant *t*-value for 95% with degrees of freedom, $df = n - 2$:

$$Cd_{SLOPE,95\%} = t_{95\%,n-2} \times SE_{SLOPE} \qquad (2.16)$$

which is calculated in E21 using:

95% Confidence deviation in slope, Cd: [E21] = T.INV.2T(0.05, C7 − 2) * E20 = 1.092

This calculation gives the **95% confidence interval** for the slope (to 1 dp) as:

$m = 1.7 \pm 1.1$

This possible range for the slope does not include $m = 0$, showing that the slope is significantly different from zero, which is consistent with the correlation and ANOVA results of $p = 0.017$.

2.1.5 **Quality of fit**

An important use of linear regression is for the calibration of experimental measurements using *known* samples to enable the analysis of *unknown* samples.

Case study: **Spectrophotometer calibration / 3. Linearity range**

—continued from 7.1.5, leading to 2.2.1

The use of a calibration line in spectrophotometry is also introduced in 7.1.1 as *related* data between absorbance and concentration, and in 7.1.5 we use the results from Minitab and SPSS to estimate the confidence in interval of an unknown concentration. We consider here the *linearity* of calibration data and, in 2.2.1, the calculation of the *confidence interval* of an unknown value using Excel.

   The data in Fig 2.5 gives the calibration data for an ICP–OES (inductively coupled plasma optical emission spectrometer), with the values of the emission intensity, I, for solutions of standard concentrations (in arbitrary units). In the interest of space, all intensity values in this case study have been divided by 1,000 and rounded to one decimal place.

   The analytical region of interest for experimental measurements is for concentrations between $C = 0$ and $C = 40$. We also know from *previous experience* that measurements of I have a standard deviation uncertainty of about $\sigma = 0.7$.

   Three replicate measurements of an *unknown* solution give an average value, $I_S = 79.2$, and we wish to calculate the best-estimate value for the concentration, C_S, of this unknown solution. However, we first assess the quality of the calibration line here by checking its linearity using residuals, and then complete the analysis in 2.2.1.

C	I_1	I_2	I_3	s
0	0	0	0	0.00
10	22.9	22.3	23.2	0.46
20	45.3	46.7	46.4	0.74
30	69	69.5	67.8	0.87
40	91.2	90.9	92.5	0.85
50	114.9	111.2	112.8	1.86
60	131.5	129.9	132.8	1.45

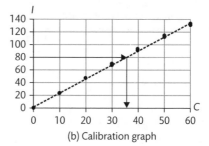

(a) Calibration data   (b) Calibration graph

Fig 2.5 Calibration data and graph for spectrophotometric calibration

We shall see in 2.2.1 that the value of $I_S = 79.2$ gives an equivalent value of $C_S = 34.6$, as shown by the arrows in the calibration graph in Fig 2.5(b). However, we are particularly concerned here with the 'quality of fit' of the calibration line.

Visually the data points in Fig 2.5(b) appear to sit on a straight line through the origin. However, we check linearity by plotting the *residuals* against C, using

Excel > Data Analysis > Regression
☑-Residual plots

to give Fig 2.6(a). Residual plots are also available from the regression options in Minitab and SPSS.

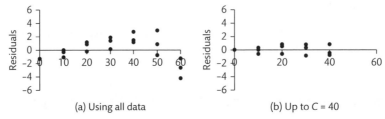

(a) Using all data   (b) Up to C = 40

Fig 2.6 Residuals around best-fit straight lines

In Fig 2.6(a) we see that there is a curvature of the line, with the upper points curving downwards. This is a common behaviour in an instrumental response which often shows decreasing sensitivity with higher concentration values.

As the principal range of interest is between 0 and 40, we investigate whether we should fit the calibration line just through these points, and we compare the statistics of the two options in Table 2.1.

Table 2.1 compares the statistics of fitting a straight line, (a) through all data points and (b) just through points up to $C = 40$. The important value is the coefficient of determination, r^2, with $r^2 = 0.9997$ for points just up to $C = 40$ which shows a better fit than for the whole range, $r^2 = 0.9986$. However, care should be taken when limiting the number of points for a best-fit line, as the fit always becomes 'better' as 'unhelpful' points are rejected. The *adjusted* r^2 value takes the change in the degrees of freedom into account, but still shows improvement for (b). Note that the p-values are both far less than 0.05 and

Table 2.1 Comparison of best-fit statistics for quality of fit

	Fit all data points:	Fit points up to C = 40:
Residual plots -	See Fig 2.6(a)	See Fig 2.6(b)
Coefficient of determination, r^2 =	0.9986	0.9997
Adjusted, r^2 =	0.9985	0.9997
Standard error of regression, SE_{REG} =	1.77	0.61
p-value for regression =	1.93×10^{-28}	3.20×10^{-24}

are of little practical value here as we are not testing whether or not there is a non-zero best-fit line.

The calculated standard error of regression, SE_{REG}, (2.1.4) is the best estimate of the experimental uncertainty in each data point, and the value of $SE_{REG} = 0.61$ for the lower 'straight line' section is consistent with our previous knowledge of the experimental standard deviation, $\sigma = 0.7$. The higher value of 1.77 for the whole data set includes the additional deviations due to the upper curvature. The analysis here has been simplified by assuming equal uncertainties for all values of I, but this does not affect the identification of the curvature in the data.

The statistical analysis of regression and the residuals indicates that it is appropriate to use the best-fit straight line just up to $C = 40$. This case study is continued in 2.2.1, where we calculate the confidence interval for the concentration of an unknown solution, for which three measurements give an average intensity value of $I_S = 79.2$.

2.2 Experimental uncertainties

Introduction

The standard error of regression, SE_{REG}, gives the best estimate of the *experimental uncertainty* around a best-fit straight line when based on the *regression data itself*, and in 2.1.4 we derived the confidence interval for the *slope* of the line. In 2.2.1 we derive the confidence interval in using a straight line for calibration and in 2.2.2 consider exact intercept values when *interpolating* or *extrapolating* the best-fit line.

In 2.2.3 we consider the experimental situation where the actual experimental uncertainty is already *known* from previous measurement, and finally in 2.2.4, we review the regression calculation in the situation where the experimental uncertainty is not the same for all points on the regression line.

The use of Minitab and SPSS for these calculations is introduced in 7.1.5.

2.2.1 Calibration uncertainty

We often use linear regression in science to provide a calibration line from which we calculate specific values using the simple straight line equation. We now introduce methods for calculating the uncertainty in those interpolated, or extrapolated, values.

Calibration uncertainty 1: Excel analysis for Fig 2.7. See also 7.1.5.
Scan here to watch the video or find it via www.oxford textbooks.co.uk/ orc/currell/

Case study: Best-fit straight line / 6. Confidence interval

—continued from 2.1.4, leading to 2.2.2

We consider the simple example of the calibration of a spectrophotometric measurement. In the Excel worksheet given in Fig 2.7, the absorbances, A (*y*-values), are entered in B4:B7 for four standard samples with known concentrations, C (*x*-values) in A4:A7. These values are also plotted, together with a 'best-fit' calibration line, in Fig 2.8 (a) and (b).

The aim of the analysis is to calculate the concentration, x_S, of an unknown solution for which three replicate measurements have recorded an average absorbance of $y_S = 0.63$.

	A	B	C	D	E	F	G	H	I	J
1	95% *CI* for the *x* value of an intercept on a free-fit (*m* and *c*) calibration line									
2	Data:			Calculations:		Uncertainties:			y_s with k	y_s
3	*x*	*y*		n =	4				replicates	exact
4	8.00	0.31		Slope, *m* =	0.037	Uncertainty, u_x =			0.746	0.496
5	13.00	0.46		Intercept, *c* =	0.010	u_x (central) =			0.737	0.482
6	18.00	0.71				* $t_{n-2,95\%}$ =			4.303	4.303
7	23.00	0.84		y_s =	0.63	Cd_x =			3.210	2.136
8				k =	3	Confidence interval (95%):				
9				x_s =	16.859	CI_x =	16.86	±	3.21	2.14
10				* SE_{REG} =	0.035					
11				Mean of *y* =	0.6	{NB: For 99% confidence, change 0.05				
12				Variance, s_x^2 =	41.67	to 0.01 in the *t*-value in I6 and/or J6}				

Fig 2.7 Confidence intervals for intercept on a 'free fit' calibration line (calculations for column J appear in 2.2.2)

We make the assumption (Beer–Lambert law) that, over our range of measurements, the relationship between *y* (A) and *x* (C) is linear.

A linear regression analysis now gives us two alternatives in calculating the slope, *m*, and intercept, *c*, of the calibration line:

1. Without any knowledge of the behaviour of the line beyond the range of measurement values we must allow a free fit for *both* slope and intercept, Fig 2.8(a), or,

2. Based on the science, we may be able to assume that the calibration line must pass through the origin of the graph, which then *only* allows uncertainty in the best-fit slope, Fig 2.8(b).

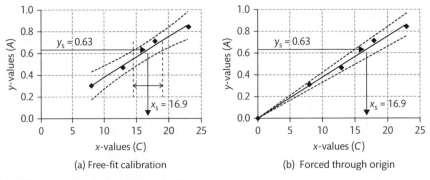

(a) Free-fit calibration   (b) Forced through origin

Fig 2.8 Fitting a calibration line to four absorbance values

In Fig 2.8 (a) and (b), the dashed lines in the diagrams show the ranges within which it is possible, given the variability of the data points, to be 95% confident of drawing the position of the 'best-fit' calibration line.

In a typical scenario, we might then make k replicate readings of the absorbance of an *unknown* solution giving an average or *mean* value of y_S. We can then derive a best-estimate value for the unknown concentration, x_S, by graphical construction as represented by the arrows in the graphs. The uncertainty, u_x, in x_S is partly generated by the range of possible x-values within the dashed lines at the y-value = y_S (shown in Fig 2.6(a)), *together with* possible uncertainty in the measured value of y_S itself.

We now develop the statistics, plus the equations in Excel, that can be used to calculate the uncertainty, u_x, in the best estimate value for x_S. In Fig 2.7 we calculate the slope, m, in cell E4 using the function SLOPE() for a free fit straight line or LINEST() if we wish the line to pass through the origin of the graph. The intercept, c, is calculated in E5 by the function INTERCEPT().

We enter the mean value, y_S, in E7, based on k measurements entered in E8, and then calculate the unknown concentration, x_S, by using the rearranged straight line equation:

$$x_S = \frac{y_S - c}{m} \qquad (2.17)$$

which is calculated in E9 as: [E9] = (E7 − E5) / E4 = 16.859.

Having calculated the best-estimate of the unknown, x_S, value, the next step is to calculate the estimate for the experimental uncertainty, u_x, in that value for a *free-fit straight line*.

$$u_x = \frac{SE_{yx}}{m} \times \sqrt{\left\{ \frac{1}{k} + \frac{1}{n} + \frac{(y_S - \bar{y})^2}{m^2 \times (n-1) \times s_x^2} \right\}} \qquad (2.18)$$

The standard uncertainty, u_x, derived in the Eqns 2.18 and 2.19, arises from two sources:

• Vertical uncertainty in the true value of y_S, based on k replicate measurements.
• Horizontal uncertainty in the interception of the calibration line.

In the derivation of the formula for u_x, we start with the standard error of regression, SE_{REG}, calculated in E10 with the function STEYX(), which gives the measure of uncertainty in the *vertical y*-direction.

The uncertainty in the *horizontal x*-direction is then given approximately by SE_{REG} / m, where m is the slope of the line. For example, for a slope, $m = 1$ at an angle of 45°, the uncertainties will be the same in both directions, but, if the slope, m, becomes *less*, the horizontal uncertainty will become *larger*.

The next factors in the equation relate to the number of replicate measurements, k, of the unknown sample and the number, n, of data points in the calibration line, which both reduce the uncertainty in proportion to their square roots (Eqn 1.21). It is assumed that the uncertainty in each y_S-value is the same as in each calibration value.

The last term in the square root of Eqn 2.18 is the factor that is responsible for the 'opening-up' of the uncertainty range at both ends of the calibration line in Fig 2.8(a).

This term becomes *zero* (and disappears) when y_S equals the mean y-value $(\bar{y})$ of the points used to generate the best-fit line, i.e. when the measured value falls in the 'centre' of the calibration values. The separate values within this term are calculated in the worksheet as follows:

$\bar{y}$ is the mean of all the y values: [E11] = AVERAGE(B4:B7) = 0.6

s_x^2 is the sample variance of x values: [E12] = VAR.S(A4:A7) = 41.67

The free-fit calculation for u_x in cell I4 using Eqn 2.18 is then given by:

[I4] = (E10 / E4) * SQRT(1/E8 + 1/E3 + (E7 − E11)^2 / (E4^2 * (E3 − 1) * E12)) = 0.746.

In a well-designed experiment, we would want the measured value, y_S, to fall close to the centre of the calibration line, i.e. close to the mean, $\bar{y}$, of the calibration y-values. In this case, the term $(y_S − \bar{y})$ is close to zero and can be ignored, giving an approximate equation, calculated in cell I5, for the *central* region of the free fit calibration line:

$$u_x \approx \frac{SE_{REG}}{m} \times \sqrt{\left\{\frac{1}{k} + \frac{1}{n}\right\}} \qquad (2.19)$$

which is calculated in I5 as:

[I5] = (E10 / E4) * SQRT(1/E8 + 1/E3) = 0.737

The scientific situation sometimes requires the best-fit straight line to pass through the origin of the graph, as in Fig 2.8(b). In this case, Eqn 2.18 for the uncertainty, u_x, would become, for a *line passing through the origin*:

$$u_x = \frac{SE_{yx}}{m} \times \sqrt{\left\{\frac{1}{k} + \frac{1}{n} + \frac{(y_s)^2}{m^2 \times (n-1) \times s_x^2}\right\}} \qquad (2.20)$$

The confidence deviation, Cd_x, is calculated in I7 using:

$$Cd_x = t_{n-2}\, u_x \qquad (2.21)$$

where the t-value, for degrees of freedom, $df = n − 2$, and with the appropriate level of confidence (typically 95% or 0.05 significance), is calculated in I6 using the equation:

[I6] = T.INV.2T(0.05, E3 − 2) = 4.303

The final result for x_S is the given as the confidence interval, CI_x

$$CI_x = x_S \pm Cd_x \qquad (2.22)$$

In the example given in Fig 2.7, the analysis shows that the 95% confidence interval using the *free-fit* uncertainty for the best estimate concentration is

$CI = 16.86 \pm 3.21$

which gives the range from 13.7 to 20.1, rounded to 1 dp.

Case study: **Spectrophotometer calibration / 4. Calibration result**

— continued from 2.1.5

We now wish to calculate the best-estimate value for the concentration, C_S, of an unknown solution, based on the calibration data for the spectrophotometer in Fig 2.5(a), where the known experimental uncertainty is given by $\sigma = 0.7$. Three replicate measurements of an unknown solution give an average value, $I_S = 79.2$.

We enter the data from Fig 2.5(a) into an Excel worksheet calculation similar to that in Fig 2.7, and the equation of the calibration line is calculated to be:

$$I = 2.29 \times C + 0.04$$

The confidence interval, *based solely on the calibration line data*, is then calculated to be:

$$CI = 34.56 \pm 0.39$$

However, we are also told that the experimental standard deviation is known to be $\sigma = 0.7$. If we replace the standard error of regression, SE_{REG}, in E10 with this value and substitute 1.96 for the *t*-value we would get a more reliable confidence interval which *includes the known experimental uncertainty* (2.2.3):

$$CI = 34.56 \pm 0.41$$

2.2.2 **Exact x/y intercepts**

The calculation in 2.2.1, and specifically Eqn 2.18, assumes that there is experimental uncertainty in the *y*-value, y_s, which is calculated as the mean of *k* measurements.

We now consider the situation where the *y*-value is known *exactly*, which removes the '$1/k$' uncertainty in the vertical direction, and we can remove the $1/k$ term from Eqn 2.18 to give:

$$u_x = \frac{SE_{yx}}{m} \times \sqrt{\left\{ \frac{1}{n} + \frac{(y_s - \bar{y})^2}{m^2 \times (n-1) \times s_x^2} \right\}} \qquad (2.23)$$

The calculations for this are performed in column J of the worksheet in Fig 2.7.

Exact x/y intercepts: Excel analysis for Figs 2.9(b) and 2.10(a). Scan here to watch the video or find it via www. oxfordtextbooks. co.uk/orc/ currell/

Case study: **Ink analysis / 3. Exact y-intercept**

—continued from 5.2.3, leading to 3.6.2 and 3.3.3

In this case study, continued from 5.2.3 and Fig 5.6, we now want to calculate the confidence intervals for the wavelengths at which each of the transmission spectra for three inks in Fig 2.9(a) cross the horizontal 50%T line.

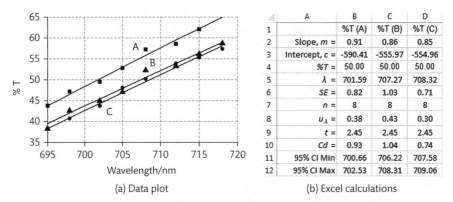

	A	B	C	D
1		%T (A)	%T (B)	%T (C)
2	Slope, m =	0.91	0.86	0.85
3	Intercept, c =	-590.41	-555.97	-554.96
4	%T =	50.00	50.00	50.00
5	λ =	701.59	707.27	708.32
6	SE =	0.82	1.03	0.71
7	n =	8	8	8
8	u_λ =	0.38	0.43	0.30
9	t =	2.45	2.45	2.45
10	Cd =	0.93	1.04	0.74
11	95% CI Min	700.66	706.22	707.58
12	95% CI Max	702.53	708.31	709.06

(a) Data plot   (b) Excel calculations

Fig 2.9 Confidence intervals for the long-wavelength transmission cut-off of black inks

Fig 2.9(b) shows the calculation steps for each of the three lines, calculating the best-estimate intercept wavelengths, λ, in row 5. The 50%T line represents an *exact* y-intercept, and we use Eqn 2.23 to calculate the uncertainty, u_λ, in row 8. The calculations of the confidence deviations in row 10 give the following results for the long wavelength cut-off, λ_{50}, for the three inks:

	A	B	C
95% confidence intervals:	701.6 ± 0.9	707.3 ± 1.0	708.3 ± 0.7

Line *A* is clearly different from *B* and *C*, but we would need to perform a two sample *t*-test (3.1.3) to decide whether the apparent difference between *B* and *C* is significant.

Further analyses:

In 3.3.3 we use an ANCOVA (ANalysis of COVAriance) to test for a *vertical* difference between *A*, *B*, and *C*.

In 3.6.2 we use repeated measures to test for *vertical* differences between *A*, *B*, and *C*.

Case study: **Best-fit straight line / 7. Standard additions**

—continued from 2.2.1

Using anodic stripping voltammetry, a solution containing an unknown quantity, q_S, of lead gives an analytical response, I_P, in Fig 2.10(a) which we plot on the y-axis in Fig 2.10(b). Additional known amounts, q, of lead are added and plotted on the x-axis, with the increasing analytical response on the y-axis.

We now extrapolate the best-fit straight line for the data back to its intercept with the horizontal q-axis, when $I_P = 0$. If we assume a linear response between I_P and q, then the intercept will occur at a value, $q = -q_S$, allowing us to calculate the value for the lead, q_S, in the original solution.

The method of standard additions is a particularly relevant example for two reasons:

- It is a further example of using an exact value of y, but, more importantly
- it demonstrates the increased uncertainty of extrapolating the calibration line beyond the range of calibration data.

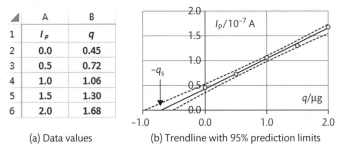

	A	B
	I_P	q
1		
2	0.0	0.45
3	0.5	0.72
4	1.0	1.06
5	1.5	1.30
6	2.0	1.68

(a) Data values

(b) Trendline with 95% prediction limits

Fig 2.10 Method of standard additions

As the I_P-value for the intercept is known exactly ($= 0$), we use Eqn 2.23 where there is no $1/k$ term in the equation.

In this analysis we are *extrapolating* beyond the range of the calibration data, and the intercept, y_S-value, is a long way from the mean value, $\bar{y}$, of the calibration values. This gives a large value for the $(y_S - \bar{y})^2$ term in Eqn 2.23, with a considerably increased uncertainty, which can be seen in the spreading 95% prediction limits drawn on either side of the trendline.

The calculation gives a confidence interval for the intercept when $I = 0$ as -0.71 ± 0.22, which then directly gives the confidence interval for q_S:

$q_S = 0.71 \pm 0.22$ μg.

The graph shows how an apparently reasonable set of calibration points can still result in a large uncertainty in the intercept with the axis. In the calculation we see that the confidence deviation when $I_P = 0$ is ± 0.22 whereas in the *central* region of the calibration data the confidence deviation would be only ± 0.08. This increased uncertainty in the final result is due to the extrapolation of the line beyond the data points, in which the second term in the square root now becomes the major factor. The most effective way of reducing this uncertainty is by increasing the variance, s_x^2, of the values along the x-axis, i.e. by making measurements over a larger range of x-values (provided that we are confident that the response remains linear).

2.2.3 Known uncertainty

In the regression analysis, developed in Section 2.1, the random uncertainty in experimental values is estimated *solely* from the sample data itself on the basis of how much the data deviates from the best-fit straight line, and it is recorded as the standard error or regression, SE_{REG}.

We now consider the situation where the experimental uncertainty for the measurement is *already known*, either from previous experience or from an estimation based on

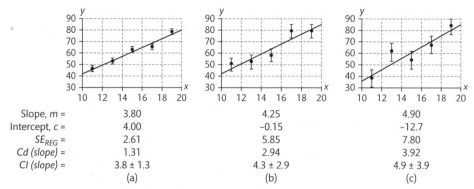

	(a)	(b)	(c)
Slope, m =	3.80	4.25	4.90
Intercept, c =	4.00	−0.15	−12.7
SE_{REG} =	2.61	5.85	7.80
Cd (slope) =	1.31	2.94	3.92
CI (slope) =	3.8 ± 1.3	4.3 ± 2.9	4.9 ± 3.9

Fig 2.11 Randomly selected sample data

the practical steps in the experimental process. In statistical terms we can now assume that population standard deviation, σ, *is a known value.* We can compare this with the situation discussed in 1.5.4.

The problem of relying *solely on sample data* is illustrated in Fig 2.11 by randomly selecting three sets of five values from a population of simulated data where the true y-values are given by

$$y = 4.0 \times x$$

and where the known standard deviation uncertainty in the y-values is given by $\sigma = 5.0$. This relatively large uncertainty has been used to improve the visualization of the differences.

The three sets of results, (a), (b), and (c), in Fig 2.11 are representative, but not extreme, examples of randomly selected experimental data sets. The lengths of the error bars drawn on either side of the data points are equal to the standard error, SE_{REG}, calculated from the data itself and represent the experimental uncertainty estimated *solely* from the five data points.

We see that, if the points happen, *by chance,* to lie in a near straight line, then the resultant residual values are small and the calculation interprets this as low overall uncertainty, e.g. in graph (a), $SE_{REG} = 2.61$. However, where the random points are more spread out away from a straight line, then this is interpreted as a large uncertainty, e.g. in graph (c), $SE_{REG} = 7.80$.

Relying on the random spread of just five data points is clearly not an accurate method of estimating experimental uncertainty, particularly when the slope and intercept of the best-fit line adjusts to minimize these deviations.

If you know *beforehand* the true standard deviation, σ, of the experimental uncertainty then you can compare your value of σ with the calculated value, SE_{REG}. If SE_{REG} appears to be much smaller than a known uncertainty, σ, then you should check that the separate measurements are truly random, but if it is much larger, then you should check for additional variations in the data (e.g. nonlinearity in 2.1.5).

If you are convinced that your value of σ is a valid description of the experimental uncertainty, then you can replace SE_{REG} with σ in the regression calculations, and Eqn 2.15 for the standard error of the slope of the straight line now becomes:

$$SE_{SLOPE} = \frac{\sigma}{\sqrt{\sum(x_i - \bar{x})^2}} \Rightarrow \frac{\sigma}{\sqrt{s_x^2 \times (n-1)}} \tag{2.24}$$

In the above example, the variance of the x-values (11, 13, 15, 17, 19) is $s_x^2 = 10.0$, which gives

$$SE_{SLOPE} = \frac{5.0}{\sqrt{10 \times 4}} = 0.7906$$

As we are now using the 'population' value for standard error rather than an estimated value, we can use the z-value (1.5.4) of 1.96 *instead of the t-value for (n – 2) degrees of freedom,* which further improves the precision of the result.

Confidence deviation of slope for a *known* standard deviation is:

$$Cd_{SLOPE,95\%} = 1.96 \times SE_{SLOPE}$$

which gives $Cd_{SLOPE,95\%} = 1.55$ for our example data.

The confidence intervals, CI, for the slopes estimated from the three replicate sets of data will now be:

A: 3.8 ± 1.6   B: 4.3 ± 1.6   C: 4.9 ± 1.6

The use of prior knowledge of experimental uncertainty provides a better overall *scientific* result.

2.2.4 Weighting uncertainties

In the regression analyses performed in previous units, we have assumed that all the data points have the same random uncertainty, i.e. they are all equally weighted in importance. However, it is not unusual to find that the uncertainty varies between data points.

It is possible to take account of these variations by 'weighting' the importance of each data value. If we can represent the uncertainty in a value by its standard deviation, u, (and variance, u^2) then the weighting factor, w, for each point is given by the proportionality relationship:

$$w \propto \frac{1}{u^2} \tag{2.25}$$

Weighting data: Excel analysis for Fig 2.12. Scan here to watch the video or find it via www.oxfordtextbooks.co.uk/orc/currell/

Case study: **Exponential decay / 2. Weighted linearization**

—continued from 2.Introduction and 2.3.4, leading to 2.4.3, 3.4.7, and 7.2.3

Fig 2.12 (same data as Fig 2.16) gives radioactive counts, N_t, in column B for times, t, in column A. We analyse this data in a variety of ways throughout the book, and in this analysis we wish to take into account the varying uncertainties in the data when we linearize the relationship (see 2.3.4) by plotting the logarithm, $\ln(N_t)$, against t, to obtain a straight line.

	A	B	C	D	E	F
1	t	N	$\ln(N)$	$\sqrt{N}$	u	$w = 1/u^2$
2	0	97	4.57	9.85	0.10	96.33
3	1	63	4.14	7.94	0.13	62.33
4	2	36	3.58	6.00	0.17	35.33
5	3	25	3.22	5.00	0.20	24.33
6	4	15	2.71	3.87	0.26	14.33
7	5	6	1.79	2.45	0.43	5.32

Fig 2.12 Weighting exponential decay values

N_t is expected to follow Eqn 2.34:

$$N_t = N_0 \times \exp(-0.693 \times t/T_{1/2})$$

where $T_{1/2}$ is the half-life and N_0 is the count rate measured when $t = 0$.
We see in 2.3.4 that we can *linearize* this equation by taking natural logarithms to get:

$$\ln(N_t) = \ln(N_0) - (0.693/T_{1/2}) \times t$$

such that the slope of a best-fit straight line of $\ln(N_t)$ against t will be $m = -0.693 / T_{1/2}$.
The values of $\ln(N_t)$ against t are plotted in Fig 2.13(a).

Initially we will assume that the uncertainties in the $\ln(N_t)$ values are all the same. If we now use the techniques of 2.1.4 to perform a linear regression on these points by calculation (or possibly a best-fit by eye), we get a 95% confidence range for the slopes of possible best-fit lines as

Slope assuming *equal uncertainties*: $m = -0.53 \pm 0.11$.

This would translate approximately into a 95% confidence interval for the half-life, $T_{1/2} \approx 1.31 \pm 0.28$.

However, the simplistic assumption that there is the same uncertainty in each point is far from true in this case, and we will now investigate how the varying uncertainty can affect the results.

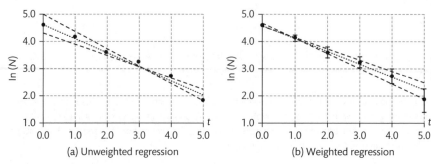

(a) Unweighted regression   (b) Weighted regression

Fig 2.13 Taking uncertainties into account

We have two factors affecting the final uncertainties:

- The uncertainty in a radioactivity count of N_t is equal to $\sqrt{N_t}$ (Eqn 1.15).
- The act of taking the log of N_t will further transform the relative uncertainties of different values.

In order to derive the uncertainty estimates, u_t, at each time, we first calculate the uncertainties, $\sqrt{N_t}$, in column D for each value of N_t. e.g.

$$[D2] = \text{SQRT(B2)} = 9.85.$$

We now need to estimate how the uncertainty $\sqrt{N_t}$ translates into the uncertainty, u_t, when we take the value of $\ln(N_t)$. A simple way of approximating this in Excel is to take logs of the *extreme values* $N_i + \sqrt{N_i}$ and $N_i - \sqrt{N_i}$ and use half the difference:

$$u_t = 0.5 \times \{\ln(N_t + \sqrt{N_t}) - \ln(N_t - \sqrt{N_t})\}$$

which we calculate in column E, e.g.

$$[E2] = 0.5 * (\text{LN}(B2+D2) - \text{LN}(B2-D2)) = 0.10.$$

The data values together with u_t plotted as error bars are shown in Fig 2.13(b). We can see that the process of linearization has dramatically increased the uncertainty in points with *lower* values of N.

We also see that the point at $t = 0$ now has relatively little uncertainty which constrains the possible spread of the best-fit lines as they pass through this point.

The simplest method of deriving *weightings* for the experimental results is to estimate the relative weighting, w, in column F using Eqn 2.25 and the uncertainties in column E, e.g.

$$[F2] = 1/E2^\wedge 2.$$

Using the weighted linear regression in SPSS or Minitab:

Minitab > Stat > Regression > Regression >
Fit Regression Model...
Response: *LnN*
Continuous predictors: *t*
> Options...: Weights: *w*
☑ Output: Gives the same values as in Fig 2.14

SPSS > Analyze > Regression > Linear...
Dependent: *LnN*
Independent(s): *t*
WLS Weights: *w*
☑ Output: Fig 2.14

Coefficients[a,b]

Model		Unstandardized Coefficients B	Std. Error	Standardized Coefficients Beta	t	Sig.
1	(Constant)	4.591	.047		98.151	.000
	t	-.487	.026	-.994	-18.894	.000

a. Dependent Variable: LnN
b. Weighted Least Squares Regression - Weighted by w

Fig 2.14 Weighted least squares regression using SPSS

The intercept in Fig 2.14 gives $\ln(N_0) = 4.591$ from which we can derive $N_0 = 98.6$.
The slope of -0.487 with a standard error of 0.026 multiplied by the t-value of 2.78, gives the 95% confidence range for the best-fit slope:

Slope using data weighting, $m = -0.49 \pm 0.07$

This would translate approximately into a 95% confidence interval for the half-life, $T_{1/2} \approx 1.42 \pm 0.21$.

In Fig 2.13 (a) and (b), we see that the data point at $t = 5.0$ causes a significant difference between the two calculations. In the *un-weighted* regression, it exerts more influence than its uncertainty should allow and forces the best-fit line into a steeper slope. Also, the reduced relative uncertainty of the point $t = 0$ in the *weighted* regression restricts the range of possible slopes that can be drawn, resulting in a more precise estimation of the half-life.

2.3 Linearization techniques

We have introduced, in 2.1 and 2.2, the statistical tools that are available to analyse data that can be expected to have a *linear* relationship. However, if we now have *nonlinear* data, then we have two broad options:

- Use a specific analytical technique to fit a nonlinear model (e.g. a polynomial) that reflects the expected system behaviour. These techniques are introduced in Section 7.2.
- Use a mathematical transformation to *linearize* the data so that it can be treated as a straight line, and then use the familiar straight line techniques.

These *linearization* methods use two main approaches:

- Change of variable.
- Taking logarithms of both sides of an equation to handle exponential and power equations.

We start with the basic 'change of variable' technique in 2.3.1, and then review some of the key properties of logarithms and exponential relationships in 2.3.2 and 2.3.3, before developing their use in the linearization of exponential equations in 2.3.4 and power relationships in 2.3.5. Finally, we introduce combined examples in 2.3.6, with a final warning about the effect of linearization on the uncertainty values of the different data points.

2.3.1 Change of variable

If we have a nonlinear relationship between variables p and q, then we can seek to rearrange the equation into the form of a straight line equation:

$$f(p) = m \times f(q) + c$$
$$\downarrow \qquad \downarrow$$
$$y = m \times x + c$$

where $f(p)$ and $f(q)$ could be nonlinear functions of p and q respectively.

If we plot the *values* of $f(p)$ on the y-axis and the *values* $f(q)$ on the x-axis, then we can measure the values of m and c in the equation from the slope and intercept of the best-fit straight line.

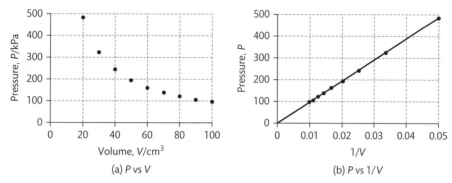

(a) *P* vs *V*   (b) *P* vs 1/*V*

Fig 2.15 Linearization of *P* vs 1/*V*

We can demonstrate this technique using the simple example of the ideal gas equation, where the pressure, P, volume, V, and absolute temperature, T, of n moles of an 'ideal' gas are given by:

$$P = n\frac{RT}{V}$$

and R is the gas constant $= 8.31\ \mathrm{JK^{-1}mol^{-1}}$.

Plotting the values of P and V for a fixed quantity of gas at a constant temperature, $T = 293\mathrm{K}$, gives the nonlinear curve in Fig 2.15(a). We can use this data to calculate the quantity of gas in moles.

We can rearrange the equation to give

$$P = nRT\left(\frac{1}{V}\right) + 0$$

$$\downarrow \quad \uparrow \quad \downarrow \quad \uparrow$$
$$y = m \times x + c$$

which is in the form of a straight line, *provided* that we plot P on the y-axis and the values of the function, $1/V$, on the x-axis. In this case, the slope and intercept of the resulting straight line in Fig 2.15(b) are given by:

Slope: $m = nRT$
Intercept: $c = 0$

Using the values from Fig 2.15(b), the calculated slope is $m = 9.74 \times 10^3$, with pressure measured in kPa and the volume in $\mathrm{cm^3}$. Converting the units of pressure to Pascals, Pa, and $1/V$ from $\mathrm{cm^{-3}}$ to $\mathrm{m^{-3}}$, the slope becomes

$$m = 9.74 \times 10^3 \times 10^3 \times 10^{-6} = 9.74$$

which allows us to calculate the number of moles, n, of the gas:

$$n = m / (RT) = 9.74 / (8.31 \times 273) = 0.004 \text{ mol}$$

Note that it is important to be careful to use the correct *units* when interpreting the results.

2.3.2 Using logarithms

Logarithms can be defined as the *inverse operation* of a power. If y is equal to b raised to a power, x, then, reversing the process, x is equal to the logarithm to base b of y, which can be written as:

$$\text{If } y = b^x \quad \text{then} \quad x = \log_b(y) \tag{2.26}$$

Although it is possible to use any value for the base, b, we generally use either logs to base 10 or logs to base e, where e is Euler's constant $= 2.71828\ldots$.

Logs to base 10 are convenient in that they relate to the standard decimal system with powers of 10.

- $\log_{10}(x)$ is normally abbreviated as $\log(x)$.

Logs to base e are also used extensively because e has a unique property in rate of change equations. The rate of change of e^x with x simply equals e^x: $\frac{d}{dx}(e^x) = e^x$.

These are also often referred to as natural or Naperian logarithms.

- $\log_e(x)$ is normally abbreviated as $\ln(x)$. Note that this is $\ln(x)$ *not* in(x).

The inverse definition of logarithms leads to the important relationships:

$$\ln(e) = 1 \quad \text{and} \quad \log(10) = 1 \tag{2.27}$$

A key effect of taking logs of an equation is that it **moves any power** (e.g. B in the equation below) **onto the equation line**:

$$\ln(A^B) = B \times \ln(A) \quad \text{and} \quad \log(A^B) = B \times \log(A) \tag{2.28}$$

which gives important relationships when there are powers of e or 10:

$$\ln(e^B) = B \times \ln(e) = B \quad \text{and} \quad \log(10^B) = B \times \log(10) = B \tag{2.29}$$

Another important property of logs in our calculations is that 'the logarithm of a *product* becomes the *sum* of the individual logarithms':

$$\ln(A \times B) = \ln(A) + \ln(B) \quad \text{and} \quad \log(A \times B) = \log(A) + \log(B) \tag{2.30}$$

and 'the logarithm of a *ratio* is the *difference* between the logarithms of the numerator and denominator':

$$\ln(A/B) = \ln(A) - \ln(B) \quad \text{and} \quad \log(A/B) = \log(A) - \log(B) \tag{2.31}$$

In respect of linearizing equations, we will investigate the effect of taking logarithms of two main types of equation:

Exponential growth and decay with a *variable* within a power (see 2.3.4):

$$N_t = N_0 \times e^{kt}$$

Equations with an unknown *constant* power (see 2.3.5):

$$E = A \times T^B$$

where B is a constant with an unknown value.

It would also be possible to linearize the gas laws equation, $P = nRT/V$, by taking logarithms of both sides of the equation, giving

$$\ln(P) = \ln(nRT) - \ln(V)$$

and then use the intercept of the straight line, with slope of -1, to calculate the value of nRT.

2.3.3 Exponential relationships

Exponential relationships occur in many branches of science, and their treatment has developed in different ways. However, it is possible to interpret their behaviour using the general equation:

$$N_t = N_0 \times e^{kt} \tag{2.32}$$

For example, the elimination of a drug of concentration, C, from the body in a time, t, can be described in pharmokinetics by

$$C_t = C_0 \times e^{-Kt}$$

where $K (= -k$ above) is the elimination constant.

We see below how a number of context-specific exponential equations are related to Eqn 2.32 by using specific expressions for k to fit the context of the problem.

Generation time, T_G, (or doubling time) is the time a population takes to double in number:

$$N_t = N_0 \times 2^{\left(t/T_G\right)}$$

For example, when $t = T_G$, the above equation gives $N_t = 2 \times N_0$.
As we know that $2.0 = e^{0.693}$ (because $\ln(2) = 0.693...$), we can substitute for '2' in the above equation to derive:

$$N_t = N_0 \times e^{0.693\left(t/T_G\right)} \tag{2.33}$$

which matches the *general* equation, with

$$k = 0.693 /T_G \quad \text{and} \quad T_G = 0.693 /k$$

Radioactivity half-life, $T_{1/2}$, is the time during which the radioactivity falls to one half of its value:

$$A_t = A_0 \times 0.5^{\left(t/T_{1/2}\right)}$$

For example, when $t = T_{1/2}$, the above equation gives $A_t = 0.5 \times A_0$.

As we know that $0.5 = e^{-0.693}$ (because $\ln(0.5) = -0.693...$), we can substitute for '0.5' in the above equation to derive:

$$A_t = A_0 \times e^{-0.693\left(t/T_{1/2}\right)}$$

(2.34)

which matches the *general* equation, with

$$k = -0.693/T_{1/2} \quad \text{and} \quad T_{1/2} = -0.693/k$$

Decimal reduction time, T_D, is the time a population takes to fall to 10% of the initial value:

$$N_t = N_0 \times 10^{-\left(t/T_D\right)}$$

For example, when $t = T_D$, the above equation gives $N_t = N_0/10$.
As we know that $10 = e^{2.30}$ (because $\ln(10) = 2.30...$), we can substitute for '10' in the above equation to derive:

$$N_t = N_0 \times e^{-2.30\left(t/T_D\right)}$$

(2.35)

which matches the general equation, with

$$k = -2.30/T_D \quad \text{and} \quad T_D = -2.30/k$$

Time constant, τ, is the time during which a value falls to $1/e = 36.8\%$ of its initial value:

$$V_t = V_0 \times e^{-\left(t/\tau\right)}$$

(2.36)

which matches the general equation, with

$$k = -1/\tau \quad \text{and} \quad \tau = -1/k.$$

2.3.4 Linearizing the exponential

We can now linearize the general exponential equation and derive the implications for the different *context* equations. Starting with Eqn 2.32:

$$N_t = N_0 \times e^{kt}$$

we choose to take natural logs (to base e) of both sides because the base in this equation is e.

Taking natural logs of both sides:

$$\ln(N_t) = \ln(N_0 \times e^{kT})$$

We then use the properties of the logarithms, from Eqns 2.30 and 2.28, to develop the right-hand side of the equation.

Log of a product is the sum of the logs:

$$\ln(N_t) = \ln(N_0 \times e^{kt}) = \ln(N_0) + \ln(e^{kt})$$

Eqn 2.28 gives $\ln(e^{kT}) = kT$:

$$\ln(N_t) = \ln(N_0) + \ln(e^{kt}) = \ln(N_0) + kt$$

With some rearrangement, we can write:

$$\ln(N_t) = k \times t + \ln(N_0)$$
$$\downarrow \quad \downarrow \downarrow \quad \downarrow$$
$$y = m \times x + c$$

(2.37)

which compares directly with the straight line equation, provided that we plot values of $\ln(N_t)$ on the y-axis and values of t on the x-axis.

We would then expect that the slope and intercept of this plot would be given by:

Slope: $m = k$

Intercept: $c = \ln(N_0)$ from which we can 'reverse' the logarithm to give, $N_0 = e^c$.

We can now interpret the slope of the equation using the values of k from the different scientific contexts discussed above.

Generation time, $T_G = 0.693 / m$

Radioactivity half-life, $T_{1/2} = -0.693 / m$

Decimal reduction time, $T_D, = -2.30 / m$

Time constant, $\tau, = -1 / m$

Case study: **Exponential decay / 3. Linearizing the exponential**

—continued from 2.Introduction, leading to 2.2.4, 2.4.3, 3.4.7, and 7.2.3.

Fig 2.16 gives radioactive counts, N_t, in column B for times, t, in column A. We aim to calculate the half-life of the radioactive decay using a linearization technique.

	A	B	C	D	E	F
1	*t*	*N*	ln(*N*)		Slope, *m* =	-0.531
2	0	97	4.57		Intercept, *c* =	4.664
3	1	63	4.14			
4	2	36	3.58		$T_{1/2}$ =	1.305
5	3	25	3.22		N_0 =	106.1
6	4	15	2.71			
7	5	6	1.79			

Fig 2.16 Radioactive decay data

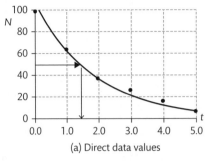
(a) Direct data values

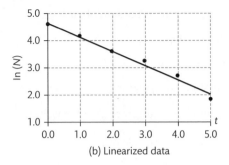
(b) Linearized data

Fig 2.17 Linearization of radioactive decay

The exponential radioactive decay of N_t with time, t, is recorded in Fig 2.17(a), and, by eye, we could *estimate* that the time taken for the radiation to fall to one half of its initial value is about 1.4 s.

In the process of linearization, we first calculate the natural log, $\ln(N_t)$, of each value of N_t in column C, e.g.

$$[C2] = LN(B2)$$

If we then plot $\ln(N_t)$ against t in Fig 2.17(b), we see that we have an approximately linear relationship, which we can analyse by calculating the slope and intercept of the best-fit straight line:

Slope, m: [F1] = SLOPE(C2:C7, A2:A7) = −0.531

Intercept, c: [F2] = INTERCEPT(C2:C7, A2:A7) = 4.664

The half-life, $T_{1/2}$, is then related to m by the equation (calculated in F4):

$T_{1/2} = -0.693 / m$: [F4] = −0.693 / F1 = 1.31

The value of N_0 can be derived from the equation $c = \ln(N_0)$ from which we can calculate:

$N_0 = e^c$: [F5] = EXP(F2) = 106.1

In this calculation, we obtained the values $N_0 = 106.1$ and $m = -0.531$, having assumed that the

• random uncertainties follow a normal distribution, and
• the uncertainties are the same for each *transformed* value, $\ln(N)$.

However, neither is true because the inherent uncertainty in N is proportional to $\sqrt{N}$ based on a Poisson distribution for random events and the process of taking the logarithm will also result in different absolute uncertainties. In 2.2.4 we use a process of data *weighting* to allow for the varying uncertainty, giving:

$N_0 = 98.6$ and $m = -0.487$

In 2.4.3 we use a process of iterative analysis to develop other possible models of analysis for nonlinear regression, and these methods are repeated using the generalized linear model in 3.4.7.

A normal distribution that assumes equal variance for the *untransformed* data points, gives

$N_0 = 97.9$ and $m(k) = -0.476$

Assuming *correctly* that the uncertainties in distribution and magnitude are given by the Poisson distribution gives

$N_0 = 99.4$ and $m(k) = -0.494$

2.3.5 Unknown power

As an example of an equation with an unknown constant power, we use a simple power equation relating variables E and T with constants A and B:

$$E = A \times T^B$$

where B is a power with an unknown value.

Again we take logs of both sides. In this case there is no preference for using either ln or log.

$$\ln(E) = \ln(A) + B \ln(T)$$

which can be rearranged to compare this directly with the straight line equation:

$$\ln(E) = B \times \ln(T) + \ln(A)$$
$$\searrow \quad \downarrow \quad \downarrow \quad \nearrow \qquad\qquad (2.38)$$
$$y = m \times x + c$$

We can see that the equation will act as a straight line provided that we treat $\ln(E)$ as the y-variable and $\ln(T)$ as the x-variable.

In this case, we would expect that the slope and intercept of this plot would be given by:

Slope: $m = B$

Intercept: $c = \ln(A)$, giving $A = e^c$

See 7.2.4 for an example of this type of calculation using the 'Rowing' case study.

2.3.6 Combined linearization

Some examples require combinations of the above processes. For example, the calculation of the activation energy, E, in the Arrhenius equation

$$k = A\exp\left(-\frac{E}{RT}\right)$$

(where k and T are the variables, R is the gas constant, A is an unknown constant) would require a plot of $\ln(k)$ against $(1/T)$, based on the rearranged equation

$$\ln(k) = \ln(A) - \left(\frac{E}{R}\right) \times \left(\frac{1}{T}\right)$$

with the slope of linear regression, $m = -E/R$.

For the Michaelis–Menten equation, which gives the initial velocity of an enzyme reaction, v, as a function of the substrate concentration, S,

$$v = \frac{v_{max}S}{K_M + S}$$

(where K_M is the Michaelis–Menten constant and v_{max} would be the maximum reaction velocity for large values of S) it is possible to rearrange the equation to give:

$$\frac{1}{v} = \left(\frac{K_M}{v_{max}}\right) \times \frac{1}{S} + \frac{1}{v_{max}}$$

In this case, a plot of $1/v$ against $1/S$ gives $1/v_{max}$ as the intercept, and then, using the slope, (K_M/v_{max}), it is possible to calculate the value of K_M.

2.3.7 Error warning

When using a linearization technique for nonlinear data, the transformation process will also act on the errors/uncertainties in the data points, and this can have the effect of distorting the *importance* of some of the points in the final regression process, resulting in slope and intercept errors. For example, refer to the linearization of an exponential decay using *weighting* in 2.2.4 and an *iterative* analysis in 2.4.3.

2.4 Iteration using Solver

Solver is an add-in in Excel which uses a process of *iteration* to fit a mathematical model to the supplied data within defined constraints. If, for example, we wish to calculate the *coefficients*, m and c, of a best-fit straight line, $y = mx + c$, the iteration:

- starts with initial *guesses* for the values for these coefficients
- calculates the value of an overall *test statistic* (e.g. sum of residuals) that is to be used to measure the goodness of fit
- uses specific algorithms *step-by-step* to change the coefficients in directions that should improve the goodness of fit
- repeats the steps until the test statistic gets *very close to* a desired (e.g. minimum) value.

Although this is not an exact *calculation* of the coefficients, the repetitive power and accuracy of modern computing can produce very accurate *estimated* coefficients.

In 2.4.1 we demonstrate this use of Solver for linear regression using the sum of the squares of the residuals (2.1.1) as the test statistic to be minimized, and in 2.4.2 we demonstrate the use of maximum likelihood estimation, MLE, for the same data. In 2.4.3 we use the flexibility of Solver to investigate how the underlying experimental uncertainty can be modelled within nonlinear regression.

2.4.1 **Operation of Solver**

We use the example of linear regression to demonstrate the operation of Solver through the identification of a best-fit straight line by minimizing the sum of squares of the residuals.

Case study: **Best-fit straight line / 8. Least squares fit using Solver**

−continued from 2.1.1, leading to 2.4.2

Fig 2.18 reproduces data from Fig 2.2, and presents the x–y values for five measurements recorded in columns B (x-data) and C (y-data), with each data pair identified by the label, i, in column A. We wish to calculate the coefficients of the best-fit straight line using the method of 'least squares'.

Using Solver: Excel analysis for Fig 2.18. Scan here to watch the video or find it via www.oxford textbooks.co. uk/orc/currell/

	A	B	C	D	E	F	G	H
1	i	x	y	y'	R	R^2	MLE	Probability
2	1	12	28	0.00	-28.00	784.00		0.00507
3	2	20	27	0.00	-27.00	729.00		0.00509
4	3	28	56	0.00	-56.00	3136.00		0.00407
5	4	40	59	0.00	-59.00	3481.00		0.00394
6	5	48	89	0.00	-89.00	7921.00		0.00260
7		Pairs, n =	5		SE_{RESID}, ΣR^2 =	16051.00	Combined:	1.08E-12
8		Slope, m =	0.000		SE_{REG} =	73.15		×10⁶
9		Intercept, c =	0.000				Objective:	1.08E-06

Fig 2.18 Starting values for a least squares fit using Solver (calculations in column H are developed in 2.4.2)

The calculations of residuals, R, and the sum of squares of residuals, $SS_{RESID} = \Sigma R^2$ in Fig 2.18 are explained in 2.1.1. In this calculation, the *starting* values for the slope, m, and intercept, c, of a best-fit straight line are entered into cells C8 and C9, with example values of 0.0 for both. Using these values, the predicted values for y' are calculated in column D, from which the values of R, R^2, and $SS_{RESID} = 16,051$ are then calculated.

The process of linear regression seeks to arrive at values of m and c that give the lowest possible value for SS_{RESID}. In 2.1.1 we used the functions, SLOPE and INTERCEPT, to calculate these values *directly*, obtaining values $m = 1.656$ and $c = 2.782$, respectively. We now demonstrate the use of Solver to arrive at the same values through the process of *iteration*.

Solver Parameters	✕
Se̲t Objective:	F7
To: ○ Ma̲x ◉ Mi̲n ○ V̲alue Of:	0
By Changing Variable Cells:	
$CS8:$CS9	

Fig 2.19 Section of the Solver dialogue window in Excel

In the Solver dialogue window in Fig 2.19, we enter F7 containing SS_{RESID} as the 'objective', and check the ⊙ **Min** option that directs the algorithm to find a *minimum* value for the 'objective' value. We then identify the cells C8 and C9 as the values of m and c that can be changed to arrive at this minimum. Solver then goes through an iterative process of changing C8 and C9, until it settles on values where any further change will begin to *increase* F7. In this example, Solver arrives at the values $m = 1.656$, $c = 2.782$, and $SS_{RESID} = 300.48$, which agree with the direct calculations in 2.1.1.

There are situations where Solver fails to end with the required answer. For complex mathematical models, it is possible that there are 'local' minima in the values of the 'objective' variable. This is similar to local dips in the side of a hill as it slopes down towards the bottom of a valley, and rain can be trapped in the local dips and fail to reach the very lowest point. It is also possible that there is a local maximum between the starting point and the true minimum in another 'valley'. If there is a possibility of such local problems, then it can be useful to try starting the iteration from a different point, e.g. with different values of C8 and C9. It is also important to check that the 'objective' values being calculated are not so *small* that no iteration occurs, and in 2.4.2 we multiply the objective by 10^6 to ensure that it does not fall below the resolution limits of Excel.

With any iteration, it is necessary to establish rules by which the process can decide when it should stop, and these involve setting minimum change levels. However, for most operations the default settings in Solver work well.

2.4.2 Maximum likelihood estimation

Maximum likelihood estimation, MLE, is an alternative to the least squares, LS, method for calculating a best-fit model to a set of data. It works by first calculating the probabilities with which the observed set of experimental values would occur for different possible models. For example, in the case of linear regression, each *model* is a specific set of values of slope, m, and intercept, c, giving a possible 'true' straight line, and, for each of these models, the process calculates the *relative probability* of the observed experimental values occurring *by chance*. The iteration then changes model values of m and c to find the model which gives the *maximum* probability for the observed data, and these values become the 'best-fit' result.

Case study: **Best-fit straight line / 9. Maximum likelihood using Solver**

—continued from 2.4.1

We use the same data as in Fig 2.18 (taken from Fig 2.2) which presents the x–y data for five measurements recorded in columns B (x-data) and C (y-data), with each data pair identified by the label, i, in column A. We wish to calculate the coefficients of the best-fit straight line using the 'maximum likelihood estimation' method.

The calculations of residuals, R, and the sum of squares of residuals, $SS_{RESID} = \Sigma R^2$, for this data are explained in 2.1.1 and 2.4.1. In addition, the standard error, SE, which is the *estimated* standard deviation uncertainty in each data point, is calculated in F8 from the value of ΣR^2 in F7 using Eqn 2.14:

Standard error, *SE*: [F8] = SQRT(F7 / (C7 − 2))

The *target* values for the slope, *m*, and intercept, *c*, of a best-fit straight line are entered into cells C8 and C9, starting with values of 0.0 for both, and then the predicted values, *y'*, are calculated using these values of *m* and *c*.

We assume that the random experimental uncertainty is given by the *normal* distribution, and the key statistic is the *probability* with which each *observed* value of *y* in column D would be randomly selected from a normal distribution with mean, *y'*, in column C and standard deviation, *SE*, in F8. For example, for the first data value:

[H2] = NORM.DIST(*D2*, C2, F$8, FALSE)

The *combined relative probability* for all five data points is then calculated by *multiplying* all the individual probabilities together in H7:

[H7] = PRODUCT(H2:H6) = 1.08×10^{-12}

It is this value that the maximum likelihood process seeks to *maximize*, but the iteration process in Solver does not work with such low values, and we choose to multiply by a factor of 10^6 to produce a more realistic 'objective' value in H9:

[H9] = H7 $* 10^6$ = 1.08×10^{-6}

Running the Solver iteration, we now seek to maximize H9 by checking the ⦿ **Max** option to change the values of C8 and C9. The result is the same as in 2.4.1 with the values $m = 1.656$, $c = 2.782$.

The underlying reason for the MLE model and the LS models giving the same result in this particular calculation is that they *both* rely on the *normal distribution* in their calculations. In 2.4.3 we can now see how the maximum likelihood model can be used for data that does not follow the normal distribution.

2.4.3 Nonlinear regression

By using maximum likelihood estimation in Solver, we can unpick the various elements in modelling regression. It is useful to use the MLE method to compare the situations when the underlying uncertainty in the data is due to a Poisson distribution and not a normal distribution, and we use the example of radioactive decay.

Nonlinear regression: Excel analysis for Fig 2.20. See also 7.2.3. Scan here to watch the video or find it via www.oxfordtextbooks.co.uk/orc/currell/

Case study: **Exponential decay / 4. Nonlinear regression using Solver**

—continued from 2.3.4, leading to 3.4.7 and 7.2.3

Fig 2.20 presents the same *x–y* data as Fig 2.16 with six measurements of radioactive decay recorded in columns B (*x*-data) and C (*y*-data), with each data pair identified by the label, *i*, in column A. We wish to calculate the best-fit coefficients for an exponential decay model.

	A	B	C	D	E	F	G	H
1	i	t	N	N'	Poisson	Normal	R	R²
2	1	0	97	90	0.0310	0.0328	-7.00	49.00
3	2	1	63	60	0.0474	0.0630	-2.67	7.14
4	3	2	36	40	0.0515	0.0518	4.44	19.71
5	4	3	25	27	0.0731	0.0657	2.11	4.44
6	5	4	15	18	0.0763	0.0602	3.17	10.05
7	6	5	6	12	0.0233	0.0388	6.18	38.19
8		Pairs, n =	6	Combined =	9.83E-09	1.64E-08	ΣR^2 =	128.53
9		Amp const, N_o =	90.00		×10⁶	×10⁶	SE_{REG} =	5.67
10		Decay const, K =	0.400	Objective =	0.0098	0.0164		

Fig 2.20 Data for nonlinear regression using Solver

Fig 2.20 gives the results of a measurement of radioactive decay, with experimental data of N counts as a function of time, t. The predicted counts, N', in column D are calculated using the mathematical model:

$$N' = N_o \times e^{-Kt}$$

where N_0 is the amplitude constant in C9 and K is the decay constant in C10, with $K = -k$ in Eqn 2.32.

The calculations for the iterative analysis are developed in 2.4.1 and 2.4.2. We start with initial values of N_0 and K equal to 90 and 0.4 respectively, which give the predicted values of N' in the table.

We now use two different statistical assumptions for the uncertainty distributions in the data:

- **Poisson distribution** for randomly generated frequency values, which gives a standard deviation (1.3.3) equal to the square root of the value of the theoretical count, $\sqrt{N'}$.

- **Normal distribution**, which, in this example, is assumed to have the same standard deviation for all data values, equal to standard error of regression in H9, calculated from the squares of the residuals.

For the **Poisson** distribution, the probability for each of the observed counts, N, being observed, given the corresponding theoretical value, N', is calculated in column E, e.g. for data, $i = 1$:

[E2] = POISSON.DIST(C2, D2, FALSE) = 0.0310

As with the calculation in 2.4.2, the product of all the probabilities is calculated in E8 and then multiplied by a factor of 10^6 to give a realistic 'objective' probability in E10 for entry into Solver.

Solver can then be used to change the values of N_0 and K to obtain a *maximum* value for the 'objective' in E10. This results in the best-fit values, using the Poisson error distribution:

$$N_0 = 99.41 \text{ and } K = 0.494 \text{ giving } k = -0.494$$

For the **normal** distribution, the probability for each of the observed counts, N, being observed, given the corresponding theoretical value, N', is calculated in column F, e.g. for data, $i = 1$:

[F2] = NORM.DIST(D2, C2, H\$9, FALSE) = 0.0328

where H9 is the estimated standard deviation in the experimental data.

Solver can then be used to change the values of N_0 and K to obtain a *maximum* value for the 'objective' in F10. This results in the best-fit values, using the normal error distribution:

$N_0 = 97.86$ and $K = 0.476$ giving $k = -0.476$

We can also use the same worksheet to perform the *least squares* analysis by using Solver to change the values of N_0 and K to obtain a *minimum* value for the sum of squares of residuals in H8. As expected, this results in the same best-fit values as above when using the normal error distribution.

This case study example is also analysed in 2.3.4, providing a review of the different options for performing nonlinear regression for an exponential decay.

3 Hypothesis testing

Introduction

The hypothesis test is a key element in the scientific method, in which a proposed hypothesis is tested experimentally by measuring the value of a *test statistic* and then using a statistical analysis to calculate the probability that the observed value could have occurred by chance. If the probability is low, typically less than one in twenty, then it may be *reasonably* safe to assume that an underlying scientific effect might be responsible. The decision logic has been introduced in Section 1.6, including the p-value, significance level, α, and the possibilities of Type I and Type II errors.

In this chapter we introduce different forms of hypothesis testing and analysis, both as a basis for analyses in later chapters and as a perspective of the variety of possible methods. The 'theoretical' basis is developed mainly through modelling with Excel and without difficult statistics, but provides the understanding necessary for the implementation of these tests using Minitab and SPSS in Part II.

Section 3.1 develops the 't' and 'z' statistics for testing differences in mean values, including the family of standard t-tests.

Section 3.2 develops the F-statistic that is at the heart of the 'analysis of variance' approach to testing and introduces the basic ANOVA.

Section 3.3 further extends the ANOVA concept to include the simultaneous testing for the effects of multiple factors.

Section 3.4 reviews the overlapping analyses of t-tests, linear regression, and ANOVAs to develop the concept of general linear models.

Section 3.5 introduces nonparametric testing with the example of the Mann–Whitney test and with reference to the range of alternative techniques.

Section 3.6 considers the additional power of making repeated measurements, either with two measurements for paired tests or more for a 'repeated measures' analysis.

Section 3.7 develops the 'chi-squared' approach to testing the frequencies with which observations fall into different categories.

Section 3.8 then considers the situation with binary outcomes of just two categories leading to tests for proportions.

Section 3.9 uses repetitive modelling in Excel to give an introduction to the developing field of 'resampling' for the calculation of p-values.

3.1 *t*-tests and *z*-tests

The *t*-test, developed by William Gosset under the pen name 'Student', is usually the first statistical test presented to science students, for testing the difference between the mean values of two data samples. In this section we start with the underlying principle of the test, demonstrating its applicability to other problems, before reviewing the 'named' *t*-tests. We also consider the *z*-test (3.1.5) which allows for the fact that we sometimes know the *experimental uncertainty* in a measurement beforehand and do not need to rely solely on the sample data to estimate uncertainty as is the case with the *t*-test.

3.1.1 **General principle of hypothesis testing**

In one of the most common types of hypothesis test (Section 1.6), we record an *observed* (*non-zero*) value for a variable, but we wish to test whether

- the *true* value is *zero* and that the observed value is just due to random experimental variations, or
- the *true* value is really *not* zero.

The test calculates a *statistic* which is the ratio between the observed *value* and the *uncertainty* in that measured value, defined as the *standard error*, *SE*:

$$t_S \text{ (or } z_S) = \frac{\text{Observed value}}{\text{Standard error, } SE}. \tag{3.1}$$

When the uncertainty is estimated from the data itself, this ratio is the *t*-statistic, t_S, but when the *SE* is already known accurately (e.g. by many previous measurements) then it becomes the *z*-statistic, z_S (1.5.4).

We first met the standard error in the calculation for the confidence interval (1.5.2) of a single replicate measurement, but we also meet calculations for standard error occurring for other characteristic values, e.g. the slope of a graph in 2.1.4.

If the *observed value* in Eqn 3.1 is *much* larger (either positive or negative) than the *standard error*, *SE*, giving a large t_S (or z_S) ratio, then we could be confident in deciding that the true value was not zero and did not occur just by random chance. However, for less extreme values, we use a set of *critical values* for t_S (and z_S) to act as reference values to help us make this decision.

Using a *critical value*, t_C, we would decide that the observed value was significant, and did not occur by chance, if

$t_S \geq t_C$ for positive values of t_S or if

$t_S \leq -t_C$ for negative values of t_S.

The value of t_C can be found from published tables or by using the function T.INV.2T(α, *df*) in Excel, which depends on

- the confidence that we want in our decision, which is defined by the significance level, α, (1.6.1) and

- the degrees of freedom, *df*, (1.5.1) which increases with the number of measurements made to calculate the standard error.

If we were to make a large number of replicate measurements (i.e. giving a large value of *df*), then we could estimate the experimental uncertainty quite accurately. Reversing this logic, if we already know the uncertainty, then this information would be equivalent to a very large value of *df*. Hence, the *t*-value decreases with larger values of *df* and becomes equal to the *z*-value for 'infinite' *df*. The *z*-value for 95% confidence is $z_C = 1.96$, and can usually be taken as 2.00 for most practical purposes.

The basic test, using the *t*-statistic, is at the core of a wide range of tests for specific variables (e.g. *t*-test in 3.1.2), but we can also use Eqn 3.1 directly. For example, the analysis of residuals for *species* given in Table 5.3 in 5.4.7 records a *skewness* value of 1.064 with a standard error, *SE*, of 0.269. As the observed value (which would be zero for a normal distribution) is greater than $\pm 2 \times SE$, then we decide that there is evidence of *significant* skewness in the data.

Difference in slopes: Excel analysis for Fig 3.1. Scan here to watch the video or find it via www.oxford textbooks. co.uk/orc/ currell/

Case study: **Bacterial growth / 2. Difference in slopes using *t*-test**

—continued from 5.2.2 (overview), leading to 3.4.3

Using data from the graph in Fig 5.5, we wish to test whether the rates of growth of the bacterial populations are *different* for different cleaner concentrations, *C2* and *C3*, between the times $t = 60$ and 85. The values of *C2* and *C3*, measured in luminescence units, are reproduced in an Excel worksheet in Fig 3.1.

	A	B	C	D	E	F
1	Data:			Calculation:	C3	C2
2	Time / t	C3	C2	Slope =	0.2277	0.2784
3	60	1.76	2.98	n =	6	6
4	65	2.78	3.68	SE(reg) =	0.1753	0.2952
5	70	3.93	5.54	SE(slope) =	0.0084	0.0141
6	75	5.06	6.8	Difference =		0.0507
7	80	6.55	8.55	SE(difference) =		0.0164
8	85	7.24	9.55	t-stat =		3.0914
9				df =		8
10				p =		0.0149

Fig 3.1 Case study: Bacterial growth / 2. Difference in slopes

The null hypothesis for testing whether the two slopes are equal is:

H₀: Difference between the slopes of the two lines, *C2* and *C3*, is zero.

We start by calculating the slopes (E2 and F2), the standard errors of regression (E4 and F4), and standard errors of the slopes (E5 and F5) using the methods introduced in 2.1.4, e.g.:

Slope: [E2] = SLOPE(B3:B8, $A3:$A8)
Standard error of regression: [E4] = STEYX(B3:B8, $A3:$A8)
Standard error of the slope: [E5] = E4 / SQRT(VAR.S($A3:$A8) * (E3-1))

We calculate the *observed value* in this analysis, which is the *difference* between the slopes:

Difference in slopes: $[F6] = F2 - E2 = 0.0507$

and the *standard error* of the difference is calculated by combining the two *SE* uncertainties, using Eqn 1.12 from 1.4.4:

Standard error of difference: $[F7] = SQRT(E5^2 + F5^2) = 0.0164$

We wish to test whether the *difference* in slopes, measured in F6, is significantly different from zero, and hence we calculate the *t*-statistic in Eqn 3.1 as the ratio of F6 divided by F7:

$$t_S = \frac{\text{Observed value}}{\text{Standard error, } SE} : \qquad [F8] = \frac{F6}{F7} = 3.09$$

The degrees of freedom for this calculation is derived from the sample sizes of the two data sets, and is given by: $df = n_1 + n_2 - 4$:

Degrees of freedom, *df*: $[F9] = E3 + F3 - 4 = 8$

The two-tailed *p*-value can then be calculated:

p-value: $[F10] = T.DIST.2T(F8, F9) = 0.0149$

As $p = 0.0149$ is less than 0.05 we conclude that there *is* a significant difference between the two slopes. We see that this agrees with the *p*-value that we calculate in 3.4.3 when using general regression with the factors: *Time*, *Conc* and *Time*Conc*.

3.1.2 **One sample *t*-test**

In a one sample *t*-test, the *n* replicate measurements of a variable with a true value, μ, give an observed sample mean, $\bar{x}$. The aim of the test is to assess whether the unknown *true* value, μ, differs from a *specified* value, μ_0.

We have already met the confidence interval (1.5.2) of a simple measurement where $\bar{x}$ is the best estimate of an unknown true value, μ,

$$CI(\mu) = \bar{x} \pm Cd \Rightarrow \bar{x} \pm (t \times SE) \Rightarrow \bar{x} \pm \left\{ t \times \frac{s}{\sqrt{n}} \right\} \qquad (3.2)$$

In this equation, the *t*-value is used to calculate the critical limit within which we have a chosen level of confidence of finding the true value, μ. The *t*-test uses the same critical level statistics but expressed in a different way.

In the one sample *t*-test, the best estimate of the true value is the sample mean, $\bar{x}$, and the 'observed value' in Eqn. 3.1 is the difference between the sample mean and the specified value $= \bar{x} - \mu_0$. The standard error in this 'value' is the standard error in $\bar{x}$, $SE = s/\sqrt{n}$ (Eqn 1.21).

The relevant statistic becomes:

$$t_S = \frac{\text{Observed value}}{\text{Standard error}} = \frac{(\bar{x} - \mu_0)}{SE} = \frac{(\bar{x} - \mu_0)}{s / \sqrt{n}} \qquad (3.3)$$

The null hypothesis of the test is that there is no difference between μ and μ_0,

$$H_0: \mu = \mu_0$$

The proposed, or alternative hypotheses, could be

- for a two-sided or two-tailed test, $H_1: \mu \neq \mu_0$
- for a one-sided or one-tailed test, $H_1: \mu > \mu_0$ or $H_1: \mu < \mu_0$

The critical values, t_C, can be obtained from published tables using degrees of freedom given by:

$$df = n - 1 \tag{3.4}$$

The reason for the '−1' is that one bit of information has been used in the calculation of the sample mean value.

Case study: **Blood alcohol / 8. One sample *t*-test**

−continued from 1.6.2, leading to 8.1.1

The data in cells A2:A6 in Fig 3.2 show five replicate measurements of alcohol level (mg/100 ml) in a blood sample. We wish to test whether the true value, μ, of alcohol in the sample is greater than $\mu_0 = 80$ mg/100 ml. Note that in this calculation (unlike in 1.6.2) we do not assume any previous knowledge of the measurement uncertainty.

The hypotheses are:

Null hypotheses, H_0: The true value, $\mu = 80$ mg/100 ml.
Alternative/proposed hypothesis, H_1: The true value, $\mu > 80$ mg/100 ml.
The specified value of 80 is entered into cell C5.

	A	B	C	D	E
1	Data:	Statistics:		Critical values:	
2	83.8	Mean =	82.82	1-tail, t_c =	2.13
3	81.2	St dev, s =	2.54	2-tail, t_c =	2.78
4	84.8	St Error, SE =	1.14	p-values:	
5	85.1	Test value =	80	1-tail, p =	0.034
6	79.2	t-statistic, t_s =	2.48	2-tail, p =	0.068

Fig 3.2 Case study: Blood alcohol / 8. One sample *t*-test

We calculate the relevant sample statistics for Eqn 3.3.

Mean: [C2] = AVERAGE(A2:A6) = 82.82
Standard deviation: [C3] = STDEV.S(A2:A6) = 2.54
Standard error (Eqn 1.21):   [C4] = C3/SQRT(5) = 1.14

The relevant test statistic, using Eqn 3.3, is:

t-statistic: $\qquad$ [C6] = (C2 − C5)/C4 = 2.48

The critical t-values are derived for a significance of 0.05, using degrees of freedom (Eqn 3.4), $df = n - 1 = 4$:

One-tailed critical value: $\qquad$ [E2] = −T.INV(0.05, 4) = 2.13

(The reason for the negative sign is that the t-value is calculated from the left-hand tail area, and an area of 0.05 gives a *negative* value for t)

Two-tailed critical value: $\qquad$ [E3] = T.INV.2T(0.05, 4) = 2.78

The p-values are then calculated directly from the t-statistic and degrees of freedom:

One-tailed p-value: $\qquad$ [E5] = T.DIST.RT(C6, 4) = 0.034

(calculates just the area of the right-hand tail, as in Fig 1.19)

Two-tailed p-value: $\qquad$ [E6] = T.DIST.2T(C6, 4) = 0.068

(calculates the area of both tails of the distribution.)

We see that, for a one-tailed test, $t_S > t_C$, suggesting that there is a significant difference, which agrees with the p-value, $p = 0.034 < 0.05$, but that, for the two-tailed test, $t_S < t_C$, suggesting that the apparent difference could have occurred by chance, which agrees with the p-value, $p = 0.068 > 0.05$. We can note that, for this symmetrical calculation, the two-tailed p-value is twice the one-tailed p-value as given in Eqn 1.27.

The *original* question was whether the blood alcohol level was *greater* than 80, and it is therefore acceptable to use the one-tailed test result. It would not be acceptable to choose the significant one-tailed test *after* seeing the results.

See 6.1.4 for the use of Minitab and SPSS for a one sample t-test.

3.1.3 **Two-sample t-test**

We can develop a similar test for a possible difference between the true means (μ_A and μ_B) of two samples, based on the best estimate values $\bar{x}_A$ and $\bar{x}_B$ from sample sizes of n_A and n_B, with standard deviations s_A and s_B.

This test is also called the independent samples t-test because each data value is measured as an *unrelated* measurement *independently* of any other value. We consider the possibility of *related* measurements in the paired t-test in 3.6.1.

We compare the difference in the two sample means with the standard error in this difference.

$$t_S = \frac{\text{Difference}}{SE(\text{Difference})} = \frac{(\bar{x}_A - \bar{x}_B)}{s' \times \sqrt{\dfrac{1}{n_A} + \dfrac{1}{n_B}}} \qquad (3.5)$$

The standard error is calculated using a *pooled* standard deviation, s', which is a weighted 'average' of the two *sample* standard deviations, s_A and s_B. This calculation assumes that the two samples are drawn from *populations* with the same standard deviation, which is then estimated by using s'.

$$s' = \sqrt{\frac{(n_A - 1) \times s_A^2 + (n_B - 1) \times s_B^2}{n_A + n_B - 2}} \tag{3.6}$$

Note that we always combine uncertainties by combining *variances* (1.4.4).
The degrees of freedom for this test is given by:

$$df = n_A + n_B - 2 \tag{3.7}$$

Two bits of information have already been used in calculating the two sample mean values.

Two sample t-test and F-test: Excel analysis for Fig 3.3. Scan here to watch the video or find it via www.oxford textbooks. co.uk/orc/ currell/

> ## Case study: River *pH* / 2. Two sample *t*-test and *F*-test
>
> —continued from 6.2.1 (overview), leading to 3.2.2 (to 3.6.1 for the paired *t*-test and to 3.5.1 for the Mann–Whitney test)
>
> The values in B2:E2 and B3:E3 in Fig 3.3 record four replicate *pH* values taken from each of two rivers, A and B. We wish to test for a difference between the true mean values for each river μ_A and μ_B respectively.
>   (The *F*-statistic referred to in L5 is developed in 3.2.1.)

The hypotheses for testing whether there is a difference in the *pH* values of the two rivers, *A* and *B*, become:

Null hypothesis, $H_0: \mu_A = \mu_B$
Alternative/proposed hypothesis, $H_1: \mu_A \neq \mu_B$

▲	A	B	C	D	E	F	G	H	I	J	K	L
1							Mean, x	StDev, s	Size, n		df	Variance
2	A	7.56	7.52	7.7	7.61		7.60	0.0776	4		3	6.03E-03
3	B	7.47	7.4	7.49	7.55		7.48	0.0618	4		3	3.82E-03
4												
5				Pooled stdev, s' =				0.0702			F statistic, F =	1.58
6				t-statistic, t_s =			2.42				p-value =	0.36
7	df =	6		t-critical (95%), t =			2.45					
8				p-value =			0.0520	T.TEST() =	0.0520			

Fig 3.3 Case study: River *pH* / 2. Two sample *t*-test and *F*-test (3.2.1)

In this calculation we make the assumptions that:

- the four recorded values for each river are independent *replicate* measurements.

- the population of measurements for each river follow a normal distribution. In this example, given the relatively small variations of the replicate results and no reason to anticipate non-normality, we are probably safe to assume a normal distribution (Section 5.4).

- the population standard deviations (and variances) for each river are *equal* (homoscedasticity). We will see in Section 3.2.1 that the *p*-value given in L6 of the worksheet shows that there is no significant difference between the variances.

Using the Excel worksheet in Fig 3.3, we calculate the sample mean values $\bar{x}_A$ and $\bar{x}_B$ in G2 and G3 and their sample standard deviations s_A and s_B in H2 and H3.

The pooled standard deviation s', is calculated using Eqn 3.6 in H5:

s': $[H5] = SQRT(((I2-1)*H2^2 + (I3-1)*H3^2)/(I2+I3-2)) = 0.0702$

The *t*-statistic is calculated using Eqn 3.5 in G6:

t_S: $[G6] = (G2-G3)/(H5*SQRT(1/I2+1/I3)) = 2.42$

The degrees of freedom are calculated using Eqn 3.7 in B7:

df: $[B7] = I2+I3-2 = 6$

The critical *t*-value (two-tailed) is calculated in G7:

t_C: $[G7] = T.INV.2T(0.05, B7) = 2.45$

The *p*-value (two-tailed) is calculated in G8:

p: $[G8] = T.DIST.2T(G6, B7) = 0.0520$

We can also calculate the *p*-value (two-tailed) *directly* from the *raw data* using the T.TEST() function in I8:

p: $[I8] = T.TEST(B2:E2, B3:E3, 2, 2) = 0.0520$

where the first '2' in the argument identifies a two-tailed test and the second '2' is for data with equal variance. We see that this agrees with the value calculated in G8.

We can base our conclusion for this hypothesis test on either of the comparisons:

- The *t*-statistic is less than the critical value: $t_S = 2.42 < t_C = 2.45$.
- The *p*-value is greater than the default significance level: $p = 0.052 > \alpha = 0.05$.

Both comparisons show that there is not enough evidence to reject the null hypothesis, and we conclude that the apparent difference could have occurred by chance.

If the *original* hypothesis had been to test whether $\mu_A > \mu_B$, we would need to calculate the one-tailed *p*-value. In this analysis the data uncertainties are symmetrical, and hence the *p*-value for a one-tailed test would be half of the two-tailed test, giving $p = 0.026$ (Eqn 1.27). For the one-tailed hypothesis, we would conclude that the *pH* of river A was indeed greater than that of river B, but it would be incorrect to decide to use the one-tailed hypothesis *after* you have seen the fact that it would suggest a significant effect in a specific direction.

In this case study, the two samples happened to be of the same size ($n = 4$), but the *t*-test can be applied in exactly the same way for samples of different sizes.

The calculation for the *t*-test assumes that the data is derived from populations with normal distributions. If the data is not normally distributed, then, in principle, it is necessary either to transform the data to a near normal distribution (5.4.7), or to use the equivalent non-parametric test: the Mann–Whitney test (3.5.1). However, the *t*-test is *robust* for minor deviations from normality, and this means that it will tend to give the correct conclusion even if the distributions are not exactly normal. It is most likely to fail if a distribution is significantly *skewed*, i.e. with a long tail.

See 6.2.5 for the use of Minitab and SPSS for the two (independent) sample *t*-test.

3.1.4 **Unequal variances**

The standard calculation for the two sample *t*-test assumes *homoscedasticity* or *homogeneity of variance*, i.e. the two populations have the same variances (and standard deviations). However we may wish to perform the *t*-test on samples that are drawn from populations with *different* standard deviations. In this case, a modified *t*-statistic by B L Welch uses both standard deviations separately and introduces a non-integer value for the degrees of freedom. With its increased complexity, Welch's modified test is normally performed in software.

In principle, we should perform a hypothesis test for the equality of variance (*F*-test or Levene's test, 3.2.1) to check whether it is necessary to use Welch's modified test. However, the variance test can be unreliable for *small* samples, and, given the fact that the standard *t*-test is *robust* in accepting such differences, we often only use a variance test when the *scientific* conditions for the two samples suggest that the variances could be different (see Section 5.4).

It is possible to test for a difference in variance using the *F*-test in Excel (3.2.1). SPSS automatically carries out Levene's test for variance when requested to perform the two samples *t*-test, and also gives the results to both types of *t*-test (6.2.5). Minitab performs either *t*-test (6.2.5) and has a separate menu option that reports the results of both the *F*-test and Levene's test (6.2.4).

3.1.5 *z*-tests

Statistical analyses using *t*-values do not use any *external* information about experimental variability, and calculate the uncertainty just from the sample data. For small samples, this estimation is itself subject to increased error and the relevant *t*-value automatically increases (via the degrees of freedom) to accommodate this wider uncertainty range. However, in many routine laboratory analyses the experimental uncertainty is actually known, either from extended previous experience or by a review of the inherent variability in the measurement process itself.

We saw in 3.1.1 that, if the uncertainty is already known, we replace the *t*-value with the equivalent value of *z* for the relevant level of confidence. The *t*-test then becomes a *z*-test, with

$z_C = 1.64$ for a one-sided test with 95% confidence, and

$z_C = 1.96$ for a two-sided test with 95% confidence

We can see the increased analytical *power* achieved through knowing the experimental uncertainty by reference to the 'Blood alcohol' case study. In 3.1.2, the one-sided *t*-test with a mean value of 82.82 for five data values gives $p = 0.034$, but by using the data in Fig 1.19 with a known standard deviation uncertainty of 2.0 we get a more significant result with $p = 0.00075$ for the same sample mean of 82.82.

3.2 Analysis of variance

As its name suggests, the analysis of variance technique tests for significant differences by analysing variances within the data. The key statistic in this process is the sum of squares, SS, (1.5.1) which measures the overall variation in a data set. We can then calculate the mean square, MS, for that data by dividing by the degrees of freedom (Eqn. 1.18):

$$MS = \frac{SS}{df}$$

The value of the mean square, MS, for a *single data set* is equal to the *variance* of the data.

We start by introducing the F-test, which is the key statistical test at the heart of the general analysis of variance (ANOVA) series of techniques.

3.2.1 *F*-test

The **F-test** tests whether there is a significant difference between the variances, s_A^2 and s_B^2, of two data samples of size n_A and n_B. We calculate the F statistic:

$$F_S = \frac{s_A^2}{s_B^2} \tag{3.8}$$

which has degrees of freedom (1.5.1) given by,

$$df_A = n_A - 1 \text{ and } df_B = n_B - 1 \tag{3.9}$$

The null hypothesis is that both samples have the same variance, but we would accept the one-tailed proposed hypothesis, that s_A^2 was greater than s_B^2 if

$$F_S \geqslant F_C$$

where F_C is the critical value available from tables or by using F.INV.RT(α, df_A, df_B).

Referring to Fig 3.3, the variances s_A^2 and s_B^2 are calculated in L2 and L3 by simply squaring the standard deviations in H2 and H3, and the F-statistic, $F = 1.58$, in L5 is derived using Eqn 3.8. The degrees of freedom for both samples in K2 and K3 are derived from the sample sizes using Eqn 3.9.

We calculate the p-value in L6 using the function: F.DIST.RT(F_S, df_A, df_B):

p-value: p = F.DIST.RT(L5, K2, K3) = 0.36

Since $p > 0.05$, we conclude that there is no evidence of a difference in variance, and the observed difference could have occurred by random chance.

Levine's test is also a test for a difference in variance, but it is a distribution-free test that does not assume the normal distribution. It is not used in the ANOVA calculations, but it is used in SPSS for testing for equality of variances between samples (6.2.5).

3.2.2 Basic principle of ANOVA calculations

The family of 'analysis of variance' calculations (ANOVAs) start by identifying the different sources of variation within the data (often referred to as partitioning the variance). The variations in a simple ANOVA are combined using the *sum of squares*:

$$SS_{TOTAL} = SS_{RANDOM} + SS_{FACTOR} \tag{3.10}$$

For example, the *factor* being tested could be a variation *between* the mean values of two samples and the *random* variations would then be due to the experimental uncertainty *within* each sample.

Taking the degrees of freedom of each component into account, we can calculate (Eqn 1.18) the *mean square* components:

MS_{RANDOM} is the variance due only to the random experimental uncertainty, and

MS_{FACTOR} is the additional variance due to the factor that is being tested.

The ANOVA process compares these *mean square* values, and if MS_{FACTOR} is much greater than MS_{RANDOM} then it would be clear that the factor must be having an effect. We use the *F*-test to test whether MS_{FACTOR} is *significantly* greater than MS_{RANDOM}:

$$F = \frac{MS_{FACTOR}}{MS_{RANDOM}} \tag{3.11}$$

MS_{RANDOM} often appears as MS_{ERROR} in ANOVA results tables.

We will use the same example as for the two sample *t*-test to illustrate a simple ANOVA calculation.

Analysis of variance: Excel analysis for Fig 3.4. Scan here to watch the video or find it via www.oxford textbooks. co.uk/orc/ currell/

Case study: River *pH* / 3. ANOVA calculations

—continued from 3.1.3, leading to 3.4.2

In Fig 3.4, the *pH* values of two rivers, A and B, are recorded in B2:E2 and B3:E3 respectively, and we wish to test if there is a significant difference in their mean values.

	A	B	C	D	E	F	G	H
1							Mean	Variance
2	A	7.56	7.52	7.7	7.61		7.60	0.0060
3	B	7.47	7.4	7.49	7.55		7.48	0.0038
4	Sample size, *n* =	4		Factor levels, *k* =		2		
5	MS(between) =	0.0288	Variance of sample means, *VSM* =			0.0072		
6	MS(within) =	0.0049	Mean of sample variances, *MSV* =					0.0049
7	*df* (between) =	1						
8	*df* (within) =	6			*F* =	5.85	*p* =	0.0520

Fig 3.4 Case study: River *pH* / 3. ANOVA calculations

The first step is to calculate the *experimental* variability separately *within* each set of replicate measurements, recording the variance values in H2 and H3, and then calculate the mean of these sample *variances*, *MSV*, in H6. This value will not be affected by any differences *between* the samples.

    Mean of the sample variances, MSV: $[H6] = \text{AVERAGE(H2:H3)} = 0.0049$

This gives directly the mean square (within), MS_W, in B6, which is the best estimate of the *random* variance in the data, and is not affected by the difference between the samples.

$$MS_W = MSV = 0.0049$$

The second step is to include the variability due to the *factor*, which, in this case, has resulted in the difference in the two mean values in G2 and G3. We calculate the *variance* of these sample mean values, *VSM*, in G5.

    Variance of the sample means, VSM: $[G5] = \text{VAR.S(G2:G3)} = 0.0072$

The factor variance due to the difference between the means is then given by mean square (between), MS_B, which is calculated in B5 as

$$MS_B = VSM \times n = 0.0288.$$

The *n* term reverses the reduction in uncertainty that had occurred when taking the mean of *n* values in each sample (Eqn 1.21).

The terms 'within' and 'between' are convenient here because they refer directly to calculations *within* and *between* the samples, and the terminology also appears elsewhere in understanding other data analysis techniques (e.g. 3.6.2). We can also relate these terms respectively to the 'random' and 'factor' terms in the ANOVA calculation:

$$MS_W = MS_{RANDOM}$$
$$MS_B = MS_{FACTOR}.$$

The *F*-test is then used to test whether MS_B is significantly greater than MS_W. If it is, then this would show that the factor being tested was also significant. We test for any significant difference by calculating the *F*-statistic in E8.

    $F_s = \dfrac{MS_B}{MS_W}:$ $[E8] = \text{B5 / B6} = 5.85$

We then calculate the associated *p*-value, with degrees of freedom in B7 and B8, using the function

    p-value: $[G8] = \text{F.DIST.RT(E8, B7, B8)} = 0.052$

We see that this gives the same value as the *t*-test in 3.1.3. There is not enough evidence at a significance of 0.05 for a difference between the *pH* values of the two rivers.

3.2.3 One-way ANOVA

In 3.2.2 we developed the working principle of the ANOVA using the same data as for the two sample *t*-test in 3.1.3. In practice however, ANOVA calculations are normally performed using a dedicated statistics software package, and, for comparison, Fig 3.5 gives the output that would be obtained using statistical software (in this example, Minitab) for the same data. The use of SPSS and Minitab for ANOVA analyses is given in Section 6.3.

```
Source  DF       SS        MS      F      P
River    1  0.02880  0.02880   5.85  0.052
Error    6  0.02955  0.00492
Total    7  0.05835
```

Fig 3.5 ANOVA results (Minitab) for the data in Fig 3.4

In addition to the same *MS* values, *F*-statistic, and *p*-value, the typical ANOVA calculation also presents the *SS* values, from which the mean squares, *MS*, are then derived by dividing by the relevant degrees of freedom (Eqn 1.18). Note that the addition of sums of squares agree with Eqn 3.10:

$$0.05835 = 0.02955 + 0.02880.$$

This introductory example was effectively testing for the difference between two *levels* of one factor, where the factor was the *choice* of water sample, but, for just two levels, there was no advantage in using an ANOVA over the simple *t*-test. The real value of the ANOVA procedure is that, unlike the *t*-test, it can be extended to test for a difference between *three or more* levels of the factor, by simply including the variations from all levels in the calculation of *MS*(between).

One-way ANOVA: Minitab and SPSS analyses leading to Fig 3.8. Scan here to watch the video or find it via www.oxfordtextbooks.co.uk/orc/currell/

Case study: **Catalyst / 1. One-way ANOVA (overview)**
—leading to 3.3.1

Fig 3.6 gives the percentage yields from a chemical reaction using three catalysts, *C1*, *C2*, and *C3*. The measurements are repeated over four days, *D1*, *D2*, *D3*, and *D4*, and we wish to test whether there is a significant difference in yield due to the choice of catalyst.

3.3.1 / 2. Two-way ANOVA: Develops the analysis to include the effect of multiple factors.

3.3.2 / 3. Interactions: Develops the two-way ANOVA to include an interaction between factors.

	A	B	C	D	E
1		D1	D2	D3	D4
2	C1	79	78	82	77
3	C2	74	76	71	72
4	C3	78	75	75	81

Fig 3.6 Percentage yields using catalyst, *C1*, *C2*, and *C3* on four days *D1*, *D2*, *D3*, and *D4*

In this initial calculation, we will treat the yields on the four days as *replicate* measurements for each catalyst, giving three samples of four replicates each. This data is said to be *balanced* because all three levels have the same number of data values.

The first step is to put the data into the *column* format that is expected in both Minitab and SPSS for unrelated samples:

Yield%	Catalyst	Day	CatN
79	C1	D1	1
74	C2	D1	2
78	C3	D1	3
78	C1	D2	1
76	C2	D2	2
75	C3	D2	3
82	C1	D3	1
71	C2	D3	2
75	C3	D3	3
77	C1	D4	1
72	C2	D4	2
81	C3	D4	3

Fig 3.7 Stacked data from Fig 3.6

In this format, every variable and factor is entered into its own column. Each data value is identified as a separate record on a unique row, and the entries in the other columns provide the information related to that particular data value.

In this analysis, the *factor* that is being tested is the choice of catalyst, with *C1*, *C2*, and *C3*, being the three *levels*. The variable *CatN* has been added to express the factor levels in numeric form, as this is required by SPSS for the basic analysis used in Fig 3.8(b). The days on which specific measurements were made are also recorded, but in this analysis they are considered to be replicate measurements for each catalyst.

The ANOVA tests the null hypothesis:

H_0: The sample means for all catalysts are equal, $\mu_{C1} = \mu_{C2} = \mu_{C3}$.

The calculation requires that the data values are collected *independently*, with the random variations following a *normal distribution*, and with *equal variance* (homoscedasticity) for the values recorded under different experimental conditions. For data recorded as a proportion or percentage, we could consider transforming the data with an 'arcsin($\sqrt{P}$)' transformation (5.4.7). However, in this example the variations in the values are small compared to the difference from either end (0% or 100%) of the data range and such a transformation is unlikely to be necessary.

Yield

Source	DF	SS	MS	F	P
Catalyst	2	69.50	34.75	5.85	0.024
Error	9	53.50	5.94		
Total	11	123.00			

(a) Minitab

	Sum of Squares	df	Mean Square	F	Sig.
Between Groups	69.500	2	34.750	5.85	.024
Within Groups	53.500	9	5.944		
Total	123.000	11			

(b) SPSS

Fig 3.8 One-way ANOVA results using the data in Fig 3.7

The ANOVA result tables using basic one-way analyses of variance in Minitab and SPSS are given in Fig 3.8. The factor variance is identified as the 'catalyst' for Minitab and 'between groups' for SPSS, and the random variance as 'error' and 'within groups' respectively. In both cases, the relevant *F*-statistic is calculated as 5.85 which gives a *p*-value = 0.024.

The *total* degrees of freedom for the ANOVA is equal to the number of data values (initial bits of information) minus one, giving 11. The degrees of freedom for the *factor* is equal to $n - 1$ where n is the number of levels in the factor. In this case, there are three catalysts which gives three levels and two degrees of freedom for the *catalyst* factor. The difference between these values then gives the degrees of freedom, 9, for the random *error* variance.

Since $p < 0.05$, we reject the null hypothesis and conclude that there is a difference between the mean values of *at least one pair* of samples. The ANOVA procedure is able to detect whether a difference occurs between several samples, but it does not identify which level is different from which other level(s). The procedures for finding *where* the differences lie is dealt with in 3.2.4 under post hoc tests.

It is reasonable to ask whether using an ANOVA to identify a difference between several samples has any advantage over using *multiple t*-tests to test for significant differences between each *pair* of samples separately. In fact, there are two problems with using multiple tests to *identify* differences:

- The numbers of pairs to be tested would increase rapidly with more samples, e.g. just four samples would require six possible tests.
- The probability of a Type I error (typically the default 0.05) occurs for *every t*-test, giving an increasing probability of an error in *at least one* of the tests as the number of tests increases, whereas the error probability for the *single* calculation ANOVA remains at the default 0.05.

The problem of interpreting multiple p-values may be addressed by using the Bonferroni correction (1.6.4) which modifies the required significance level, reducing the Type I error probability:

Bonferroni significance $= \alpha / n$

where n is the number of multiple tests.

However, given the problems with multiple t-tests, it is far simpler just to use a single ANOVA.

The uses of Minitab and SPSS for ANOVA calculations are developed in 6.3.5, where we see that it is possible to perform ANOVA calculations using the general linear model (Section 3.4) which gives greater flexibility to the analysis.

3.2.4 Post hoc comparison tests

The ANOVA analysis detects whether there is a significant difference between one or more of the sample mean values, but it does not identify which sample(s) may be different from the others.

There exist a range of procedures called *post hoc* (from Latin 'after this') tests that can be used to identify where any differences lie, *after* the ANOVA has confirmed that they do exist. These tests are also called *comparison* tests. A common test is the Tukey test which compares each pair of samples to identify differences. Although this sounds like multiple t-tests, the Tukey procedure uses variance data from *all* the samples in assessing the differences between pairs and is less likely to make errors than using multiple t-tests.

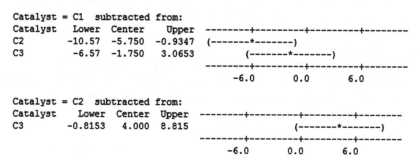

```
Grouping Information Using Tukey Method and 95.0% Confidence
Catalyst  N    Mean  Grouping
C1        4   79.00  A
C3        4   77.25  A B
C2        4   73.25    B
Means that do not share a letter are significantly different.


Tukey 95.0% Simultaneous Confidence Intervals
All Pairwise Comparisons among Levels of Catalyst


Catalyst = C1  subtracted from:
Catalyst   Lower   Center    Upper  --------+---------+---------+--------
C2        -10.57   -5.750  -0.9347  (-------*-------)
C3         -6.57   -1.750   3.0653         (-------*-------)
                                    --------+---------+---------+--------
                                         -6.0       0.0       6.0


Catalyst = C2  subtracted from:
Catalyst    Lower   Center   Upper  --------+---------+---------+--------
C3        -0.8153   4.000   8.815                   (-------*-------)
                                    --------+---------+---------+--------
                                         -6.0       0.0       6.0
```

Fig 3.9 Tukey test output (Minitab 16) for the data in Fig 3.7 (Minitab 17 provides the same information in a rearranged format)

The Minitab 16 output from a Tukey test for the data in Fig 3.7 (ignoring the effect of different days) is given in Fig 3.9. The 'grouping information' section shows that the catalysts C1 and C2 are significantly different from each other because C1 is just in group A and C2 is just in group B, but C3 appears in both groups A and B and is therefore not significantly different from C1 or C2. The 'pairwise comparison' sections show the *confidence intervals* of the *differences* between catalysts both numerically and also graphically within the brackets. The fact that the confidence interval for the difference between C1 and C2, from −10.57 to −0.9347, does not include 0 shows that this is a significant difference. However, the confidence interval of C1 compared with C3 *overlaps* 0 and hence shows no significant difference. The third plot of C2 compared with C3 shows that the confidence interval of the difference also *overlaps* 0, and thus there is no significant difference between these catalysts. Minitab 17 uses a graphical plot to display the confidence intervals.

Multiple Comparisons

Dependent Variable: Yield
Tukey HSD

(I) CatN	(J) CatN	Mean Difference (I-J)	Std. Error	Sig.	95% Confidence Interval	
					Lower Bound	Upper Bound
1.00	2.00	5.75000*	1.72401	.021	.9365	10.5635
	3.00	1.75000	1.72401	.586	-3.0635	6.5635
2.00	1.00	-5.75000*	1.72401	.021	-10.5635	-.9365
	3.00	-4.00000	1.72401	.104	-8.8135	.8135
3.00	1.00	-1.75000	1.72401	.586	-6.5635	3.0635
	2.00	4.00000	1.72401	.104	-.8135	8.8135

*. The mean difference is significant at the 0.05 level.

Fig 3.10 Tukey test output (SPSS) for the data in Fig 3.7

The equivalent output from SPSS is given in Fig 3.10, where we see the same conclusions. The three catalysts have been coded by scale values 1, 2, and 3 respectively, and the same confidence intervals have been calculated between pairs of samples. SPSS also calculates a *p*-value for the significance of the difference between each pair, with:

1.00 and 2.00 (i.e. between *C1* and *C2*) $p = 0.021$ indicating a *significant* difference

1.00 and 3.00 (i.e. between *C1* and *C3*) $p = 0.586$ indicating *no* significant difference

2.00 and 3.00 (i.e. between *C2* and *C3*) $p = 0.104$ indicating *no* significant difference

which is in agreement with the result from Minitab.

There is a wide range of possible post hoc tests. Most will set a significance level, sometimes called the *family rate* that takes into account the 'family' of multiple tests being made, i.e. the maximum probability of one Type I error in *all* of the comparisons. See, for comparison, consideration of the Bonferroni correction in 1.6.4.

Tests available in Minitab and/or SPSS include:

Table 3.1 Post hoc tests

Tukey HSD	Compares all samples pairwise. It is a common default choice. HSD (honestly significantly different).
Fisher LSD	Compares all samples pairwise. Uses *individual* error rate which is the maximum probability of a Type I error in *every* comparison. LSD (least significant difference).
Dunnett	Compares each sample with one *control* sample which needs to be identified.
Hsu's MCB	Compares each sample with the sample which has either the highest or lowest mean (to be selected). MCB (multiple comparisons with the best).
Scheffe	Allows all possible combinations of sample means to be tested, and tends to be more conservative.
Bonferroni	Based on multiple *t*-test but with adjusted significance level.
Sidak	As for Bonferroni, but producing tighter bounds for the confidence intervals.

3.3 Multiple factors ANOVA

We saw in Section 3.2 the use of a one-way ANOVA to perform a one factor, multi-level, analysis, but we now extend the concept of the ANOVA to test more than one factor. Each factor will have two or more levels, and each combination of different levels is sometimes called a *treatment*. The ANOVA is said to be balanced if every treatment has the same number of replicate measurements. In this section, we are only considering *univariate* data, i.e. we are measuring the *same* output variable for different factor levels.

3.3.1 Two-way ANOVA

We start with a two-way ANOVA by illustrating a variant of the analysis performed in 3.2.3.

Case study: **Catalyst / 2. Two-way ANOVA**

—continued from 3.2.3, leading to 3.3.2

Fig 3.11 gives similar data to that in Fig 3.6, with the percentage yields from a chemical reaction, but using three *new* catalysts, *C4*, *C5*, and *C6*. The measurements are repeated over four days, *D1*, *D2*, *D3*, and *D4*. Again, we wish to test whether there is a significant difference in yield due to the choice of catalyst.

Multi-factorial ANOVA (Minitab): Analysis for data in Figs 3.11 and Fig 3.15. Scan here to watch the video or find it via www.oxford textbooks. co.uk/orc/ currell/

	A	B	C	D	E
1		D1	D2	D3	D4
2	C4	76	77	74	74
3	C5	75	74	72	71
4	C6	77	75	75	72

Fig 3.11 Reaction yields as functions of *catalyst* and *day*

Using a *one*-way ANOVA to test for differences between the new catalysts, *C4*, *C5*, and *C6* in Fig 3.11, gives the results in Fig 3.12 with $p = 0.236$, which suggests that there is no significant difference.

```
Source    DF    SS     MS     F     P
Catalyst   2   11.17  5.58  1.70  0.236
Error      9   29.50  3.28
Total     11   40.67
S = 1.810    R-Sq = 27.46%    R-Sq(adj) = 11.34%
```

Fig 3.12 One-way ANOVA output (Minitab) for the data in Fig 3.11

Multi-factorial ANOVA (SPSS): Analysis for data in Figs 3.11 and Fig 3.15. Scan here to watch the video or find it via www.oxford textbooks. co.uk/orc/ currell/

However, if we now plot the individual yields for each of the three catalysts for the four days we see a pattern emerging in Fig 3.13 (see 6.4.3 for deriving this 'interaction' plot).

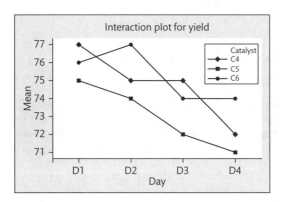

Fig 3.13 Interaction plot (Minitab 16) for the data in Fig 3.11

There appears to be a general downward trend in the yield over the four days. This may, for example, be due to the degradation over time of chemical reagents that were produced at the beginning of the investigation.

From a statistical perspective, the one-way ANOVA is unable to tell the difference between the *systematic* changes over the four days and possible *random* variations, and can only interpret the changes as an increased *uncertainty* in the data giving $SS_{ERROR} = 29.5$. With this apparent increase in uncertainty, the calculated *F*-statistic will be reduced, making it less likely that any differences between the catalysts can be seen as significant.

However, we can identify the effect of the different days as a *second* factor in the analysis, and perform a *two*-way ANOVA, giving the results in Fig 3.14 which reports *p*-values for both of the *Catalyst* and *Day* factors. For the analysis of multiple factors we use the general linear model options for performing ANOVAs, as described in Sections 3.4 and 6.4.

```
Source     DF   Seq SS   Adj SS   Adj MS      F      P
Catalyst    2  11.1667  11.1667   5.5833   6.93  0.028
Day         3  24.6667  24.6667   8.2222  10.21  0.009
Error       6   4.8333   4.8333   0.8056
Total      11  40.6667
S = 0.897527    R-Sq = 88.11%    R-Sq(adj) = 78.21%
```

Fig 3.14 Two-way GLM/ANOVA output (Minitab) for the data in Fig 3.11

The *Day* factor has $p = 0.009$ showing that *Day* is indeed a significant effect, and we also see that the ANOVA has been able to separate the variance due to the *Day* from the *Error*, reducing the remaining random uncertainty, from $SS_{ERROR} = 29.5$ in Fig 3.12 to $SS_{ERROR} = 4.8$ in Fig 3.14. The goodness of fit of the model (4.4.1) is also shown to increase from $R^2(\text{adj})$ from 11.3% to 78.2%. With this decreased uncertainty and improvement in fit, the analysis is now able to confirm, with $p = 0.028$, that there is a significant difference between the different levels of *Catalyst*.

3.3.2 Interactions between the different factors

We can take the complexity of the ANOVA one stage further by looking at possible *interactions* between the different factors.

Case study: **Catalyst / 3. Interactions**

—continued from 3.3.1

In a further investigation, Fig 3.15 gives the yields of two *new* catalysts C6 and C7, measured at three increasing temperatures, *T1*, *T2*, and *T3*. It is important to note that, in this data set, there are two *replicate* measurements made at each combination of factor levels, giving *two rows of data* for each catalyst.

	A	B	C	D
1		T1	T2	T3
2	C6	77	73	74
3	C6	74	75	78
4	C7	69	74	78
5	C7	70	76	78

Fig 3.15 Reaction yields as functions of *catalyst* and *temperature*

Performing a two-way ANOVA calculation with just the *catalyst* and *temperature* as possible factors gives the results in Fig 3.16, which appear to show that neither the choice of catalyst or the temperature have any significant effect on the yield of the reaction.

```
Source      DF  Seq SS   Adj SS   Adj MS     F      P
Catalyst     1   3.000    3.000    3.000  0.44  0.527
Temp         2  40.667   40.667   20.333  2.96  0.109
Error        8  55.000   55.000    6.875
Total       11  98.667
S = 2.62202   R-Sq = 44.26%   R-Sq(adj) = 23.35%
```

Fig 3.16 Two-way GLM/ANOVA output (Minitab 16) for the data in Fig 3.15

However, it is again useful to present the data visually, as in Fig 3.17, where we have plotted the *mean of each pair* of replicate yields against the increasing temperatures. The points from the different catalysts are linked for clarity.

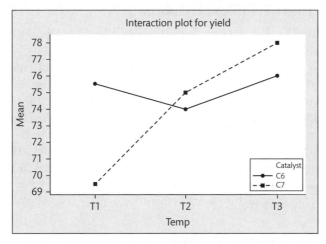

Fig 3.17 Interaction plot (Minitab 16) for the data in Fig 3.15

From the graph it appears that there may be a difference between the ways in which the *level* of temperature influences the *effect* of each catalyst. The yield for *C6* appears to be unchanged with temperature, whereas the yield for *C7* increases with temperature. This interlinked behaviour is described as an *interaction* between the two factors.

We can include the interaction as an additional factor to be tested in the ANOVA, giving the output in Fig 3.18 for Minitab and Fig 3.19 for SPSS, both using the general linear model analyses.

```
Source         DF  Seq SS   Adj SS   Adj MS     F      P
Catalyst        1   3.000    3.000    3.000  1.06  0.343
Temp            2  40.667   40.667   20.333  7.18  0.026
Catalyst*Temp   2  38.000   38.000   19.000  6.71  0.030
Error           6  17.000   17.000    2.833
Total          11  98.667
S = 1.68325   R-Sq = 82.77%   R-Sq(adj) = 68.41%
```

Fig 3.18 Two-way GLM/ANOVA output (Minitab) with *interaction* for the data in Fig 3.15

The important conclusion is that the *interaction* between catalyst and temperature is significant, with $p = 0.030$. In this respect, *both* catalyst and temperature are significant factors in the overall model, even though the *p*-value for the 'average' catalyst effect is greater than 0.05. By including interaction as an additional factor, the R^2(adj) value has increased from 23.35% to 68.41% demonstrating the better overall fit of the model to the experimental values.

Dependent Variable: Yield

Source	Type III Sum of Squares	df	Mean Square	F	Sig.
Corrected Model	81.667[a]	5	16.333	5.765	.027
Intercept	66901.333	1	66901.333	23612.235	.000
Catalyst	3.000	1	3.000	1.059	.343
Temp	40.667	2	20.333	7.176	.026
Catalyst * Temp	38.000	2	19.000	6.706	.030
Error	17.000	6	2.833		
Total	67000.000	12			
Corrected Total	98.667	11			

a. R Squared = .828 (Adjusted R Squared = .684)

Fig 3.19 Two-way GLM/ANOVA output (SPSS) with *interaction* for the data in Fig 3.15

An *interaction* is frequently labelled as product of the two (or more) factors, e.g. *Catalyst** *Temp*. The use of the *product* in labelling an interaction also reflects the way in which an interaction can be modelled in a GLM of the system (Eqn 3.12).

It is important to note that an ANOVA can only identify an *interaction* if the data includes *replicate measurements* under the same combinations of factor levels. The ANOVA needs to estimate the *experimental uncertainty* before it can separate the interaction from experimental error. Comparing the two sets of results, we can see that, in Fig 3.16, the variance due to the interaction is 'hidden' within the *Error* term with $SS_{ERROR} = 55.0$, but in Figs 3.18 and 3.19 we have a partitioning of this variance to give $SS_{INTERACTION} = 38.0$ plus $SS_{ERROR} = 17.0$.

3.3.3 Analysis of Covariance, ANCOVA

The factors considered in previous sections each have a *limited* number of levels, typically defined by the experiment, e.g. three levels of temperature in Fig 3.15. We now introduce a factor that is a *continuous* variable whose value we do not necessarily control in the selection of our subjects for analysis, but which might show some correlation with the measured variable. This is referred to as a *covariate*.

ANCOVA
Excel and Minitab analyses for Fig 3.20. See also 6.4.8. Scan here to watch the video or find it via www. oxfordtextbooks. co.uk/orc/ currell/

Case study: Ink analysis / 5. ANCOVA analysis 1

—continued from 5.1.6 (overview), 5.2.3, and 3.6.2

The data in Fig 3.20(a), plotted in Fig 3.20(b), shows the percentage transmission, %*T*, for two inks *A* and *B* measured at *different* wavelengths, and we wish to test for a *vertical* difference in %*T* between the lines. For a related problem in 3.6.2, we are able to use a 'repeated measures' analysis which requires that measurements are made for each line at the *same* wavelengths. This is not the case here, but we will assume that the %*T* values for each line varies *linearly* with wavelength, as appears in Fig 3.20(b).

	A	B	C	D	E	F	G
1	Ink	Wavelength	%T	Best-fit	Residuals		
2							
3	A	698.00	47.16	43.98	3.18	Best-fit line:	
4	A	705.00	52.84	50.21	2.63	Slope =	0.89
5	A	712.00	58.63	56.45	2.18	Intercept =	-577.95
6	B	695.00	38.48	41.30	-2.82		
7	B	702.00	45.37	47.54	-2.17		
8	B	708.00	52.56	52.89	-0.33		
9	B	715.00	56.45	59.12	-2.67		
10	Two-Sample t-test results:						
11		p =	0.432	p =	0.001		

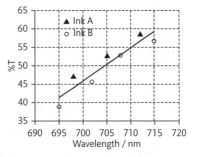

(a) Excel calculations   (b) Plotting data from (a)

Fig 3.20 Case study: Ink analysis / 5. ANCOVA analysis

We wish to test for a difference in %T between lines A and B, but a two sample t-test based on the values in column C gives $p = 0.432$ (in C11) which fails to identify a significant difference. However, in Fig 3.20(b), we can see that there may be a difference between the A and B samples in that the A points tend to be more to the *top left* of the graph and the B points to the *bottom right*. However, the t-test *only* aims to test for a difference between the two groups *in the vertical direction*, and the effect of the wavelength covariate has been to spread out both samples vertically, giving the appearance of a greater uncertainty.

In the analysis of covariance, ANCOVA, we first use a *linear regression analysis* to calculate the slope (G4) and intercept (G5) values for the best-fit straight line that relates %T to wavelength. We then calculate the points on the best-fit line (column D) for each wavelength, giving the trendline in Fig 3.20(b). Then, in column E, we take the differences between the values in columns C and D to get the *residuals* for each wavelength. These residuals are plotted in Fig 3.21, where we can see that there is a clear difference in vertical *residuals* between lines A and B. If we now perform a two sample t-test between the residual values we get $p = 0.001$ (E11), reporting a highly significant difference between the two lines.

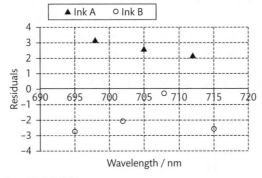

Fig 3.21 Residual values from Fig 3.20(b)

For comparison, we use the GLM/ANCOVA analysis on the three lines A, B, and C, using the data from Fig 3.40(a):

Minitab

Minitab > Stat > ANOVA > General Linear Model > Fit General Linear Model...

Reponses: $\%T$ **Factors:** *Ink*   **Covariates:** *Wavelength*

$\rightarrow$ Output: Similar to Fig 3.22 (a)

and then for the post hoc tests and data plots:

Minitab > Stat > ANOVA > General Linear Model > Comparisons...

Response: $\%T$ **Type of comparison:** ∇ *Pairwise*

Select a post hoc test (3.2.4), e.g. ☑-*Tukey*

Choose terms for comparison: Double click on *ink* to see: C Ink

> Graphs... ☑-**Interval plot for difference in means**

> Results...

☑-**Grouping information**

☑-**Tests and confidence intervals**

$\rightarrow$ Output: Same information as in Fig 3.22 (b)

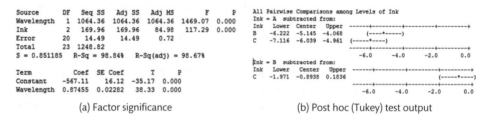

(a) Factor significance   (b) Post hoc (Tukey) test output

Fig 3.22 ANCOVA output for data from Fig 3.40(a) (Minitab 16) (Minitab 17 uses a graphical plot to display the confidence intervals)

The p-values in the ANCOVA test output in Fig 3.22(a) confirm that the wavelength effect is significant and also that there is a significant difference somewhere between the inks. It also reports the linear regression calculation to find the best-fit straight line for the three lines *A*, *B*, and *C*, giving slope, $m = 0.875$ and intercept, $c = -567$, which closely matches the values of 0.89 and -578 that we obtained above using reduced data for just lines *A* and *B*.

From the Tukey test results in Fig 3.22(b), we see that neither of the confidence intervals (given by the bracketed ranges) for the differences between *A* and *B* and *A* and *C* overlap 0, so we conclude that there is a significant difference between *A* and both *B* and *C*. This agrees with the Excel result obtained for *A* and *B* in Fig 3.20(a). However, the confidence interval, -1.97 to 0.184, for the difference between *B* and *C* does overlap 0, which means that we are unable to detect the difference between *B* and *C*.

The analysis of this data using SPSS is carried out in 6.4.8.

3.4 General linear model

The general linear model (GLM) uses the techniques of linear regression to develop a mathematical model to describe the factor effects, interactions, and uncertainties in a system. The historical advantage of the ANOVA was perhaps that it gave a more intuitive understanding of the analysis and that small problems could be analysed by hand. However, with modern software the GLM provides a more comprehensive analytical approach that includes regression, ANOVAs, and t-tests.

The GLM develops a regression analysis, as in Chapter 2, and assumes that the dependent variable is derived from a normal distribution with a constant variance, which is consistent with a wide range of other familiar techniques, e.g. t-test, ANOVA, linear regression. Not only does it duplicate these standard analyses, but it also provides a greater flexibility in their implementation for different analytical problems. The general*ized* linear model (3.4.7) provides more flexible options permitting a choice in the underlying distribution and integrated transformations.

3.4.1 General linear model

We saw in (2.1.1) that the behaviour of y as a simple function of x can be written as

$$y = mx + c \text{ or } y = b_0 + bx.$$

If the variation of y is dependent on more factor variables, x, then we can express it using a linear combination of x_A, x_B, etc.:

$$y = b_0 + b_A x_A + b_B x_B + b_{AB} x_A x_B + b_C x_C \ldots \tag{3.12}$$

where the $x_A x_B$ term represents an *interaction* (3.3.2) between the factors x_A and x_B.

It is also possible to introduce a *power* term into the analysis by using 'interaction' terms through products of the *same* variable: $x_A^2 = x_A x_A$ and $x_B^3 = x_B x_B x_B$, etc.

We can investigate the effect of the interaction term if we choose example values for the coefficients, $b_0 = 0.0$, $b_A = 0.5$, $b_B = 0.5$ and $b_{AB} = 4.0$, producing the equation:

$$y = 0.5 \ x_A + 0.5 \ x_B + 4.0 \ x_A x_B.$$

We calculate the observed values of y for values of $x_A = 0.5$ and 2.0 and $x_B = 0.5$ and 3.0, giving:

Table 3.2 Interaction calculation

	$x_B = 0.5$	$x_B = 3.0$
$x_A = 0.5$	$y = 0.25 + 0.25 + 1.0 = 1.5$	$y = 0.25 + 1.5 + 6.0 = 7.75$
$x_A = 2.0$	$y = 1.0 + 0.25 + 4.0 = 5.25$	$y = 1.0 + 1.5 + 24.0 = 26.5$

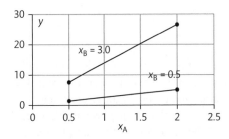

Fig 3.23 Graphical illustration of an interaction

We can plot the values from Table 3.2 using the graph in Fig 3.23, in which we see that, due to the interaction term, the two lines have different *slopes*, and the *effect* of x_A on y is *dependent* on the value of x_B.

3.4.2 **GLM, ANOVA, and the *t*-test**

It is useful to use a simple example to demonstrate how the linear regression approach can provide the same analytical results as the familiar *t*-test.

General linear model: Excel analysis for Fig 3.24. Scan here to watch the video or find it via www. oxfordtextbooks. co.uk/orc/ currell/

Case study: **River *pH* / 4. GLM, ANOVA, and the *t*-test**

—continued from 3.1.3 and 3.2.2

The independent samples of *pH* data for two rivers, A and B, were introduced in 3.1.3, and are reproduced here in Fig 3.24. A *t*-test and ANOVA calculation (3.2.2) gave the same *p*-value = 0.052 for a significant difference in their mean values. We now use a linear regression analysis to retest for this difference.

For consistency with linear regression in Chapter 2, we identify the measured *pH* values in Fig 3.24 as the '*y*' response values, and identify the two levels, *A* and *B*, with dummy '*x*' values, 1 and 2. We can now plot the data on the *x–y* graph in Fig 3.25.

	A	B	C	D	E	F	G	H	I
1		'x'	pH						
2	A	1	7.56	n =	8				
3	A	1	7.52	SE =	0.0702				
4	A	1	7.7						
5	A	1	7.61		SS	df	MS	F	p
6	B	2	7.47	Regression (B):	0.0288	1	0.0288	5.85	0.052
7	B	2	7.4	Residual (W):	0.0296	6	0.0049		
8	B	2	7.49	Total:	0.0584	7			
9	B	2	7.55						

Fig 3.24 Case study: River *pH* / 4. GLM, ANOVA, and the *t*-test

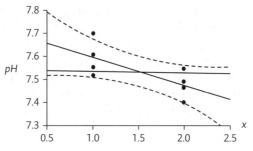

Fig 3.25 *pH* values for two rivers, coded as '1' and '2'

A hypothesis test for a significant *difference* between the mean values of the data at $x = 1$ and $x = 2$ becomes equivalent to showing that the line between their true mean values will *not* have a zero slope. Hence the null hypothesis for this test becomes:

H_0: The linear relationship between *pH* and '*x*' has a *zero* slope.

Using the techniques in Section 2.2, we can draw the 95% confidence limits, given as dashed lines in Fig 3.25, within which possible best-fit lines may lie. We see immediately that it is just possible, within 95% confidence, to draw a horizontal best-fit line that does have a *zero slope* at a constant value of about 6.53 *pH*. We cannot therefore reject the null hypothesis, and this implies that we cannot be confident at 0.05 significance that there is a *difference* between the samples *A* and *B*. This agrees with the *t*-test conclusion in 3.1.3.

Using the same techniques from 2.1.2 and 2.1.4, we derive the values for a hypothesis test:

SE_{REG}:	[E3] = STEYX(C2:C9, B2:B9)	= 0.0702
SS_{RESID}:	[E7] = E3^2 * (E2 − 2)	= 0.0296 (using Eqn 2.14)
SS_{TOT}:	[E8] = VAR.S(C2:C9) * (E2 − 1)	= 0.0584 (using Eqn 2.8)
$SS_{REGRESS}$:	[E6] = E8 − E7	= 0.0288 (using Eqn 2.9)

where the *factor* in the analysis is the regression fit of the data to a straight line and the *random* uncertainty is given by the residual values.

Dividing by the relevant degrees of freedom (Eqn 1.18), we can calculate the *MS* values, 0.0288 in G6 and 0.0049 in G7, and then obtain the *F*-statistic in H6 using Eqn 3.8:

$$F = \frac{MS_{REGRESS}}{MS_{RESID}} = \frac{0.0288}{0.0049} = 5.85$$

The *p*-value for a non-zero slope can then be calculated:

p-value: [I6] = F.DIST.RT(H6, F6, F7) = 0.052

This is exactly the same value as calculated using the two sample *t*-test (3.1.3) and ANOVA (3.2.2). Hence, we see the parallel between the techniques of linear regression and the analysis of variance.

Historically, statistical packages have grouped their menu options based on specific *tests*, but the use of GLM calculations has created overlaps between these tests, resulting in different approaches to analysis. In Minitab 16, GLM options are available under both *ANOVA* and *Regression* headings. In SPSS v20 the individual tests can still be accessed through the legacy dialogs, but the main menu option under *Analyze* is *General Linear Model* followed by the identification of the data structure (univariate, multivariate, repeated measures) before presenting a range of possible analyses.

The practical use of the general linear model to conduct ANOVA 'style' *factor* analyses is developed in Chapter 6. Section 3.4.3 uses the general regression option for analysing a 'regression' problem.

3.4.3 General regression

The following case study gives a further example of the overlap, from a *regression* perspective, between regression and ANOVA techniques.

General regression (Minitab): Analysis for Fig 3.26 data. See also 7.2.5. Scan here to watch the video or find it via www. oxfordtextbooks. co.uk/orc/ currell/

> ## Case study: **Bacterial growth / 3. Difference in slopes as an interaction**
> —continued from 3.1.1
>
> In 3.1.1 we used a *t*-statistic analysis which identified a difference between the *slopes* of two bacterial growth curves with a *p*-value = 0.0149. We now analyse the same data, reproduced in Fig 3.26, using a general regression model.

We wish to test whether there is a difference in the slopes of the two (nearly) straight lines, C3 and C2, in Fig 5.5, over the *t*-range from 60 to 85. The same data from Fig 3.1 is reproduced in Fig 3.26, but with the output variable, *Data*, entered into the single column, B.

	A	B	C	D	E
1	*Time*	*Data*	*Conc*	*ConcN*	*Interact*
2	60	1.76	C3	1	60
3	65	2.78	C3	1	65
4	70	3.93	C3	1	70
5	75	5.06	C3	1	75
6	80	6.55	C3	1	80
7	85	7.24	C3	1	85
8	60	2.98	C2	2	120
9	65	3.68	C2	2	130
10	70	5.54	C2	2	140
11	75	6.8	C2	2	150
12	80	8.55	C2	2	160
13	85	9.55	C2	2	170

Fig 3.26 Bacterial growth as a function of time under difference conditions, C3 and C2

A general linear model gives *Data* as the linear function:

$$Data = b_0 + b_1 \times Time + b_2 \times Conc + b_{12} \times Time \times Conc$$

The 'average' slope of the two lines with time will be given by the coefficient, b_1, but the *difference* in slope between the two lines will be generated by the *interaction* term, b_{12}.

The null hypothesis for the test of a difference between the two slopes is then:

H_0: The interaction term, b_{12}, is zero.

It is important to note that the concentrations are only identified here as *nominal* values, *C3* and *C2*, but this is not a problem as we do not need to calculate a numeric *value* for b_{12}. We only need to test whether it is zero or not.

Minitab

Minitab > Stat > Regression > Regression > Fit Regression Model...

Responses: *Data*   **Continuous predictors:** *Time*   **Categorical predictors:** *Conc*

> Model: Highlight *Time* and *Conc* in **Terms in the model,** then

Cross predictors and terms in the model - Add

Delete any cross terms that are not required.

The terms in the model should now include *Time, Conc, Time*Conc*

→ Output: similar to Fig 3.27

```
Source        DF    Seq SS    Adj SS    Adj MS        F         P
Regression     3   64.5545   64.5545   21.5182  365.105  0.000000
Time           1   56.0205   56.0205   56.0205  950.517  0.000000
Conc           1    7.9707    0.1723    0.1723    2.924  0.125661
Time*Conc      1    0.5632    0.5632    0.5632    9.557  0.014858
Error          8    0.4715    0.4715    0.0589
Total         11   65.0260
```

Fig 3.27 General regression calculation for data from Fig 3.26 (Minitab 16)

This gives $p = 0.0149$ for the interaction term which agrees with the calculation in 3.1.1, again showing that there is a significant difference between the two slopes.

As a matter of interest, the difference, for *Conc*, between the 'sum of squares' values *seqSS* = 7.907 and *adjSS* = 0.1723 occurs because *seqSS* is calculated *before* the interaction term is taken into account, and the variance contribution then becomes less *after* including the interaction (*adjSS*). This topic is discussed in 3.4.5.

It is also useful to note that, if we were using a standard ANOVA calculation, it would not be possible to analyse for an *interaction* term, because this data does not include any replicate measurements from which the ANOVA could assess the experimental uncertainty (3.3.2). However, the GLM regression calculation is able to assess this uncertainty by measuring the *residual values* between the data values and the best-fit linear model.

It is possible to derive the same results by performing a standard regression calculation in Minitab 16, but to do this it is necessary to code the concentrations with the dummy *numeric* variable, *ConcN*, in column D, and introduce a new dummy variable, *Interact*, in column E whose value is simply the product *Time*ConcN*.

Minitab 16

Minitab> Stat > Regression > Regression...

Response: *Data*

Predictors: *Time ConcN Interact*

→ Output: Fig 3.28

```
The regression equation is
Data = - 9.90 + 0.177 Time - 2.05 ConcN + 0.0507 Interact

Predictor      Coef   SE Coef       T      P
Constant     -9.903     1.895   -5.23  0.001
Time        0.17691   0.02595    6.82  0.000
ConcN        -2.049     1.198   -1.71  0.126
Interact    0.05074   0.01641    3.09  0.015
```

Fig 3.28 Standard regression calculation in Minitab 16 for data from Fig 3.26

The values of the coefficients, b_0, b_1, b_2, and b_{12}, in the regression equation are not numerically relevant because they depend on our choice of *dummy* variables for the concentration. However, the important value is $p = 0.015$ for the *Interact* term, which again agrees with the other hypothesis tests.

3.4.4 Fixed and random factors

With the increased flexibility in the GLM analyses, we need to be more aware of exactly how the statistical analysis matches the objectives of the scientific investigation, and we start by looking at the difference between *fixed* and *random* factors.

Using GLM in software, we are able to define which of the input factors might be 'random' as opposed to 'fixed'. The important difference is the scope of any conclusions that can be drawn from the analysis.

A **fixed factor** has specific levels defined by the experiment, e.g. choice of water samples when measuring *pH*, and, for the fixed factor, the conclusion is relevant *only* for the levels chosen from that factor. For example, when testing two water samples, a conclusion of 'no significant difference' only applies to those samples, and cannot be extended to any other water samples.

A **random factor** has its levels selected at random from a *population* of possible levels, and a conclusion of 'no significant difference' then applies to the whole population of levels.

3.4.5 Sequential and adjusted sums of squares

The concept of sum of squares, *SS*, was introduced in 2.1.2 and in 3.2.2 for the different factors in an ANOVA. An important feature of attempting to fit a *number* of factors into a mathematical model is that it is possible to consider different *sequences* with which the different factors are introduced into the model.

Sequential sum of squares, *seqSS*, for a given factor, describes the amount of variation that is explained by that factor after the *previous* terms have been taken into account. This value could also include some variations that might be explained by a new factor that is to be introduced *later*.

Adjusted sum of squares, *adjSS*, for a given factor describes the amount of variation that is explained by that factor after *all other* terms have been taken into account. Unlike *seqSS*, the *adjSS* term will not include any variations that can be explained by a factor that is to be introduced later (e.g. in Fig 3.27). The *adjSS* value can be interpreted as indicating how much *new information* is provided by this factor separately from other factors.

In the following examples we see how the *adjSS* value calculated for a factor can be less than the *seqSS* for the same factor, because a *later* factor in the sequence explains some of the variation that was previously included in the *seqSS* value. We also see that, for the *last* factor in the model, the *adjSS* value will equal the *seqSS* value because the *previous* terms are the same as *all other* terms in the model.

A choice can be made as to whether to use the *seqSS* or *adjSS* values to calculate the *MS* values in the *p*-value calculations. The **Type I sum of squares** calculation uses *seqSS* but the normally default **Type III sum of squares** calculation uses *adjSS* values.

Case study: **Correlated variables / 2. Sums of squares**

—continued from 4.1.4

The data in Fig 3.29 presents an output variable, R, measured together with two input variables, v1 and v2, reproduced from Fig 4.7(a). An additional variable, v3, has been added, which is the same as v2 except for two small changes in rows 3 and 5, and this effect is discussed in 3.4.6. The discussion in 4.1.4 differentiates between bivariate and partial correlation for related variables, and we now investigate the effect of these relationships between input variables when developing a regression model. The analysis in 3.4.6 demonstrates the difference between 'pure error' and 'lack of fit'.

	A	B	C	D
1	R	v1	v2	v3
2	10.00	2.24	0.94	0.94
3	12.30	2.39	1.11	1.12
4	9.60	1.71	0.76	0.76
5	8.00	1.85	0.89	0.91
6	9.80	1.65	0.85	0.85
7	9.60	2.18	1.12	1.12
8	8.20	2.08	0.81	0.81
9	9.50	2.30	0.91	0.91
10	7.40	1.69	0.72	0.72
11	11.90	2.50	1.25	1.25

Fig 3.29 Response variable, R, with possible predictor variables, v1, v2, and v3

We start with a general regression analysis of R against $v1$, with the result in Fig 3.30.

```
Source       DF   Seq SS    Adj SS    Adj MS       F          P
Regression    1   10.1809   10.1809   10.1809   6.80989   0.0311508
      v1      1   10.1809   10.1809   10.1809   6.80989   0.0311508
Error         8   11.9601   11.9601    1.4950
Total         9   22.1410
S = 1.22271        R-Sq = 45.98%            R-Sq(adj) = 39.23%
```

Fig 3.30 Regression analysis of R vs v1

The analysis shows an apparent 'dependence' on $v1$, with $p = 0.031$. However, in Fig 3.31 we include $v2$ in the analysis.

Source	DF	Seq SS	Adj SS	Adj MS	F	P
Regression	2	13.6980	13.6980	6.84901	5.67845	0.034241
v1	1	10.1809	0.0890	0.08904	0.07382	0.793695
v2	1	3.5171	3.5171	3.51714	2.91603	0.131461
Error	7	8.4430	8.4430	1.20614		
Total	9	22.1410				

S = 1.09824   R-Sq = 61.87%   R-Sq(adj) = 50.97%

Fig 3.31 Regression analysis of R vs v1 and v2

The regression model in Fig 3.31 is again significant with $p = 0.034$, and we can see the increased 'fit' with the data through the increase in the R^2(adj) from 39.23% to 50.97%.

Notice the dramatic effect of introducing the v2 term on the SS values for v1. Before considering the effect of v2, we have $seqSS(v1) = 10.1809$ which would include the variations due to the *following v2* term, but after including the effect of v2, the value becomes $adjSS(v1) = 0.0890$. This suggests that it is v2 that is responsible for much of the variance in the data.

However, why has the p-value increased (from 0.031 to 0.034) when we have introduced an important term, which we might expect to reduce the p-value? The answer is that by including another factor, we increase the probability (p-value) of obtaining the observed data through random chance when using *two* variables rather than just one.

Examining the data in Fig 3.31, we might also ask whether there is an apparent contradiction in that, although the overall regression is significant, the p-values for both v1 and v2 are greater than 0.05. The answer is that the analysis is using the Type III method where the *individual* p-values are based on the *adjSS* values, which provide the significance of the individual factor *after* the effect of the other factors have been taken into account. We saw in the previous analysis that v1 *alone* already provides evidence of a significant effect, so that the *additional* information from v2 is then less significant with $p = 0.131$, and hence there is no contradiction, and similarly for v1. The individual p-values for v1 and v2 correspond to their individual *partial correlations* (4.1.4) with R, which measure the remaining variation after the other factors have been taken into account.

Source	DF	Seq SS	Adj SS	Adj MS	F	P
Regression	2	13.6980	13.6980	6.84901	5.67845	0.034241
v2	1	13.6090	3.5171	3.51714	2.91603	0.131461
v1	1	0.0890	0.0890	0.08904	0.07382	0.793695
Error	7	8.4430	8.4430	1.20614		
Total	9	22.1410				

S = 1.09824   R-Sq = 61.87%   R-Sq(adj) = 50.97%

Fig 3.32 Changing the order of v1 and v2

If we now consider that v2 might be more relevant than v1, and *change the order* in which the model is developed, we find that the conclusions in Fig 3.32 are the same as the previous analysis. The order of v1 and v2 does not make any difference to the p-values, because, using the Type III method, the p-values are calculated using *adjSS* values which are independent of factor order.

Source	DF	Seq SS	Adj SS	Adj MS	F	P
Regression	1	13.6090	13.6090	13.6090	12.7604	0.0072707
v2	1	13.6090	13.6090	13.6090	12.7604	0.0072707
Error	8	8.5320	8.5320	1.0665		
Total	9	22.1410				

S = 1.03272   R-Sq = 61.47%   R-Sq(adj) = 56.65

Fig 3.33 Regression analysis of R vs v2

Finally, with the belief that *v1* is not an important factor, we *remove* it from the analysis, and we now see, in Fig 3.33, a very significant effect. The use of the single factor, *v2*, in our model has considerably reduced the likelihood of the observed data occurring *by chance* to $p = 0.0073$. You should also note that the degrees of freedom, *df*, for regression equals the number of factors involved, and the reduction from two to one similarly affects the calculated *p*-value.

3.4.6 Lack of fit and error

It is possible to illustrate a further aspect of using the general linear model. In Fig 3.29 the data set *v3* is the same as *v2*, *except* that the value in row 3 is changed from 1.11 to 1.12 and the value in the row 5 from 0.89 to 0.91.

```
Source          DF   Seq SS   Adj SS   Adj MS      F          P
Regression       1  13.4215  13.4215  13.4215  12.3140  0.007972
  v3             1  13.4215  13.4215  13.4215  12.3140  0.007972
Error            8   8.7195   8.7195   1.0899
  Lack-of-Fit    6   3.9495   3.9495   0.6582   0.2760  0.907072
  Pure Error     2   4.7700   4.7700   2.3850
Total            9  22.1410
S = 1.04400      R-Sq = 60.62%        R-Sq(adj) = 55.70%
```

Fig 3.34 Regression analysis of *R* vs *v3*

Performing the general regression analysis for the single factor *v3* we get the results in Fig 3.34. The small change in the *v2* data has made only a minimal difference in the *p*-value results compared with Fig 3.33. However, there are now *two pairs* of *replicate* measurements in the data which have the same *v3* values of 1.12 and 0.91. By using these replicate values, the analysis is able to make an estimate of the inherent *experimental uncertainty* in the measurements, and this variation is calculated in the output as the 'pure error'. Then, by removing 'pure error' from the total error variations, the analysis can attribute the remaining variation to a 'lack of fit' between the calculated model and the experimental data.

3.4.7 Generalized linear model

The general linear model considered so far is based on the assumption that the data is normally distributed and that the underlying link between input and output data is a linear response. The general*ized* linear model (GsdLM) takes the analysis further to include systems that do not necessarily satisfy those two assumptions, and thereby seeks to encompass a wider range of analyses within one unified approach.

The GsdLM analysis starts by specifying the *three main elements* of the system being analysed:

- The experimental **uncertainty distribution** which describes the inherent random distribution of experimental data, e.g. normal, Poisson.
- A **link function** which compensates for nonlinear behaviour, either as a result of a theoretical relationship in the system (e.g. using a log function to linearize an

exponential, 2.3.4) or in the type of variables being analysed (e.g. using the logit function for a binary or ordinal variable, 8.3.4).

• Identifying the types of input variable(s) as **factor(s)** and/or **covariate(s)**.

Although the details of the process are beyond the scope of this book, we can demonstrate the use of the GsdLM by analysing a simple exponential radioactive decay.

Generalized linear model (Poisson loglinear): SPSS analysis leading to Fig 3.35. See also 6.4.7. Scan here to watch the video or find it via www.oxfordtextbooks.co.uk/orc/currell/

> ## Case study: **Exponential decay / 5. Generalized linear model**
> —continued from 2.3.4 and 2.4.3, leading to 7.2.3
>
> The linearization of radioactive decay to calculate half-life was described in 2.3.4, based on the data in Fig 2.16. We now use the *GsdLM* to repeat the analysis.

The counts, N_t, in radioactivity as a function of time, t, are given by the exponential decay:

$$N_t = N_0 \times e^{kt}$$

with $k = -0.693/T_{1/2}$, where $T_{1/2}$ is the half-life.
By taking logarithms of both sides of the equation we get:

$$\ln(N_t) = \ln(N_0) + k \times t$$

To calculate $T_{1/2}$ we need to measure the slope in the straight line defined by the covariates $\ln(N_t)$ and t. The *link function* between N_t and t is therefore a logarithm.
The *uncertainty* in radioactivity is based on the very low random probability of each atom decaying, and therefore follows a Poisson distribution (1.3.3).
We can see how this is implemented in **SPSS** using the data from Fig 2.16:

SPSS > Analyze > Generalized Linear Models > Generalized Linear Models...
Type of Model: Either select a standard option: ⊙-*Poisson loglinear*
   or: ⊙-*Custom* and select **Poisson** distribution and **Log** link function
Response: Dependent variable: N
Predictors: Select t as a **Covariate**
Model: Specify t for the **Model**
→ Output: Fig 3.35

Parameter Estimates

Parameter	B	Std. Error	95% Wald Confidence Interval		Hypothesis Test		
			Lower	Upper	Wald Chi-Square	df	Sig.
(Intercept)	4.599	.0859	4.431	4.768	2863.529	1	.000
t	-.494	.0460	-.584	-.403	115.002	1	.000
(Scale)	1ᵃ						

Dependent Variable: N
Model: (Intercept), t
  a. Fixed at the displayed value.

Fig 3.35 *SPSS* output using the GsdLM

The output in Fig 3.35 confirms that the model fits the data with p-values less than 0.0005, and provides best estimate values for the coefficients, B, of the equation:

$$\ln(N_t) = B_{Intercept} + B_t \times t = 4.599 - 0.494 \times t$$

We can then derive values for N_0 and k:

$\ln(N_0) = 4.599$ from which we can derive $N_0 = 99.4$, and

$k = -0.494$.

These results agree with the result obtained in 2.4.3 using Solvcr with the Poisson distribution.
We could repeat the analysis (incorrectly) using the normal distribution of uncertainty with the log link function and obtain:

$\ln(N_0) = 4.584$ from which we can derive $N_0 = 97.9$, and $k = -0.476$

which also agrees with the similar analysis in 2.4.3.

3.5 Nonparametric analyses

Nonparametric tests do not make any assumptions about the distribution of the random variations in the data. They use the values only to establish their relative *rank order*, with the magnitude of the *difference* between adjacent values having no significance. As these differences are not relevant, the calculations are not affected if particular sections of data are more or less 'spread out' in value. They are independent of the 'shape' of the distribution of values, and are therefore called 'distribution-free' tests.

We normally need to use nonparametric tests if we have

- continuous data that does not have a known (e.g. normal, Poisson) distribution
- ordinal data (e.g. Likert scale data) or
- data that has already been given ranked values.

The common nonparametric tests are based on similar principles, in which a test statistic is derived from the distribution of ranks, and is then either compared with a critical value or used to calculate a p-value. There is no need to describe the process underlying each test, and we will just consider the example of the Mann–Whitney test which is the nonparametric equivalent of the independent samples t-test. Section 3.5.2 reviews other nonparametric tests in relation to their parametric equivalents.

3.5.1 Mann–Whitney example

The Mann–Whitney test is a hypothesis test for a difference between the *medians* of two independent samples. The typical calculation first works out the *rank values* of all of the data, and then derives a test statistic that describes the distribution of these ranks between the samples. Then, assuming that the null hypothesis were true, it calculates the probability of that distribution, or one more extreme, occurring by chance. As with other tests, the final stage can be completed either by comparing the calculated test statistic with a critical value or, in modern software, by calculating a p-value directly.

Case study: River pH / 5. Mann–Whitney test

—continued from 3.1.3, leading to 3.9.2

Fig 3.36 gives the pH (in row 2) from two samples relating to two rivers, A and B, identified in row 1.

For the nonparametric test we test for a difference in the *median* values, given the null hypothesis:

H$_0$: The two population medians are equal, $m_A = m_B$.

	A	B	C	D	E	F	G	H	I	J	K
1	River		A	A	A	A		B	B	B	B
2	Data / pH		6.56	6.52	6.7	6.61		6.47	6.4	6.49	6.55
3	Rank		3	5	1	2		7	8	6	4
4					$W_A =$	11				$W_B =$	25
5			$n_A =$	4	$U_A =$	1		$n_B =$	4	$U_B =$	15

Fig 3.36 Case study: River pH / 5. Mann–Whitney test

The procedure is to rank all of the data values across *both* samples in row 3. It does not matter whether we rank from 'high to low' or 'low to high' as long as we are consistent. When ranking *negative* values, it is useful to interpret the statement that '*a* is greater than *b*' as actually stating that '*a* is more *positive* than *b*', e.g. $-3 > -4$ is interpreted as '-3 is *more positive* than -4'. Where two or more data have the same values, the *average* rank should be returned for each of the possible values.

In Fig 3.36, we calculate the rank of each value in row 3 out of all the values from C2 to K2, using the function (for example):

[C3] = RANK.AVG(C2, $C2:$K2).

The '$' signs lock the column values when we copy this function to other cells in row 3 to generate the other ranks in C3 to K3.

We then calculate the rank totals, W_A, and W_B, in F4 and K4, for *each* sample,

$$W_A = 3+5+1+2 = 11$$
$$W_B = 7+8+6+4 = 25.$$

The two possible *extreme* situations for two samples each with four values would be:

- If the two sample median values were *very different* with *no overlap* in sample values, then W_A would be equal to 10 ($=1+2+3+4$) and W_B equal to 26 ($=5+6+7+8$), or vice versa.

- If the median values were the *same* (null hypothesis), then W_A and W_B would become closer in value.

We see in 3.9.2 how we can use the Monte Carlo method to test whether the W values are more extreme than would be expected by chance, and calculate a p-value. However, when using the critical value approach we need to derive a new statistic, U_i, for each sample, i, (where i is A or B):

$$U_i = W_i - n_i(n_i + 1)/2 \qquad (3.13)$$

which give values for each sample of

$$U_A = W_A - n_A(n_A + 1)/2 = 11 - 4\times5/2 = 1$$
$$U_B = W_B - n_B(n_B + 1)/2 = 25 - 4\times5/2 = 15.$$

These values can then be compared with a table of critical values, U_C, where we **reject the null hypothesis** if either:

$$U_A \leq U_C \text{ or } U_B \leq U_C.$$

In our example, for two samples each of size four, the critical values are $U_C = 0$ for two-tailed test and $U_C = 1$ for a one-tailed test. For a *two-tailed test*, neither U_A nor U_B are less than or equal to $U_C = 0$, hence we *do not reject* the null hypothesis. However, if we had originally decided to perform a *one-tailed test*, $U_C = 1$, and, because $U_A = U_C$, we would accept that the median for river A was a significantly higher than for river B. These conclusions are consistent with the parametric t-test result in 3.1.3.

3.5.2 Nonparametric and parametric test equivalents

Table 3.3 gives the nonparametric alternatives equivalent to the most common parametric tests, together with links for examples of their operation.

Table 3.3 Parametric and nonparametric test equivalents

Parametric test	Link	Nonparametric test	Link
One sample t-test (mean)	6.1.4	One sample Wilcoxon test (median)	6.1.5
Two sample t-test (means)	6.2.5	Mann–Whitney test (medians)	6.2.6
Paired t-test (means)	6.2.7	Paired Wilcoxon test (medians)	6.2.8
1-way ANOVA (means)	6.3.5	Kruskal–Wallis test (medians)	6.3.7
2-way ANOVA (means)	6.4.4	Friedman test (medians)	6.4.6
Pearson's r (linear correlation)	7.1.4	Spearman's p (monotonic correlation)	7.1.4
F-test (variance)	6.2.4	Levene's test (variance)	6.2.4

The following tests use *ranking* as a means of quantifying the relationships within the data set.

Wilcoxon signed rank test (6.1.5) is the nonparametric equivalent of the **one sample t-test**, and tests whether the sample has been drawn from a population that has a *median* value that is different from a specific value, m_0. The calculation derives a test statistic, W, based on the *ranking* of values on either side of the test value, m_0.

Wilcoxon paired test (6.2.8) is the nonparametric equivalent of the **paired *t*-test**, and tests whether there is a significant difference between the median values of two *related* samples.

Kruskal–Wallis test (6.3.7) is the nonparametric equivalent of the **one-way ANOVA,** in that it tests for a difference between the median values of *k*-samples (where $k > 2$). All data values are given their rank value, and the test develops a test statistic, *H*, which is a measure of how *unevenly* the ranked values are spread between the samples. A large *H* value indicates a difference between the median values of the samples, and can be compared with a critical value using the chi-squared distribution.

	A	B	C	D	E
1		D1	D2	D3	D4
2	C4	76	77	74	74
3	C5	75	74	72	71
4	C6	77	75	75	72

Fig 3.37 Same data as Fig 3.11 for a two-way analysis of catalyst yields

Friedman test (6.4.6) is the nonparametric equivalent of the **two-way ANOVA,** in that the analysis of the effect of one factor is tested while 'blocking' the effect of the other. For the data in Fig 3.37, a Friedman test would be testing for a difference in median values due to the *day* factor, *D1, D2, D3,* and *D4,* while taking into account (blocking) the catalyst levels *C4, C5,* and *C6.* The test for the significance of the *catalyst* factor would require the data to be *transposed* with the columns defined by the catalyst and the rows by the day.

The next tests use *binomial probability* as a basis for their analysis.

Sign test is another nonparametric equivalent of the **one sample *t*-test**, but, unlike the Wilcoxon test, it only *counts* the numbers of values in the sample on either side of the test value, m_0, and does not rank the differences. It is therefore less powerful than the Wilcoxon test.

Runs test (6.1.6) is a simple test for the randomness with which values above and below a particular value appear in a data set, and can be used as a check on the expected randomness of experimental data.

Fisher's exact test (4.2.3) tests the numbers of events falling into *two categories* (i.e. proportions) with the frequencies that might be expected by chance.

Some measures of relative relationship uses *concordant and discordant data pairs* (4.3.5). For example, **Kendall's test for concordance** (4.4.3) can be used to test for the *agreement* between several variables.

The other major family of nonparametric analyses (developed in Section 3.7) is based on the **chi-squared** *probability distribution*. These analyse categorical data and have no parametric equivalents.

3.6 Repeated measurements

An important option in experimental design is called 'repeated measures' and involves making repeated measurements of the *same subject*, but under different conditions. A common example is where the same test is repeated 'before' and 'after' an intervention on each one of a number of different subjects, and consequently this experimental arrangement is often called a 'within subject' design. The term 'subject' follows the common use of this type of analysis in questionnaires asking related questions on human 'subjects'. However, in general, the term 'subject' only represents a unique 'link' that identifies related data values, and can easily represent an inanimate connection, e.g. the specific *pH* meters in the case study in 3.6.1 below.

When the linked measurements are between just *two* samples, the analysis is called a 'paired' test. Where repeated measures *within* subjects are used together with tests *between* subjects it is called a *mixed* design.

3.6.1 Paired samples

Pairing between two data samples occurs when each data value in one sample shares a unique 'link' with one data value in the other sample. In this section, we will develop the *parametric* analysis of the paired *t*-test for testing for a difference in mean values. The *non-parametric* equivalent is the paired Wilcoxon test for a difference in median values. How to use Minitab and SPSS for the two tests is given in 6.2.7 and 6.2.8 respectively.

Case study: River *pH* / 6. Paired *t*-test

—continued from 3.1.3

We saw in 3.1.3 that when using the independent samples *t*-test for four *pH* measurements from two rivers, A and B, there was no significant difference between the mean *pH* values. However, we now include the additional information that four *pH* meters, M1, M2, M3, and M4, were used, with each meter making one measurement of each river. Hence each pair of data values are now linked by a unique meter, as in rows 2 to 5 in Fig 3.38.

⊿	A	B	C	D	E	F	G	H	I
1		A		B		Diff, *d*		Size	4
2	M1:	6.56	-	6.47	=	0.09		*df* =	3
3	M2:	6.52	-	6.4	=	0.12			
4	M3:	6.7	-	6.49	=	0.21		*t*-statistic, t_s =	3.70
5	M4:	6.61	-	6.55	=	0.06		*t*-critical (95%), *t* =	3.18
6								*p*-value =	0.034
7				Mean		0.120			
8				StDev		0.065		TTEST() =	0.034

Fig 3.38 Case study: River *pH* / 6. Paired *t*-test

In terms of experiment design we call this a 'repeated measures' design or a 'within subjects' design, where the different 'subjects' are identified as the different *pH* meters, and the paired values are 'related'.

To analyse the data, it is appropriate to take the differences, 'within subjects', between the values of each linked pair, calculated as *Diff*, *d*, in column F. The sample mean and standard deviation of these differences are calculated in F7 and F8 respectively. If there is no difference in the true mean values of *A* and *B* then we would expect that the true mean value, $\bar{d}$, of *Diff* would be zero.

Hence the *paired* test has now become a *one sample t*-test (3.1.2), testing whether the mean of *Diff* is significantly different from zero. We can derive the relevant *t*-statistic in I4 using Eqn 3.3:

$$t_s = \frac{(\bar{d} - 0)}{s / \sqrt{n}}: \qquad [I4] = F7 / (F8 / SQRT(I2))$$

The degrees of freedom for the test are given by Eqn 3.4, $df = n - 1$, and calculated in [I2]

df: $[I2] = I1 - 1 = 3$

We then calculate the *p*-value (two-tailed) from the *t*-statistic:

p-value: $[I6] = T.DIST.2T(I4, I2) = 0.034$

or, from the raw data directly:

p-value: $[I8] = T.TEST(B2:B5, D2:D5, 2,1) = 0.034$

where the '2' in the argument identifies a two-tailed test and the '1' is for paired data.

With $p = 0.034 < 0.05$, this paired test shows a significant difference between the *pH* values of the two rivers, whereas the unrelated *t*-test in 3.1.3 failed to detect the difference.

We can use graphical representations of the data in Fig 3.39 to see why the *paired t*-test identified a difference that was missed by the *independent* samples *t*-test. Fig 3.39(a) is drawn using a 'line graph' in Excel to plot each data value against the meter used, and it can be seen that there is a pattern of variation between the different meters, with M2 giving lower readings than the other meters. This may be due to poor initial calibration of the meters, but, whatever the physical cause, there is a *bias* (systematic error) between the meters, which is ignored by the *independent* samples *t*-test and appears as increased random uncertainty in the data. The independent samples *t*-test is then unable to identify the difference in mean values in the presence of the greater apparent random uncertainty. The *paired t*-test performs the difference calculation 'within' each pair of meter readings where the random variation 'between' the meters can have no effect.

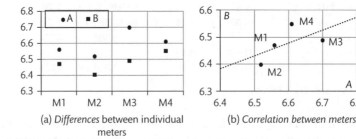

(a) *Differences* between individual meters

(b) *Correlation* between meters

Fig 3.39 Identifying bias between *pH* meters

It is also useful to plot each *pair* of meter readings from the two rivers on an *x–y* graph, as in Fig 3.39(b) where each dot represents a separate *pH* meter. Any *bias* between the meters will result in higher reading meters towards the top right and the lower reading meters towards the bottom left of the graph, and we would expect to see a positive slope for a best-fit straight line. In this case, a test for correlation between the data values gives $r = 0.588$ which does show a degree of positive correlation, although it is not statistically significant with $p = 0.412$.

3.6.2 Repeated measures

When the number of 'linked' samples exceeds two it becomes a *repeated measures* (within subjects) design. As an example, we use the 'Ink analysis' case study data to test for a difference between three inks: *A, B,* and *C*.

Case study: Ink analysis / 4. Repeated measures

—continued from 5.1.6 and 5.2.3, leading to 3.3.3

The data in Fig 3.40(a) gives the percentage transmission, *%T*, for three inks as a function of wavelength, and we wish to test whether there is a significant difference between them. We can use a repeated measures analysis because there is a unique wavelength link between measurements made with each ink.

Repeated measures 1: SPSS analysis leading to Figs 3.41, 3.42, and 3.43. See also 6.3.8. Scan here to watch the video or find it via www.oxford textbooks. co.uk/orc/ currell/

	Wavelength	A	B	C
1	695.00	44.03	38.48	38.47
2	698.00	47.16	42.93	40.93
3	702.00	49.45	45.37	43.86
4	705.00	52.84	47.49	48.05
5	708.00	57.27	52.56	50.23
6	712.00	58.63	53.42	53.95
7	715.00	62.04	56.45	55.50
8	718.00	65.33	58.89	57.45

(a) SPSS data

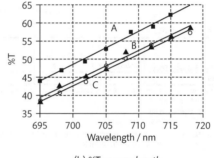

(b) *%T vs wavelength*

Fig 3.40 'Repeated' *%T* values for three inks at each of eight wavelengths

---

We analyse this data using the *parametric* 'repeated measures' analysis in SPSS, and in Minitab we can treat the 'within-subjects' factor and 'between-subjects' factor as factors in a two-way ANOVA. The *nonparametric* analysis of repeated measures uses the Friedman test introduced in 6.4.6.

SPSS

SPSS > Analyze > General Linear Model > Repeated Measures...
Within Subjects Factor Name: *Inks* (we enter a name to describe the 'repeated' factor)
Number of levels: 3 (corresponding to the three inks)
> Add
> Define: Transfer *A, B,* and *C* to **Within Subjects Variables (Inks)**
> Model... ⊙-Custom Transfer *Inks* to **Within Subjects Model**

> **Options...** Transfer *Inks* **to Display Means for**
  ☑-**Compare Main Effects**
  Confidence Interval Adjustment: Select Bonferroni or Sidak
→ Output: Figs 3.41, 3.42, and 3.43

As with the two sample example, the analysis works by taking differences between the possible pairs of each subject, and an important requirement is that the variances of these values should be the same between all possible pairs. A test for this equality of variance, albeit with some limitations for both large and small samples, is Mauchly's test of sphericity. SPSS performs this test as a default (Fig 3.41), and also gives values for epsilon, which is a multiplication factor that can be applied to the degrees of freedom for the F-test if the sphericity condition is not met.

We see in Fig 3.41 that Mauchly's test of sphericity gives, $p = 0.571$, which allows us to assume that the equality of variance condition has been met. SPSS then gives p-values in Fig 3.42 for significant differences in the sample mean value, both on the assumption that the sphericity condition is met and also for other corrected tests if it is not: Greenhouse–Geisser, Huynh–Feldt, and Lower-bound (most conservative).

Mauchly's Test of Sphericity[a]

Measure: MEASURE_1

Within Subjects Effect	Mauchly's W	Approx. Chi-Square	df	Sig.	Epsilon[b] Greenhouse-Geisser	Huynh-Feldt	Lower-bound
Inks	.830	1.121	2	.571	.854	1.000	.500

Tests the null hypothesis that the error covariance matrix of the orthonormalized transformed dependent variables is proportional to an identity matrix.

  a. Design: Intercept
    Within Subjects Design: Inks

  b. May be used to adjust the degrees of freedom for the averaged tests of significance. Corrected tests are displayed in the Tests of Within-Subjects Effects table.

Fig 3.41 Test for equality of variance in a repeated measures analysis

This is a 'mixed' design which tests for differences both between wavelengths and between inks. The analysis gives $p = 0.000$ (not given in figures here) for the 'between-subjects' test confirming that there is a significant difference in %T for different *wavelengths*. However, we wish to test, 'within-subjects', for a difference between the *inks*, A, B, and C, and this is reported in Fig 3.42 showing a significant difference between at least two of the inks.

Tests of Within-Subjects Effects

Measure: MEASURE_1

Source		Type III Sum of Squares	df	Mean Square	F	Sig.
Inks	Sphericity Assumed	169.964	2	84.982	164.845	.000
	Greenhouse-Geisser	169.964	1.709	99.461	164.845	.000
	Huynh-Feldt	169.964	2.000	84.982	164.845	.000
	Lower-bound	169.964	1.000	169.964	164.845	.000

Fig 3.42 Significance results for the 'within subjects' test

Although not required in this analysis, we can see that the corrected tests in Fig 3.42 use the same F-statistic, 164.8, but the degrees of freedom are adjusted by the value of epsilon. Note for example that the corrected degree of freedom for the Greenhouse–Geisser test is equal to the initial degree of freedom multiplied by the relevant epsilon value, 0.854, from Fig 3.41:

$$2 \times 0.854 = 1.709 \text{ (within rounding errors)}$$

The direct post hoc option is for locating differences between 'between-subjects' factors. However, we wish to locate the difference 'within-subject', i.e. between A, B, and C, and we use the 'compare main effects' for inks under options as a means for applying a post hoc test (LSD, Bonferroni, or Sidak) to the repeated measurements. Using the Bonferroni test we get the comparisons in Fig 3.43.

Pairwise Comparisons

Measure: MEASURE_1

(I) Inks	(J) Inks	Mean Difference (I-J)	Std. Error	Sig.[b]	95% Confidence Interval for Difference[b]	
					Lower Bound	Upper Bound
1	2	5.145*	.275	.000	4.284	6.006
	3	6.039*	.390	.000	4.819	7.258
2	1	-5.145*	.275	.000	-6.006	-4.284
	3	.894	.398	.179	-.352	2.140
3	1	-6.039*	.390	.000	-7.258	-4.819
	2	-.894	.398	.179	-2.140	.352

Based on estimated marginal means
*. The mean difference is significant at the .05 level.
b. Adjustment for multiple comparisons: Bonferroni.

Fig 3.43 Post hoc tests for 'within-subjects'

In Fig 3.43 the inks, A, B, and C, are labelled as 1, 2, and 3, corresponding to the order of entry into the software, and the Bonferroni comparison then identifies significant differences ($p = 0.000$) between A and B and between A and C, but not ($p = 0.179$) between B and C, which is consistent with the results in 3.3.3.

3.7 Chi-squared analyses

The chi-squared, χ^2, test is a hypothesis test that compares frequencies with which 'events' are *observed* to occur in different 'categories' with the frequencies that could have been *expected* based on the science involved. The experimental data can appear first as a simple list of individual observations or events, and the process of adding up the *numbers* of observations/events that fall into specific categories is called 'tabulation' (putting the data into tables). We deal with tabulation and cross-tabulation in 8.1.7 and 8.2.4 respectively, but in this section we just analyse the resultant frequency values.

The principle calculation for the chi-squared statistic, χ^2, uses Pearson's formula introduced in Eqn 3.19, but can also use the Yates continuity correction in Eqn 3.21 or the likelihood ratio in Eqn 3.22.

3.7.1 **Tabulated data**

Figs 3.44 and 3.45 show the two main forms of tabulated data, in which the data values are frequencies (simple counts) recorded as integer values.

	A	B	C	D
1	Category, i	1	2	3
2	Observed, O	23	26	11
3	Expected, E	15	30	15

Fig 3.44 one-way frequency table

	A	B	C	D
1	Gender \ Subject	*Science*	*English*	*History*
2	*M*	11	18	17
3	*F*	18	20	6

Fig 3.45 two-way contingency table

For the one-way table in Fig 3.44, the chi-squared test compares the *observed* distribution of frequencies in row 2 with a distribution of *expected* frequencies in row 3, and tests whether any observed difference between the distributions could have occurred by chance, or whether the difference is statistically significant. This is often called a 'goodness of fit' test.

In the two-way 'contingency' table in Fig 3.45, the rows and columns are identified by different levels of two factors: *Gender* and *Subject*. The chi-squared analysis compares the *distributions* of frequencies between different rows and between different columns, and tests whether these differences are statistically significant. If, for example, the distribution *across the columns* changes significantly between the rows, then there is said to be an *association* between the two factors. Such an association between the factors would also cause a difference in the distribution *down the rows* for different columns.

The chi-squared test, applied to a contingency table, is a hypothesis test which identifies whether any apparent association between the two factors could have occurred by chance, and is a *symmetric* measure (4.3.2) in that it identifies a relationship between the two factors, but does not identify one as being dependent on the other. In Sections 4.2 and 4.3, we consider a range of other statistics that can be used to measure the *strength* of association.

3.7.2 **One-way 'goodness of fit'**

We can illustrate the principle of the chi-squared, χ^2, test by using the simple one-way table of observed frequencies, O, and expected frequencies, E, in Fig 3.44.

The null hypothesis for the test is:

H_0: The distribution of observed values is the same as the expected ratios (in this case given directly by expected values).

The relevant formula for the Pearson's chi-squared statistic is:

$$\chi^2 = \sum_i \frac{(O_i - E_i)^2}{E_i} \tag{3.14}$$

This formula is essentially a set of instructions which says that for each observed value, i, divide the square of the difference between the observed and expected value by the expected value, and then add up these results for all the values, i. For the example data, this gives:

$$\chi^2 = \frac{(23-15)^2}{15} + \frac{(26-30)^2}{30} + \frac{(11-15)^2}{15} = 4.27 + 0.53 + 1.07 = 5.87$$

It is clear that, as the difference between observed and expected frequencies increases, then the value of χ^2 will also increase. However, we would expect that, just by random chance, there will be differences and χ^2 will not be zero. As with other test statistics, we can use tables of critical values, χ_C^2, to decide how large χ^2 must be before we decide that the difference is significant. We accept that the difference is significant and reject the null hypothesis if:

$$\chi^2 \geq \chi_C^2$$

The significance of the calculated χ^2 value also depends on the degrees of freedom, df, which, for the one-way test is given by:

$$df = n - 1 \tag{3.15}$$

where n is the number of categories.

For the example with $n = 3$, $df = 2$, the critical chi-squared value, $\chi_C^2 = 5.99$. Since $\chi^2 < \chi_C^2$ there is not enough evidence to reject the null hypothesis and we conclude that the observed distribution is not significantly different from the expected values.

In practice, it is usually necessary to calculate the expected frequencies based on expected ratios or probabilities, as is demonstrated in the following case study.

Case study: **Chi-squared / 2. One-way 'goodness of fit' test**

—continued from 8.1.1 (overview), leading to 3.9.3

It is expected that genotypes *AB*, *Ab*, *aB*, and *ab* will be observed in the ratio 9:3:3:1.

In an experimental measurement, the number of genotypes observed in these categories were 125, 28, 39, and 8 respectively. We wish to test whether the observed frequencies show a significant deviation from the expected ratios.

Chi-squared 'goodness of fit': Excel analysis for Fig 3.46. See also 8.1.5. Scan here to watch the video or find it via www.oxford textbooks. co.uk/orc/ currell/

The null hypothesis for this test is:

H_O: The observed distribution of frequencies occurred randomly with probabilities based on the given theoretical ratios.

	A	B	C	D	E	F	G
1	Category	Expected ratios	Expected proportions	Observed numbers	Expected numbers	Difference	Chi-squared
3	i	R_i	P_i	O_i	E_i	$O_i - E_i$	$(O_i - E_i)^2 / E_i$
4	1 (AB)	9	0.5625	125	112.5	12.5	1.39
5	2 (Ab)	3	0.1875	28	37.5	-9.5	2.41
6	3 (aB)	3	0.1875	39	37.5	1.5	0.06
7	4 (ab)	1	0.0625	8	12.5	-4.5	1.62
8	$\Sigma_i R_i =$	16	$N = \Sigma_i O_i =$	200		$\chi^2 =$	5.48
9						$\alpha =$	0.050
10	$p =$	0.140	= CHISQ.DIST.RT()			$df =$	3
11	$p =$	0.140	= CHISQ.TEST()			$\chi^2_c =$	7.81

Fig 3.46 Case study: Chi-squared / 2. One-way 'goodness of fit' test

In Fig 3.46, the four categories are identified by the values of i in A4:A7 with the observed frequencies, O_i in D4:D7 and the expected ratios, R_i, in B4:B7. The chi-squared calculation works by comparing the observed frequencies, O_i, for each category i, with the frequencies, E_i, that would be observed if all the observations were distributed *exactly* according to the expected ratios.

We start by converting expected *ratios*, R_i, into expected *proportions*, P_i, by first calculating the sum of all ratios in B8

$$[B8] = SUM(B4:B7) = 16$$

and then, for each category, i, using the equation:

$$P_i = \frac{R_i}{\sum_i R_i} \tag{3.16}$$

e.g. $[C4] = B4 / B\$8 = 9 / 16 = 0.5625$ or directly $[C4] = B4 / SUM(B\$4:B\$7)$

The expected *frequency* for each category is then calculated by *multiplying* the total number of observed events, N, (calculated in D8) by the expected proportion, P_i, for that category:

$$E_i = N \times P_i \tag{3.17}$$

e.g. $[E4] = SUM(D\$4:D\$7) * C4 = 200 \times 0.5625 = 112.5$

Note that the theoretical expected values can have *non-integer* values, whereas all observed frequencies must have integer values.

Now that we have the observed and expected frequencies for each category, we can use the Pearson's formula (Eqn 3.14) for the chi-squared statistic:

$$\chi^2 = \sum_i \frac{(O_i - E_i)^2}{E_i}$$

The formula is essentially a set of instructions, which are implemented as below:

1. For each category, i, calculate $O_i - E_i$ in column F:
   e.g. $[F4] = D4 - E4 = 12.5$

2. Square this value and divide by E_i in column G:
   e.g. $[G4] = F4^2 / E4 = 1.39$

3. Take the sum of the values for all categories, i, to calculate the chi-squared value:
   $[G8] = SUM(G4:G7) = 5.48$

The degrees of freedom, calculated from Eqn 3.15, are $df = 4 - 1 = 3$, which is entered into G10.

The critical chi-squared value for a specific significance and degrees of freedom can be found in look-up tables, or it can be calculated using the function, CHISQ.INV.RT(), with the significance level, $\alpha = 0.05$ (in G9). This is the value in the chi-squared distribution which defines the upper, or right-hand, 5% tail of the distribution:

$$[G11] = CHISQ.INV.RT(G9, G10) = 7.81$$

As the experimental value of $\chi^2 = 5.48$ is less than 7.81, we decide that there is not enough evidence to indicate that the distribution is significantly different from the null hypothesis.

We can also calculate a p-value using Excel (or other software), using the function, CHISQ.DIST.RT(), based on the values of χ^2 and df,

[B10] = CHISQ.DIST.RT(G8, G10) = 0.140

or the function, CHISQ.TEST(), to calculate directly from the observed and expected values.

[B11] = CHISQ.TEST(D4:D7, E4:E7) = 0.140

The fact that the value of $p = 0.140$, calculated in B10 and B11, is greater than the significance level of 0.05 is consistent with the decision that there is not enough evidence that the distribution is significantly different from the expected ratios.

3.7.3 Low value of chi-squared

We are almost always looking for the chi-squared statistic being *greater* than a critical value to show that there is a variation in the data that is greater than would be expected by chance. However, it is useful to note that it also possible to test whether the observed variations are significantly *less* than would be expected by random chance. This would be done by calculating the *left-hand* critical value of the distribution using the function CHISQ.INV(). For example, for $\alpha = 0.05$ and $df = 3$, χ^2 would have to be less than the critical value of 0.35 to show a significant *lack* of random variation. Such a situation might point towards a problem with the data collection, possibly with the data values not being truly independent.

3.7.4 Contingency table

The other main use of the chi-squared calculation is in the analysis of a contingency table, and in this section we develop the basic test for an *association* between factors. Further analyses are presented in Section 4.2, and the analysis of contingency tables using Minitab and SPSS is then developed in Section 8.2.

A typical contingency table is defined in two dimensions by two factors. Each individual observation can be 'placed' in one of the cells defined by the two axes, and the total *number* in each cell is recorded as a 'frequency'.

Contingency table test for association: Excel analysis for Fig 3.47. See also 8.2.4. Scan here to watch the video or find it via www.oxford textbooks. co.uk/orc/ currell/

Case study: **Association / 2. Contingency table**

—continued from 8.2.1

In this example, a total of 90 children, 46 boys and 44 girls, each identify their favourite subject, *Science*, *English*, or *History*, resulting in the numbers given in Fig 3.47.

The null hypothesis for the test is:

H$_0$: The distribution of choices for boys is the same as for girls.

	A	B	C	D	E	F
1	Observed values:					
2	Gender \ Subject	Science	English	History	Totals	
3	M	11	18	17	46 $= R_1$	
4	F	18	20	6	44 $= R_2$	
5	Totals	29	38	23	90 $= T$	
6		$= C_1$	$= C_2$	$= C_3$		
7	Expected values:					
8	Gender \ Subject	Science	English	History	Totals	
9	M	14.82	19.42	11.76	46 $= R_1$	
10	F	14.18	18.58	11.24	44 $= R_2$	
11	Totals	29	38	23	90 $= T$	
12		$= C_1$	$= C_2$	$= C_3$		
13	Chi-squared values:					
14	Gender \ Subject	Science	English	History		
15	M	0.986	0.104	2.340		
16	F	1.030	0.109	2.446		
17				Sum =	7.015	
18	p-value (CHISQ.DIST.RT) =	0.030		df =	2	
19	p-value (CHISQ.TEST) =	0.030		Critical value =	5.991	

Fig 3.47 Contingency table / 2. Tests for association

The observed data values are in shaded cells B3:D4, and the first step is to calculate the total frequencies for each row, R_1 and R_2, for each column, C_1, C_2, and C_3, and the total $T = 90$.

The column totals, $C_1:C_2:C_3$, give the *average* ratios with which the events are distributed between the columns. If there is no difference between the rows, then we could expect the events in each row to be distributed in this same proportion.

If we wish to calculate the expected frequencies in the first row in the table, we need to distribute the R_1 events ($= 46$) in the ratio $C_1:C_2:C_3$, where $C_1+C_2+C_3 = T$. This is similar to the calculation that we performed in the one-way table in 3.7.2 using Eqn 3.16,

$$E_{1j} = R_1 \times \frac{C_j}{T} = \frac{R_1 \times C_j}{T}$$

In general, for row i, we can write

$$E_{ij} = \frac{R_i \times C_j}{T} \tag{3.18}$$

We use Eqn 3.18 to calculate the expected values in the shaded cells, B9:D10. For example, the expected value, E_{11}, in B9 is calculated:

$$E_{11} = \frac{46 \times 29}{90}: \qquad [B9] = \$E3 * B\$5 / \$E\$5 = 14.82$$

Note that the theoretical *expected* values can be noninteger. The use of the dollar signs in the Excel expression allows us to copy the same formula into all cells in B9:D10, because $E locks the reference to column E (for the row totals) and $5 locks the reference to row 5 (for the column totals).

We next calculate the contributions $(O - E) / E$ in shaded cells B15:D16:

e.g. $[B15] = (B3 - B9)^2/B9$

The calculated chi-squared contributions are now summed over all rows, i, and columns, j:

$$\chi^2 = \sum_{ij} \frac{(O_{ij} - E_{ij})^2}{E_{ij}} \qquad (3.19)$$

In our analysis we need to calculate, in E17, the sum of all values from B15:D16:

Chi-squared, χ^2: $\qquad$ $[E17] = SUM(B15:D16) = 7.015$

The critical chi-squared value will depend on the degrees of freedom for a contingency table, which are given by

$$df = (r - 1)(c - 1) \qquad (3.20)$$

where r and c are the numbers of rows and columns respectively.

In our example, $df = (2 - 1)(3 - 1) = 2$, which is entered into E18.

The critical chi-squared value for a specific significance and degrees of freedom can be found in look-up tables, or it can be calculated using the function, CHISQ.INV.RT(α,df), with the significance level, $\alpha = 0.05$. This is the value in the chi-squared distribution which defines the upper, or right-hand, 5% tail of the distribution:

$[E19] = CHISQ.INV.RT(0.05, E18) = 5.991$

As the experimental value of $\chi^2 = 7.02$ is greater than 5.99, we decide that there is a significant association between the choice of subjects and the gender of the child for the population from which the 90 children were a representative random sample.

We can also calculate a p-value using Excel, using the function, CHISQ.DIST.RT(), based on the values of χ^2 and df,

$[B18] = CHISQ.DIST.RT(E17, E18) = 0.030$

or the function, CHISQ.TEST(), to calculate directly from the observed and expected values,

$[B19] = CHISQ.TEST(B3:D4, B9:D10) = 0.030$

The fact that the value of $p = 0.030$, calculated in B18 and B19, is less than the significance level of 0.05 is consistent with the above decision that there is a significant association between the choice of subjects and the gender of the child.

3.7.5 Yates continuity correction

We need to note that there is a specific issue when using the standard chi-squared formula in Eqn 3.19, for problems where the degrees of freedom, df, $= 1$. Due to the fact that we are comparing a chi-squared value based on integer values with a continuous distribution, the standard calculation tends to *overestimate* the value of χ^2, which means it is more likely to produce Type I errors in borderline cases. This can be corrected by using the Yates continuity correction which modifies the formula slightly:

$$\chi^2 = \sum_{ij} \frac{\{|(O_{ij} - E_{ij})| - 0.5\}^2}{E_{ij}} \tag{3.21}$$

This formula states that for every category cell we must take the *positive* (or absolute) value of the difference between O and E before subtracting 0.5 and then squaring the result and dividing by E.

The need for this correction occurs for 2×2 contingency tables, which have just two levels for each factor and hence degrees of freedom, $df = 1$. However, the problem is equivalent to testing for a difference between two *proportions*, and it is also possible to use other tests including Fisher's exact test. See Sections 3.8 and 4.2.3.

3.7.6 Likelihood ratio

An alternative method of calculating a chi-squared value is based on comparing the likelihoods (probabilities) of obtaining the observed distribution of frequencies under the two possible hypotheses set for the test. This likelihood ratio is calculated as:

$$LR = 2 \sum_{ij} O_{ij} \times ln(\frac{O_{ij}}{E_{ij}}) \tag{3.22}$$

The statistic can also be described as G or G^2 and the test as a G-test.

The calculated value for the likelihood ratio chi-squared is usually slightly larger than the Pearson's chi-squared, χ^2, but they become closer for larger sample sizes.

3.7.7 Sample size limitations

An important limitation with the standard chi-squared test occurs with a limited number of observations and/or a large number of categories, resulting in low expected frequencies giving unreliable statistical conclusions. Cochran's criterion for the minimum reliable sample size is that all of the cells must have *expected* frequencies of at least one and at least 80% of the cells must have *expected* frequencies of five or over. Note that that the criterion refers to *expected* frequencies and it is possible that some *observed* frequencies could even be 0.

A common reason for low expected frequencies is that too many category levels have been included in the model for the amount of available experimental data. The options available when the minimum criteria are not met are:

- Collect more experimental data, although this is not always possible.
- Combine category levels (8.2.7).
- Use a resampling technique (3.9.3, 8.2.7).

3.8 Frequency and proportions

When recording frequencies in just two categories we are actual measuring *proportions*. There are different methods for testing proportions, and we will develop the use of:

- binomial theory
- normal approximation of the binomial theory
- chi-squared analysis of a 2 × 2 contingency table.

3.8.1 Probability distribution

In arriving at a frequency proportion we would normally count the number of measurements or observations that fall into each of two categories, and, for convenience, we can define the two categories here as 'Y' and 'N'. In statistics terminology, each observation is often called a *trial*, and a specific outcome, e.g. 'Y', an *event*. We can define the individual event probability, p, as:

p = probability that a *single randomly* chosen observation will give the outcome Y.

As there are just two options, we can use binomial probability theory from 1.3.3 to calculate the probability $p(r)$ that, out of n observations (trials), there will be r events in the 'Y' category:

Eqn 1.4: $p(r) =_n C_r \times p^r \times (1-p)^{(n-r)}$

Eqn 1.5: Population mean, $\mu = p \times n$

Eqn 1.6: Population standard deviation, $\sigma = \sqrt{\{np(1-p)\}}$

Given that the **proportion** of r events is defined by

$$P = \frac{r}{n}$$

we can derive expressions for the true value, Π (Greek capital *pi*), and standard deviation, σ_P, for the *proportion* in the population:

$$\textit{Proportion true value } \Pi = \frac{\mu}{n} = p \tag{3.23}$$

$$\textit{Proportion standard deviation } \sigma_P = \frac{\sigma}{n} = \frac{\sqrt{\{np(1-p)\}}}{n} = \sqrt{\frac{p(1-p)}{n}} = \sqrt{\frac{\Pi(1-\Pi)}{n}} \tag{3.24}$$

3.8.2 One proportion test

In a one proportion hypothesis test we have a sample of values which give an experimentally observed proportion, P, which is a *best estimate* of the true population proportion, Π. We wish to test whether the true proportion is equal to a specific test proportion, Π_0, which gives the null hypothesis:

H_0: $\Pi = \Pi_0$

It is possible to

- use the binomial theory to perform an 'exact' calculation of the p-value
- use the normal distribution approximation to calculate a confidence interval and perform a z-test
- use a chi-squared test for a 'goodness of fit' for the two frequencies compared to the test ratios.

These three approaches are demonstrated using the 'Frogs' case study.

One proportion: Excel and Minitab analysis for Fig 3.48. See also 6.1.7. Scan here to watch the video or find it via www.oxford textbooks. co.uk/orc/ currell/

Case study **Frogs / 2. One proportion test**

—continued from 6.1.7 (overview), leading to 3.8.3

The aim of the investigation is to test whether the proportion of female frogs in a given large lake is *greater than* 0.6. Randomly selecting a sample of 50 it is found that 37 are female, and we wish to test whether this proportion, $P = 37/50 = 0.74$, is significantly greater than the expected proportion of 0.6.

We start with the null hypothesis:

H_0: The true proportion of female frogs, $\Pi = \Pi_0 = 0.60$

Binomial test

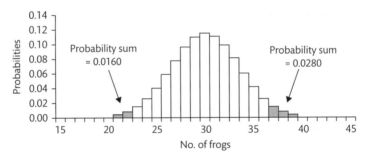

Fig 3.48 Binomial probabilities

Using the binomial equation (Eqn 1.4) for a random sample of 50 frogs ($n = 50$ trials), we can calculate the probabilities of observing r females, given that the probability of each individual being female, $p = 0.060$. The distribution of these probabilities is given in Fig 3.48, and we see that the probability of observing 37 or more female frogs is obtained simply by adding the individual shaded probabilities on the right, $p(r \geq 37) = 0.028$.

This is the probability, given that the *null hypothesis is true*, that we will observe 37 or more female frogs by chance. This is then the p-value for the one-tailed test:

p-value (1-tailed) = 0.028

Since p-value = 0.028 < 0.05, we would reject the null hypothesis and conclude that the female proportion was indeed greater than 0.6.

We can now also consider the calculation of the two-tailed p-value, but the distribution in Fig 3.48 is not exactly symmetrical, and the equation (Eqn 1.27) stating that the two-tailed p-value is twice the one-tailed value does not apply in this case. We see in Fig 3.48 that the two tails include '37 and above' with the probability of 0.028 and '22 and below' with the probability of 0.016, giving

$$p\text{-value (2-tailed)} = 0.028 + 0.016 = 0.044$$

which would show a significant difference between the observed proportion and 0.6.

Using a normal distribution approximation

We saw in 1.3.3 that the normal distribution can be used as an approximation for the binomial distribution, provided that $np(1-p) \geq 5$. In this case, $n = 50$ and the experimental value is $P = 0.74$, such that $nP(1-P) = 9.6$ which suggests that it may be a reasonable approximation.

Using the test value, the proportion standard deviation for the null hypothesis becomes

$$\sigma_P = \sqrt{\frac{0.60(1-0.60)}{50}} = 0.0693 \quad.$$

As we only make one measurement of the proportion, the standard error in the measurement is the same as the standard deviation (Eqn 1.21), and we can derive an expression for the z-statistic:

$$z = \frac{P - \Pi_0}{\sigma_P} = \frac{0.74 - 0.60}{0.0693} = \frac{0.14}{0.0693} = 2.02$$

The critical value for a one-tailed test is 1.64, and since $2.02 > 1.64$ we conclude that the measured proportion is significantly greater than 0.6. The test would also be significant for a two-tailed hypothesis where the critical value would be 1.96.

We can also derive an estimate for the confidence interval (Eqn 1.23) for the true proportion, by using the normal approximation. The experimentally measured proportion, P, is the best estimate for the true proportion, and using σ_P as the standard error and the t-value given by $z = 1.96$, we get:

$$CI = P \pm 1.96 \times \sqrt{\frac{P \times (1-P)}{n}} = 0.74 \pm 1.96 \times \sqrt{\frac{0.74 \times (1-0.74)}{50}} = 0.74 \pm 0.12$$

This is a symmetrical confidence interval, from 0.62 to 0.86, which is an *approximation* to the true confidence interval. This calculation puts the test value of 0.6 just outside lower limit of the confidence interval, again showing a significant difference.

Using Minitab to perform the same calculations (the use of SPSS is given in 6.1.7):

Minitab
Minitab > Stat > Basic Statistics > 1 Proportion...
▼ **Summarized data** (for directly entering observed frequencies)

Number of Events: 37   Number of Trials: 50

☑-**Perform hypothesis test**

Hypothesized proportion: 0.6

> **Options...** Under options we can choose one- or two-tailed tests and whether to use the normal distribution approximation

→ Output: Fig 3.49

Fig 3.49 gives the result for a one-tailed binomial test and a two-tailed test based on the normal distribution, recoding the same values as calculated above.

```
Test of p = 0.6 vs p > 0.6                Test of p = 0.6 vs p not = 0.6

                      95% Lower    Exact    Sample   X   N  Sample p        95% CI          Z-Value  P-Value
Sample   X   N  Sample p   Bound   P-Value    1        37  50  0.740000  (0.618419, 0.861581)   2.02     0.043
1        37  50  0.740000  0.618736  0.028
                                           Using the normal approximation.
```

(a) One-sided binomial test   (b) Two-sided test using the normal approximation

Fig 3.49 One proportion test (Minitab)

Performing a chi-squared test

We can perform a one-way chi-squared test with observed values of 37:13 for female and male frogs to be compared with an expected ratio of 0.6:0.4. The expected ratio of 0.6:0.4 for a total of 50 trials becomes equal to expected frequencies of 30:20, which are given as the one-way table of frequencies in Fig 3.50(a).

◢	A	B	C
1		F	M
2	Observed	37	13
3	Expected	30	20

◢	A	B	C
1		F	M
2	Lake 1	37	13
3	Lake 2	30	20

(a) One-way table of frequencies   (b) Contingency table of frequencies (see 3.8.3)

Fig 3.50 One and two proportion test data

Using the Yates continuity correction (Eqn 3.21), we can calculate the chi-squared value

$$\chi^2 = \frac{(37-30-0.5)^2}{30} + \frac{(13-20-0.5)^2}{20} = \frac{42.25}{30} + \frac{42.25}{20} = 3.52$$

As the chi-squared value of 3.52 is less than the critical value, 3.84, for $df = 1$, this test is unable to detect a significant difference between the observed proportion of 0.74 and 0.60 when based on 50 trials.

Reviewing the three different tests, we see that they all gave different results, although both the binomial test and the normal approximation z-test identified a significant difference, while the chi-squared test failed to detect a difference.

3.8.3 **Two proportions test**

> **Case study: Frogs / 3. Two proportions test**
>
> —continued form 3.8.2
>
> In 3.8.2, we compared the proportion of frogs, 37/50, to a *specific* test proportion of 0.6, but we now consider comparing the results of 37 out of 50 for one lake with the *experimental* results of 30 out of 50 for a second lake. We can express this situation using the *contingency* table in Fig 3.50(b).

Two proportions: Minitab analysis leading to Fig 3.51. See also 6.2.9. Scan here to watch the video or find it via www.oxford textbooks. co.uk/orc/ currell/

Although the two tables in Fig 3.50 have the same *values*, they represent different experimental situations. The second row in Fig 3.50(a) gives the *specific expected* values calculated for the one proportion calculation, but the second row in Fig 3.50(b) gives the *observed experimental* values for the second lake in a test between *two* proportions.

In a two proportion analysis we compare two pairs of experimentally measured frequencies from two data samples. This gives us the 2×2 contingency table, in which we test whether the distribution of frequencies (the proportion) in one row is significantly different from the distribution (the proportion) in the other row.

H_o: The proportions are the same for different rows and for different columns.

It is possible to use a chi-squared test with the Yates continuity correction, but the Fisher exact test for the 2×2 table is usually the preferred analysis as it is an exact test based on the binomial distribution.

It is also possible to use the normal approximation to the binomial distribution to test for a difference between proportions, using the test statistic:

$$z = \frac{P_A - P_B}{\sqrt{P'(1-P') \times (1/n_A + 1/n_B)}}$$

where P' is a pooled proportion for the two samples

$$P' = \frac{n_A P_A + n_B P_B}{n_A + n_B}$$

Using Minitab for a two proportions test:

Minitab

Minitab > Stat > Basic Statistics > 2 Proportions...
▼ **Summarized data** (for directly entering observed frequencies)
First: Events 37   **Trials** 50
Second: Events 30   **Trials** 50
→ Output: Fig 3.51

The output in Fig 3.51 gives $p = 0.132$ for the normal approximation and $p = 0.202$ for the Fisher's exact binomial test which both report that there is not enough evidence to claim a significant difference. The reason that the one proportion test for the data in Fig 3.50(a) finds a significant difference but the two proportion test for the same data in Fig 3.50(b) does not is that the two proportion test compares two *experimental* values both with inherent uncertainties, whereas, in the one proportion calculation, the test proportion of 30/50 is an *exact* value.

```
Difference = p (1) - p (2)
Estimate for difference:  0.14
95% CI for difference:  (-0.0422661, 0.322266)
Test for difference = 0 (vs not = 0):  Z = 1.51   P-Value = 0.132

Fisher's exact test: P-Value = 0.202
```

Fig 3.51 Two proportion test (Minitab)

The use of SPSS for a two proportion test is given in 6.2.9, and, if we use SPSS for the above calculation, it gives $p = 0.202$ for the Yates corrected chi-squared test as well as $p = 0.202$ for the two-tailed Fisher's exact test.

3.9 Resampling techniques

There is a developing range of techniques (e.g. *Monte Carlo*, *bootstrapping*) which use the experimental results observed for a system as a basis for *randomly* regenerating possible system values (typically 10,000 times), and then by analysing the *distribution* of these values it is possible to derive estimates of p-values. We can only give an introduction to the techniques here, starting with the general approach and then with examples of possible applications.

We will use the two sample 'River pH' case study to illustrate the method for calculating p-values for the t-test and Mann–Whitney test, and we will use the 'Chi-squared' case study to illustrate the calculation for a goodness of fit test. This allows us to compare the Monte Carlo results with those obtained by other methods. However, the real value of the technique is for analyses when alternative methods are not available, for example using the Monte Carlo method when low expected frequency values make the chi-squared analysis unreliable (8.2.7).

3.9.1 General approach to resampling

The basic procedure has the following steps:

Step 1. An initial analysis of the experimental data calculates the value of a relevant test statistic, e.g. t-statistic for a two sample t-test, W-statistic for a Mann–Whitney test, or the probabilities in a chi-squared test (3.7.2).

Step 2. A mathematical model is then developed, based on the null hypotheses, but including random variations in values. The resampling process then randomly generates a large number (e.g. 10,000) of possible values in the model.

Step 3. The value of the relevant test statistic is calculated for every generated data set.

Step 4. The final step calculates the *proportion* of resample data sets that give a test statistic value that is equal to, or more extreme (i.e. further from the null hypothesis) than the test statistic calculated from the *experimental* data. This proportion gives the *probability* that the null hypothesis could give an experimental result at least as extreme as the values observed—this is the *p*-value.

Because of the inherent randomness of the resampling process, there is also an inherent uncertainty in the calculated *p*-value. For this reason, the result is expressed as a *confidence interval* for the *p*-value.

3.9.2 *t*-test and Mann–Whitney test

Case study: River *pH* / 7. Monte Carlo analysis

—continued from 3.1.3 and 3.5.1

Four replicate *pH* measurements have been made of the water in each of two rivers, *A* and *B*, with the intention of performing a hypothesis test for a difference in the *pH* of the two rivers. The data is given in Fig 3.3.

Resampling *t*-test and Mann–Whitney test: Excel analysis for Fig 3.52. Scan here to watch the video or find it via www.oxford textbooks. co.uk/orc/ currell/

Step 1

The *pH* values already appear in Fig 3.3 and initial calculations have derived the *t*-test statistic, $t_s = 2.42$, and the pooled standard deviation for the two samples, $s' = 0.0702$, in 3.1.3, and the Mann–Whitney upper rank total, $W_U = 25$, in 3.5.1. These values have been entered into the Excel worksheet in Fig 3.52 in cells B4, D2, and B17 respectively.

Step 2

We develop a mathematical model of two samples with four values each. The first sample *A* is in cells F2:F5 and sample *B* in cells F6:F9. The null hypothesis assumes that there is no difference between the means of the populations from which the samples are drawn, and we will generate both samples with a mean value = 0. We assume that the two samples are drawn from populations which have the same standard deviation, and for this we use the pooled value, $s' = 0.0702$, in D2. In fact we could use any value for standard deviation in this calculation, but it is useful to use a model which matches the observed data.

To generate the random data values we use the Excel function in F2:

```
[F2] = NORM.INV(RAND(),0,$D$2)
```

which randomly selects a value from the normal distribution with mean of 0 and standard deviation given by the value in D2. This function may then be copied to other cells, F3:F9, to generate other randomly selected values for samples *A* and *B*. The '$' signs lock the reference to the row and column of cell D2.

We have now generated four random values for each of samples *A* and *B* in column F. The next step is to generate a total of 10,000 pairs of samples, all with independently selected data values. We do this by simply copying the formulae in F2:F9 to column NTU, which produces a total of 10,000 randomly generated data sets that are all based on the same model.

	A	B	C	D	E	F	G	H	NTT	NTU
1	t-test:				Rivers	1	2	3	9999	10000
2	Pooled stdev =			0.0702	A	0.007	0.125	0.176	0.079	-0.031
3					A	0.009	-0.038	0.043	0.080	0.040
4	t-statistic =	2.42			A	0.073	0.024	0.022	0.090	0.120
5					A	0.053	-0.085	-0.001	-0.054	0.091
6	Proportion =	0.0527			B	-0.029	-0.009	0.010	-0.057	0.082
7	StDev =	0.0022			B	0.013	-0.031	-0.097	0.030	-0.070
8					B	0.063	0.004	-0.074	0.004	0.055
9	p-range =	0.048	to	0.057	B	-0.043	0.158	-0.048	0.001	-0.053
10										
11					Pooled st dev =	0.041	0.088	0.065	0.055	0.071
12					t-statistic =	1.20	-0.38	2.45	1.39	1.02
13					Comparison =	0	0	1	0	0
14	Mann-Whitney test:									
15					Ranks	A	6	2	1	3	6
16					A	5	7	2	2	5
17	W-statistic =	25			A	1	3	3	1	1
18					A	3	8	5	7	2
19	Proportion =	0.0588			B	7	5	4	8	3
20	StDev =	0.0024			B	4	6	8	4	8
21					B	2	4	7	5	4
22	p-range =	0.054	to	0.063	B	8	1	6	6	7
23										
24					W_A =	15	20	11	13	14
25					W_B =	21	16	25	23	22
26					Comparison =	0	0	1	0	0

Fig 3.52 Case study: River pH / 7. Monte Carlo analysis (columns I to NTS are hidden)

Steps 3, 4: t-test

For the t-test, we first use Eqn 3.6 in row 11 to calculate the pooled standard deviation for each sample pair, e.g.

$$[F11] = SQRT((3*VAR.S(F2:F5) + 3*VAR.S(F6:F9)) / 6)$$

and then Eqn 3.5 in row 12 to calculate the t-statistic, e.g.

$$[F12] = (AVERAGE(F2:F5) - AVERAGE(F6:F9)) / (F11*SQRT(1/4+1/4))$$

For each of the generated 10,000 samples, we wish to test whether the t-value is *greater* (i.e. more extreme) than is observed for the experimental data. We do this by comparing the positive value (using the ABS() function) with the experimental value in B4, using

$$[F13] = IF(ABS(F12) > \$B4, 1, 0)$$

which returns a '1' only if the positive value of the generated t-statistic (in F12) is *greater* than the experimental value (in B4). The formulae in F11, F12, and F14 are copied to all 10,000 data sets.

We then calculate in B6 the *proportion* of data sets that record a '1' in row 13. This is the proportion of random samplings where the null hypothesis generates a test statistic greater than the experimental value. This is the *p*-value:

p-value: $[B6] = \text{SUM(F13:NTU13)} / 10000 = 0.0527$

The standard deviation uncertainty in this proportion is calculated in B7, using Eqn 3.24:

$[B7] = \text{SQRT(B6} * (1 - B6) / 10000) = 0.0022$

The confidence interval limits for the calculated *p*-value are then given in

$[B9] = B6 - 1.96 * B7$ and $[D9] = B6 + 1.96 * B7$

For the *t*-test in this particular example of randomly chosen data we get a *p*-value in the range 0.048 to 0.057, and this is consistent with the directly calculated *p*-value of 0.052 in 3.1.3.

Note that, every time the 'Enter' or 'F9 function' key is pressed, 10,000 new calculations will be performed giving new calculated confidence interval ranges for the *p*-value. We should find that, if we repeat the analysis many times, 95% of the confidence intervals should include the value 0.052.

Steps 3, 4: Mann–Whitney test

To perform the Mann–Whitney test on the two samples in F2 to F9, we first calculate the rank of each data value. For example the rank of the number in F2 (within the range of numbers in F2:F9) is calculated and recorded in F15 using the equation:

$[F15] = \text{RANK.AVG(F2,F\$2: F\$9)}$

This formula is then copied down to all cells F15:F22 to generate the ranks for the values in both data samples.

In F24 and F25 we calculate the values of W_A and W_B by simply calculating the total ranks for each of the samples, A and B. The formulae in F15 to F25 are copied to all columns up to NTU, giving 10,000 resamples of the Mann–Whitney calculation.

A pair of samples A and B will only give a value equal to, or more extreme than, the observed samples, if either W_A or W_B has a value equal to, or greater than, the experimental upper rank total, W_U, which is in B17. We can identify the data sets that fall into this category by using the comparison condition in row 26, for example:

$[F26] = \text{IF(OR(F24} >= \$17, \text{F25} >= \$B17), 1, 0)$

which returns a '1' only if the generated sets show a more extreme variation from the null hypothesis than the experimental data.

As with the *t*-test, we then calculate in B19 the *proportion* of data sets that record a '1'. This is the proportion of random samplings where the null hypothesis will record a test statistic greater than the experimental value. This is the *p*-value:

p-value: $[B19] = \text{SUM(F26:NTU26)} / 10000 = 0.0588$

The standard deviation uncertainty in this proportion is calculated in B20, using Eqn 3.24:

[B20] = SQRT(B19*(1−B19)/10000) = 0.0024

The confidence interval limits for the calculated p-value are then given in

[B22] = B19 − 1.96*B20   and   [D22] = B19 + 1.96*B20

For the Mann–Whitney test in this example calculation we get a p-value in the range 0.054 to 0.063, and this is consistent with the analysis in 3.5.1 where we did not reject the null hypothesis.

3.9.3 Chi-squared probabilities

The use of the Monte Carlo method has a distinct advantage over the standard chi-squared test when dealing with *low expected frequencies* in one or more categories (3.7.7). We develop the chi-squared case study to illustrate this application.

Resampling chi-squared: Excel analysis for Fig 3.53. Scan here to watch the video or find it via www.oxford textbooks. co.uk/orc/ currell/

Case study: Chi-squared / 3. Monte Carlo analysis

—continued from 3.7.2

The chi-squared test was used in 3.7.2 to assess whether the distribution of four experimental frequencies was significantly different from the expected ratios 9:3:3:1 predicted by Mendel's theory.

In this previous problem there was an expected frequency of at least five in each category. We now consider the same problem, but with a new set of experimental frequencies of 7, 8, 3, and 2 respectively in the four categories, which gives three categories with expected frequencies of less than five.

With the low data numbers, the standard chi-squared calculation is no longer reliable, and we use the Monte Carlo method as outlined in 3.9.1.

Step 1

In Fig 3.53, the experimentally observed numbers are entered in D3:D6, giving a total frequency of [D7] = 20.

From the expected ratios, 9:3:3:1 in B3:B6 we can calculate these ratios as probabilities in C3:C6 giving 0.563, 0.188, 0.188, 0.063 (to 3 dp), e.g.

[C3] = B3 / SUM(B$3:B$6)

We then calculate the expected values in E3:E6, e.g.

[E3] = D$7*C3

The chi-squared statistic, χ^2, for the experimental data is calculated in D8 using Eqn 3.14.

[D8] = (D3−$E3)^2/$E3+(D4−$E4)^2/$E4+(D5−$E5)^2/$E5+(D6−$E6)^2/$E6

giving $\chi^2 = 7.022$

Three of the expected frequencies in C3:C6 are less than five, which prevents us from comparing the chi-squared value with critical values from tables or using the standard calculation of p-value, but we can continue with the Monte Carlo method.

	A	B	C	D	E	F	G	H	I	J	NTU	NTV
1	Category	Expected	Expected	Observed	Expected		1	2	3	4	9999	10000
2		ratios	proportions	numbers	numbers							
3	AB	9	0.5625	7	11.25		11	9	13	9	7	12
4	Ab	3	0.1875	8	3.75		3	7	3	6	3	6
5	aB	3	0.1875	3	3.75		2	3	3	5	7	1
6	ab	1	0.0625	2	1.25		4	1	1	0	3	1
7			Total =	20								
8			Chi-squared Statistic =	7.022	Re-sampled =	7.022	3.467	0.622	3.467		7.022	3.467
9			p =	0.081	Comparison =	1	0	0	0		1	0
10			StDev =	0.0027								
11	p-range (99% CI) =		0.074	to	0.088							

Fig 3.53 Case study: Chi-squared / 3. Monte Carlo analysis (columns K to NTT are hidden)

Step 2

We now generate the first random sample of values in G3:G6, based on the null hypothesis that a total of D7 = 20 individual events are randomly allocated to the four categories with the probabilities in C3:C7. The first value in G3 generated by using the binomial distribution to randomly allocate a number out of a possible 20 based on the probability in C3:

[G3] = BINOM.INV($D7, $C3, RAND()) = 11

We now have 20 – G3 events left to allocate to the remaining three categories. Hence we allocate a number to G4, out of a possible D7-G3, using the binomial distribution but with a new probability calculated by dividing C4 by the sum of the remaining probabilities C4:C6:

[G4] = BINOM.INV(D7 – G3,$C4 / SUM($C4 : $C6),RAND()) = 3

We perform a similar allocation for G5 based on the numbers not allocated in G3 or G4:

[G5] = BINOM.INV(D7 – G3 – G4,$C5 / SUM($C5 : $C6),RAND()) = 2

Then the remaining numbers from the original 20 in D7 are entered into G6

[G6] = $D7 – G3 – G4 – G5 = 4

The formulae in G3:G6 are then copied to column NTV to give 10,000 randomly generated samples which are all based on the null hypothesis that the observed frequencies occurred randomly according to the probability ratios of 9:3:3:1.

Step 3, 4

For every resampled data set we now calculate the chi-squared statistic in row 8 by copying the calculation in D8 to all cells G8:NTV8.

The next step is to test in row 9 whether the chi-squared value for each sample in G8:NTV8 is greater than, or equal to, the experimentally observed value in D8, using for example:

[G9] = IF(G8 >= $D8, 1, 0)

The IF() function returns a '1' if the resampled data gives a chi-squared value that is equal to, or more extreme, than the observed.

The *proportion* of '1s' in row 9 equals the probability that the observed (or greater) chi-squared value could have occurred by chance from the null hypothesis, and is therefore equal to the *p*-value for the test. This proportion is calculated in D9:

p-value: [D9] = SUM(G9:NTV9) / 10000 = 0.081

The standard deviation uncertainty in this proportion is calculated in D10, using Eqn 3.24:

[D10] = SQRT(D9 * (1− D9) / 10000) = 0.0027

In this example, we calculate the 99% confidence interval limits for the calculated *p*-value (to compare with SPSS below), for which the relevant *z*-value is 2.58:

[C11] = D9 − 2.58 * D10   and   [E11] = D9 + 2.58 * D10

In this *particular* calculation, the 99% confidence interval range for the *p*-value is from 0.074 to 0.088, and, since these values are all greater than 0.05, we conclude that there is not enough evidence for a significant difference from the specified frequency ratios.

We can use SPSS to perform the same Monte Carlo analysis.

SPSS

Starting with the data as in Fig 3.54, it is necessary to first *weight* the cases (8.1.3, 8.2.4) to give the correct frequency weighting to each category:

SPSS > Data > Weight Cases....
⊙-Weight cases by:
  Frequency variable: *Freq*

and we then use the chi-squared analysis:

SPSS > Analyze > Nonparametric tests > Legacy Dialogs > Chi-Square...
Test variable list: *GTypeN* (categories must be defined by numeric values)
> Exact... ⊙-Monte Carlo
Expected values: Either accept all categories as equal *or*
  ⊙-Values: Add expected frequency values in ascending order of the numeric categories, e.g. enter 9 **Add** 3 **Add** 3 **Add** 1 **Add** if the values of *GTypeN*, 1, 2, 3, 4 describe the categories *AB*, *Ab*, *aB*, *ab*.

The 99% confidence interval in Fig 3.54(b) is 0.070 to 0.084 which is consistent with the results given by the Excel model above, allowing for the fact that each new set of random

Test Statistics

	GTypeN
Chi-Square	7.022[a]
df	3
Asymp. Sig.	.071
Monte Carlo Sig. Sig.	.077[b]
99% Confidence Interval Lower Bound	.070
Upper Bound	.084

	GTypeT	GTypeN	Freq
1	AB	1.00	7.00
2	Ab	2.00	8.00
3	aB	3.00	3.00
4	ab	4.00	2.00

a. 3 cells (75.0%) have expected frequencies less than 5. The minimum expected cell frequency is 1.3.

b. Based on 10000 sampled tables with starting seed 2000000.

(a) Frequency data entry   (a) Chi-squared results

Fig 3.54 Monte Carlo analysis for a 'goodness of fit 'test in SPSS

numbers will generate a slightly different result. SPSS also reports the *starting seed* used to generate the particular set of random numbers used.

Comparing data

Introduction

Many scientific investigations arrive at the need to *compare* sets of experimental data, either testing for the *existence* of some relationship or measuring the *strength* of that relationship. Chapter 3 introduced methods for hypothesis testing, and now this chapter brings together a range of different techniques which provide measures of association and agreement between theoretical models and experimental data, and also between different experimental measurements of the same quantities.

Section 4.1 further develops the statistics of *parametric correlation* from Section 2.1, and also introduces methods of *nonparametric correlation*.

Section 4.2 further develops the statistics for testing *association* from Section 3.7, and also introduces *Fisher's exact test* and the ability to test for *progression* in factors.

Section 4.3 considers the *strength* of the association between factors, and reviews a range of possible methods of measurement.

Section 4.4 reviews the concept of *agreement* in various contexts, including the 'goodness of fit' of analytical models, and agreements between variables and within contingency tables.

4.1 Correlation

The parametric statistics of correlation were introduced through the analysis of the straight line in 2.1.3, and these are now reviewed in relation to testing for the *existence* of, and measuring the *strength* of, *linear* correlation. We now also introduce nonparametric methods for correlation that are appropriate for ordinal data and which do not specifically assume that the relationship between the variables is that of a straight line.

4.1.1 Linear correlation

Linear correlation is a measure of the extent to which one variable increases in the same *ratio* as the increase in a second variable. The **correlation coefficient, r,** between two variables x and y can be defined as *the proportion of the variation in y that is predicted by the variation in x.*

The variation in x can be represented by its standard deviation, s_x. If the slope of the line of regression of y against x is m, then the variation in x translates into a *predicted* variation in y

given by $m \times s_x$. The *actual* variation in y is given by s_y, and thus the correlation coefficient, r, is given by the ratio:

$$r = \frac{\text{Variation in } y \text{ predicted by } x}{\text{Actual variation in } y} = \frac{m \times s_x}{s_y} \tag{4.1}$$

Pearson's correlation coefficient, r, (or product moment coefficient) is the standard parametric statistic for linear correlation (2.1.3), and the square of its value is the **coefficient of determination**, r^2, which assesses the 'goodness of fit' (4.4.1) of the data to a straight line.

The related hypothesis test calculates whether the best-fit slope is significantly different from zero, with the null hypothesis:

H_0: The best-fit straight line for the data has a *zero* slope.

The value of the correlation coefficient can range from $+1$ when all the data values fall exactly on a straight line with a *positive* slope to -1 (perfect negative correlation) where they all fall on a straight line with a *negative* slope.

The *value* of the correlation coefficient, r, is NOT dependent on the slope, m, of the best-fit straight line except that:

- if the slope of the best-fit straight line is zero, $m = 0$, then $r = 0$.
- if m is positive then r is also positive, and if m is negative then r is also negative.

Fig 4.1(a) shows examples of two data sets (circles and squares) with perfect *positive* correlation, $r = 1.000$, but with *different slopes*. A third data set (diamonds) has perfect *negative* correlation, $r = -1.000$. In all these cases the data points all lie exactly on the best-fit line. A fourth data set (triangles) has a best-fit line with a positive slope and with a correlation coefficient, $r = 0.935$.

Fig 4.1(b) shows a data set with an obvious correlation between x and y, but the correlation is not *linear*, and the calculated linear correlation coefficient actually has a zero value, $r = 0.000$. It is important to remember that r refers specifically to the correlation of data along a *straight* line.

The calculation of the correlation coefficient is *symmetrical* between x and y, and there is no suggestion that one variable is *dependent* and the other *independent* (2.1.1). Correlation between variables does not imply *causation*, and, in fact, we will see (4.1.4) that a bivariate

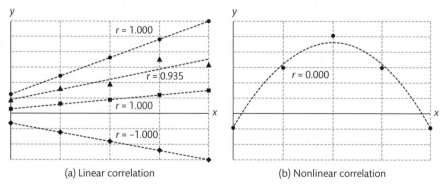

(a) Linear correlation   (b) Nonlinear correlation

Fig 4.1 Examples of correlation in x–y scatterplots

correlation between two variables can appear because they are both dependent on the variation of third variable.

The *significance* of a calculated value of r is a measure of whether the apparent correlation could have occurred by chance. It depends on the number of data pairs, n, and can be assessed by comparing r with published sets of critical values. It is also possible to calculate an equivalent p-value in Excel, via a t-value, using the equation:

$$p = \text{T.DIST.2T}(X, n-2) \tag{4.2}$$

where the value of X is calculated as

$$X = r \times \sqrt{\frac{n-2}{1-r^2}} \tag{4.3}$$

We will use the following case study to demonstrate the calculation of the parametric correlation coefficient using Eqns 4.1, 4.2, and 4.3, and use the same data to demonstrate the nonparametric correlation coefficients introduced in 4.1.2.

Case study: Toxicity assays / 2. Correlation

—continued from 7.1.1, leading to 4.4.2

Two types of assay, *B1* and *B2*, have been used to measure the percentage of cell deaths due to exposure to different drug concentrations, *C* mM, with results given in Fig 4.2. The *ranked* values for *C*, *B1*, and *B2* are also given respectively as *CR*, *BR1*, and *BR2*.

Parametric and non-parametric correlation: Excel analysis for Fig 4.2. Scan here to watch the video or find it via www.oxford textbooks. co.uk/orc/ currell/

⊿	A	B	C	D	E	F	G	H	I	J	K
1	C/Mm	LogC	B1 (x)	B2 (y)					CR	B1R	B1R
2	0.01	-2.00	2.0	4.0					1	1	1
3	0.1	-1.00	15.0	16.0		m =	1.407		2	2	2
4	1	0.00	36.0	50.5		s_x =	19.157		3	5	3
5	5	0.70	33.0	54.5		s_y =	28.176		4	3	4
6	10	1.00	35.0	65.0		r =	0.957		5	4	5
7	20	1.30	57.5	75.5		p =	0.003		6	6	6
8											

Fig 4.2 Cell death and drug concentration. The ranked values are analysed in 4.1.2.

We wish to compare the level of agreement between the two assay methods, *B1* and *B2*, in Fig 4.2 by calculating the *correlation* between the numbers of deaths using assay *B1* with those recorded using assay *B2*. Fig 4.3(a) plots *B2* as *y* against *B1* as *x* on a scatterplot using the *interval* values directly. If the two assays were in perfect *agreement* we would expect equal values at each concentration giving a straight line with a slope, $m = 1.0$, intercept, $c = 0$ and with a correlation coefficient, $r = 1.00$, but the actual trendline, with $m = 1.41$, suggests that *B2* gives *proportionately* higher results than *B1*. However, in this analysis we are just testing for *correlation* and the extent to which *B1* and *B2* values fall close to a straight trendline, and we use Pearson's r to measure the *strength* of this linear correction. The p-value tests whether there is *any* true correlation, i.e. whether the slope is *significantly* different from 0.

Calculating the slope, m, between $B1$ and $B2$ using the Excel function:

m: $[G3] = SLOPE(D2:D7, C2:C7) = 1.407$

and the sample standard deviations, s_x and s_y, for the two samples,

s_x: $[G4] = STDEV.S(C2:C7) = 19.157$

s_y: $[G5] = STDEV.S(D2:D7) = 28.176$

Using Eqn 4.1, Pearson's product moment correlation, r, is

r: $[G6] = (G3 * G4)/G5 = 1.407 * 19.157 / 28.176 = 0.957.$

It is also possible to get this result directly in Excel by simply using the function CORREL() or PEARSON().
The p-value can be calculated using Eqns 4.2 and 4.3:

p-value: $[G7] = T.DIST.2T(G6 * SQRT((6-2)/(1-G6^2)),4) = 0.003.$

Since the p-value is less than the default significance level of 0.05, we accept that there is significant evidence of linear correlation between $B1$ and $B2$.

Section 2.1.3 develops an alternative method of calculating Pearson's correlation coefficient, together with the use of Minitab and SPSS.

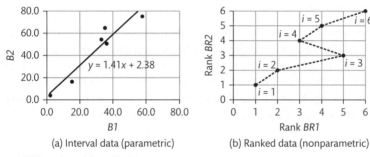

(a) Interval data (parametric)   (b) Ranked data (nonparametric)

Fig 4.3 $B1$ and $B2$ data plotted from Fig 4.2

4.1.2 **Nonparametric correlation**

The two main nonparametric correlation statistics are Spearman's rho, ρ or r_S, and Kendal's tau-b, τ, in which it is only the relative *ranking* (1.1.2) between two data values that is important, and the magnitude of the *difference* between them is irrelevant.

Fig 4.3(b) plots the *ranked* values, $BR1$ and $BR2$, of the data from Fig 4.2 as the x and y coordinates, with the points joined by a dotted line to indicate the *order* of increasing concentration. Two data points would be *positively* correlated if, when moving from one point to the other, the ranking changes in the *same way* for both values in the data pair, giving a *positive slope* in a scatterplot. They would be negatively correlated if their ranks changed in opposite directions giving a negative slope in the scatterplot. For example, in Fig 4.3(b) the point labelled $i = 4$ is negatively correlated with $i = 3$, but it is positively correlated with all other data points.

Measures of correlation for the data record the extent to which *all* the data pairs show the *same direction* of change. Perfect correlation occurs if the change in one variable is always in the same direction (either positive or negative) with respect to a change in the other variable.

Spearman's rank correlation coefficient is calculated by first identifying the rank of the values in each sample, calculating the difference, d_i, in these ranks for each data pair, and then using the equation

$$r_s = 1 - \frac{6 \sum_i d_i^2}{n(n^2 - 1)} \tag{4.4}$$

where n is the number of data pairs.

As with the Pearson's correlation coefficient, the values of the Spearman's correlation coefficient range between -1 and $+1$, and their significance can be tested by comparing the positive value with a table of critical values.

Kendall's tau, τ, is an alternative nonparametric correlation coefficient which uses the same ranked data as in Spearman's calculation. It calculates the numbers of concordant, C, and discordant, D, pairs of values within samples (4.3.5). A concordant pair of values has both ranks changing in the same direction, giving a positive slope in Fig 4.3(b), and a discordant pair change in opposite directions, giving a negative slope. For example, points 3 and 6 would be a concordant pair but 3 and 4 would be discordant.

$$\tau = \frac{C - D}{0.5n(n-1)} = \frac{C - D}{C + D} \tag{4.5}$$

The expression for τ in Eqn 4.5 assumes that there are no *tied* ranks with equal pair values, which can result in some pairs of values being neither concordant nor discordant. This situation is addressed using a modification of the simple formula to calculate Kendal's tau-b, which is introduced in 4.3.5, but we use the simpler formula here to understand the basic principles.

Using the data from Fig 4.2, the ranked values for *BR1* and *BR2* have been transposed into rows 2 and 3 in Fig 4.4. Each of the six data pairs has been identified by the value of the label, *i*, in row 1.

	A	B	C	D	E	F	G	H	I
1	*i*	1	2	3	4	5	6		
2	x_i (B1R)	1	2	5	3	4	6	*n* =	6
3	y_i (B2R)	1	2	3	4	5	6		
4	d_i	0	0	2	-1	-1	0		
5	d_i^2	0	0	4	1	1	0	$\Sigma d_i^2 =$	6
6								$r_s =$	0.829
7	*a*	*b =*	2	3	4	5	6		
8	1		C	C	C	C	C	$n_C =$	13
9	2			C	C	C	C	$n_D =$	2
10	3				D	D	C		
11	4					C	C	$\tau =$	0.733
12	5						C		

Fig 4.4 Calculation of Spearman's and Kendal's tau correlation coefficients

For **Spearman's correlation coefficient**, we first calculate, in row 4, the differences d_i in rank for each data pair, i, and then, in row 5, the square of these differences d_i^2.

The sum of the d_i^2 values is calculated in I5, $\Sigma d_i^2 = 6$, and then the correlation coefficient is calculated in I6 using Eqn 4.4.

p or r_s: $[I6] = 1 - 6 * I5 / (I2 * (I2^2 - 1)) = 0.829$

For **Kendal's tau** we consider every possible pair of values identified by the different i values, and decide whether they are concordant, with ranks changing in the *same* direction, or discordant with ranks changing in *opposite* directions.

For example, if we take the values $i = 3$ and 4 in columns D and E:

$x_3 = 5$ and $x_4 = 3$, giving $x_3 > x_4$, but

$y_3 = 3$ and $y_4 = 4$, giving $y_3 < y_4$

In this case, the x-rank does not change in the same direction as the y-rank, and this is then a *discordant* pair. In the 'results' section between A7 and G12 we then enter a D in E10 to record that the pair defined by $a = 3$ and $b = 4$ is discordant. By comparison, for the values $i = 3$ and 6, the x-rank and y-rank both change in the *same* direction, and thus form a *concordant* pair, recorded as C in G10, defined by $a = 3$ and $b = 6$.

We simply record the total numbers of concordant, $n_C = 13$, and discordant pairs, $n_D = 2$, in I8 and I9 respectively. Kendal's tau is then calculated using Eqn 4.5

τ: $[I11] = (I8 - I9) / (0.5 * I2 * (I2 - 1)) = 0.733$

SPSS performs the calculations for nonparametric correlation coefficients, as above, and their associated p-values directly:

SPSS > Analyze > Correlate > Bivariate.... Variables: *BR1 BR2*
☑**-Kendall's tau-b** ☑**-Spearman's**
→ Output: Fig 4.5

Correlations

			BR1	BR2
Kendall's tau_b	BR1	Correlation Coefficient	1.000	.733*
		Sig. (2-tailed)	.	.039
		N	6	6
	BR2	Correlation Coefficient	.733*	1.000
		Sig. (2-tailed)	.039	.
		N	6	6
Spearman's rho	BR1	Correlation Coefficient	1.000	.829*
		Sig. (2-tailed)	.	.042
		N	6	6
	BR2	Correlation Coefficient	.829*	1.000
		Sig. (2-tailed)	.042	.
		N	6	6

*. Correlation is significant at the 0.05 level (2-tailed).

Fig 4.5 SPSS Nonparametric correlation

Fig 4.5 gives the correlation coefficients of 0.829 and 0.733 as calculated above, and also gives the equivalent two-tailed *p*-values recorded as 'Sig'.

Spearman's correlation coefficient in Minitab can be obtained by using the contingency table statistics via crosstabs (8.2.4):

Minitab > Stat > Tables > Cross Tabulation and Chi-Square...
Rows: *BR1*   **Columns:** *BR2*
> Other Stats...: ☑-Correlation coefficients for ordinal categories

4.1.3 Scientific context of correlation

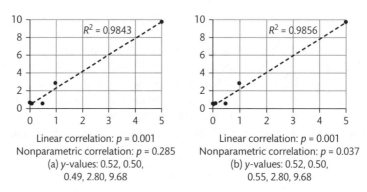

Linear correlation: *p* = 0.001
Nonparametric correlation: *p* = 0.285
(a) *y*-values: 0.52, 0.50,
0.49, 2.80, 9.68

Linear correlation: *p* = 0.001
Nonparametric correlation: *p* = 0.037
(b) *y*-values: 0.52, 0.50,
0.55, 2.80, 9.68

Fig 4.6 Effect of bunched measurements at one end of a correlation

It is useful to compare the conclusions from Pearson's and Spearman's correlation tests using some example data shown in Fig 4.6. It is not uncommon to see student data with multiple values bunched towards one end of the graph, and we use this to calculate the *p*-values for the two forms of correlation. The only difference between graphs (a) and (b) is that the *y*- value of the middle data point has a very small change from 0.49 to 0.55. We see that, for the *parametric* analysis, the R^2 value is almost unchanged between the graphs and the *p*-value remains a highly significant 0.001. However, the small change in the actual *y*-value of the middle point changes the *nonparametric p*-value from a non-significant 0.285 to a significant 0.037. The small experimental difference makes very little difference to the parametric calculation, but, because the data values are so close together in the bunch the difference changes the *rank* order between the second and third point (from 0.50 > 0.49 to 0.50 < 0.55) giving a very different nonparametric result. It is important to be aware of how experimental variations may affect the statistical interpretation of your data.

4.1.4 Bivariate and partial correlation

In situations where there are more than two related samples, it is possible to measure the correlations, pairwise, between each possible pair of samples. A *bivariate* correlation relates to the simple correlation calculated solely between each pair, without taking into account the variations of any other sample. However, it is possible that an *apparent* correlation exists between two samples because they are both also correlated with a third sample. A *partial* correlation between two samples takes into account (or controls for) the third sample.

> ## Case study: **Correlated variables / 1. Bivariate and partial (overview)**
> —leading to 3.4.5
>
> Fig 4.7(a) gives a response variable, R, and three related input variables, v1, v2, and v3. We start here by investigating the correlations between R, v1, and v2, and in Section 3.4 we compare these results with the data produced by a general regression analysis.
>
> 3.4.5 / 2. Sums of squares: Demonstrates the difference between sequential and adjusted sums of squares, leading to 3.4.6 and the difference between 'pure error' and 'lack of fit' in a regression model.

	A	B	C	D
	R	v1	v2	v3
1	R	v1	v2	v3
2	10.00	2.24	0.94	0.94
3	12.30	2.39	1.11	1.12
4	9.60	1.71	0.76	0.76
5	8.00	1.85	0.89	0.91
6	9.80	1.65	0.85	0.85
7	9.60	2.18	1.12	1.12
8	8.20	2.08	0.81	0.81
9	9.50	2.30	0.91	0.91
10	7.40	1.69	0.72	0.72
11	11.90	2.50	1.25	1.25

(a) Data

Correlations

		R	v1	v2
R	Pearson Correlation	1	.678[*]	.784[**]
	Sig. (2-tailed)		.031	.007
	N	10	10	10
v1	Pearson Correlation	.678[*]	1	.818[**]
	Sig. (2-tailed)	.031		.004
	N	10	10	10
v2	Pearson Correlation	.784[**]	.818[**]	1
	Sig. (2-tailed)	.007	.004	
	N	10	10	10

*. Correlation is significant at the 0.05 level (2-tailed).
**. Correlation is significant at the 0.01 level (2-tailed).

(b) SPSS output

Fig 4.7 Bivariate correlations between R, v1, and v2

SPSS calculates the direct bivariate correlations in Fig 4.7(b) using:

SPSS > Analyze >Correlate>Bivariate...
Variables: *R v1 v2*
☑-**Pearson**

The bivariate correlation suggests that there is a significant correlation between *v2* and *R* with $p = 0.007$ and between *v2* and *v1* with $p = 0.004$. As *v2* shows significant correlation with both *R* and *v1*, it is not surprising that there is also correlation between *v1* and *R* with $p = 0.031$.

Correlations

Control Variables			R	v1
v2	R	Correlation	1.000	.102
		Significance (2-tailed)	.	.794
		df	0	7
	v1	Correlation	.102	1.000
		Significance (2-tailed)	.794	.
		df	7	0

(a) Partial correlation between R and v1, controlling for v2

Correlations

Control Variables			R	v2
v1	R	Correlation	1.000	.542
		Significance (2-tailed)	.	.131
		df	0	7
	v2	Correlation	.542	1.000
		Significance (2-tailed)	.131	.
		df	7	0

(b) Partial correlation between R and v2, controlling for v1

Fig 4.8 Partial correlations between R, v1, and v2

SPSS performs the partial correlation between R and $v1$, while controlling for $v2$:

SPSS > Analyze >Correlate > Partial...
Variables: R $v1$
Controlling for: $v2$

The *partial* correlation between R and $v1$ in Fig 4.8(a)measures the variability in R related to $v1$ *after* the effect of $v2$ has been taken into account. We see that this now gives $p = 0.794$ which suggests that most of the variation in R can be predicted by $v2$. Similarly R also showed significant correlation with $v1$, so that the partial correlation between R and $v2$, controlling for $v1$, leaves less variation to be described by $v2$ giving $p = 0.131$. These results do not specifically decide whether there is a *scientific* cause and effect link between R and $v1$ or between R and $v2$, although the relative magnitudes of the correlation coefficients do indicate the strongest direct relationships.

We can compare these results with those in 3.4.5 where we perform a general regression analysis to model R on $v1$ and $v2$, and find the same partial p-values in Fig 3.31.

4.2 Tests for association

The concept of association between two factors in a contingency table of frequencies was introduced in 3.7.4, and the concept of an ANOVA interaction between factors in 3.3.2. In this section we:

- investigate the similarities and differences between the analytical techniques that test for association and interaction.
- develop further methods of analysis for contingency tables: Fisher's exact test, Mantel–Haenszel linear by linear association chi-squared test, and Eta for nominal-interval association.

The practical procedures for using Minitab and SPSS for association within a contingency table are demonstrated in Section 8.2.

4.2.1 Association and interaction

We start by reviewing the difference between the terms *association* and *interaction* and their underlying analytical processes.

	A	B	C	D	E	F	G	H
1		B1	B2	Totals			B1	B2
2	A1	31	19	50		A1	35.7, 25.7	16.8, 20.6
3	A2	20	30	50		A2	16.3, 22.7	32.2, 26.4
4	Totals	51	49	100				
5		Data set (a) Association					Data set (b) Interaction	

Fig 4.9 Comparing association and interaction

Fig 4.9 shows two sets of data in shaded cells. Set (a) in B2:C3 gives *frequency* values, and set (b) gives two *replicate interval* values in each cell in G2:H3. Both sets of data are obtained under conditions defined by factors *A* and *B*, each with two levels.

For set (a), 100 observations/trials have been made, each one of which *falls* into one of the cells defined by the *observed values* for *A* and *B*, giving a total count or frequency for each cell. For set (b), a total of eight experiments/observations have been conducted under conditions *defined* by *A* and *B*, and the *observed result* recorded in the relevant cell.

The data is presented in Fig 4.10 (a) and (b) respectively, in which the response values are plotted against the values for *A*, with the values for *B* differentiated by circular and triangular markers. The two lines, one for *B1* and one for *B2*, join the *values themselves* in set (a) and the *mean* of each data pair in set (b). The lines provide an indication of the change in response due to the change in level for *A*.

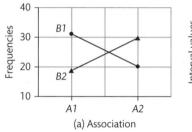 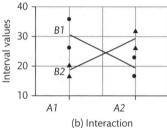

(a) Association   (b) Interaction

Fig 4.10 Factor plots for association and interaction

It certainly appears that the *response to a change in the value (level) of A depends on the value (level) of B*. The fact that the slopes of the two lines are not parallel suggests that there is an *association* in (a), and an *interaction* in (b), between the factors A and B.

The relevant tests for an *association* in (a) give results:

   Corrected (Yates, 3.7.5) chi-squared = 4.002 giving $p = 0.045$

   Fisher's exact test (4.2.3) gives $p = 0.045$.

The relevant GLM/ANOVA test (3.3.2) for an *interaction* in (b) gives $p = 0.034$.

Both analyses show a significant effect ($p < 0.05$), but it is useful to reflect on the underlying analytical processes. We see in 3.3.2 that the identification of an interaction in the GLM/ANOVA calculation requires the replicate measurements under identical conditions, such that it is possible to derive an estimate of the true experimental uncertainty in the measurement, assuming a normal uncertainty distribution. Without the replicate values in set (b) it would be impossible to distinguish between any interaction and experimental error. However, set (a) does not have replicates, but, with frequency values, the analysis uses the probabilities of the binomial distribution to estimate the uncertainty directly from the values themselves. Using Eqn 1.21, the standard deviation uncertainty in a frequency, *N*, is equal to $\sqrt{N}$, giving a standard deviation uncertainty in each of the values in (a) of approximately $\sqrt{25} \sim 5.0$.

The relatively large data uncertainties in (a) and variations in (b) explain why the association and interaction, which appear very obvious from the lines in Fig 4.10, are only just statistically significant, with p-values just less than 0.05.

A major difference between the two analytical methods is in the assumptions of the underlying statistical distribution. Analysis of (a) assumes a binomial/Poisson distribution, whereas the GLM for (b) assumes a normal distribution. The aim of the unifying *GsdLM* (3.4.7) is to incorporate the option of different distributions.

4.2.2 Tests for association

Table 4.1 lists possible hypothesis tests for the *existence* of an association, and Table 4.2 lists analyses that measure the *strengths* of association. The most common tests, e.g. Pearson's chi-squared analysis, treat the levels of the factors as nominal categories, with no sense of progression from one category to the next. However, many situations involve ordinal data, e.g. giving an opinion on a Likert scale, and the linear-by-linear test can be useful in that it takes into account this sense of progression between the levels of the two factors.

A common problem encountered, particularly in student experiments, is insufficient time to collect enough data to avoid having low expected frequencies in some cells of the table. We consider possible solutions in 8.2.7, either combining factor levels to reduce the number of cells or by using the Monte Carlo resampling approach.

Table 4.1 Tests for association in a contingency table

Statistic	Comment	Link
Pearson's chi-square	Standard statistic, but the test conclusions can be unreliable for low category numbers	3.7.4
Yates continuity correction	Correction when degrees of freedom = 1	3.7.5
Likelihood ratio	Alternative to Pearson's chi-squared value, usually giving larger values for smaller sample sizes	3.7.6
Fisher's exact test	Exact calculation based on binomial distribution. Tests for a difference in proportions in 2 × 2 tables	4.2.3
Linear-by-linear test	Test for an association between ordinal/linear factors, using a correlation analysis to identify a progression between factor levels.	4.2.4
McNemar's test	Tests for a difference in the off-diagonal elements of a 2 × 2 table. McNemar–Bowker test is similar to McNemar's test, but for larger tables.	4.4.5
Monte Carlo analysis	The Monte Carlo resampling techniques is useful for situations with low expected frequencies.	3.9.3

4.2.3 Fisher's exact test

R A Fisher addressed the contingency table problem by considering the probabilities with which different combinations of frequencies could occur while keeping the *same* row totals and column totals. In this way, he was able to develop an 'exact' test that calculated the cumulative probability that the observed distribution of values, or any more extreme distribution, could have occurred by chance.

Computer software can handle large contingency tables, but we can demonstrate the principle by analysing the simple 2×2 table between factors X and Y in Fig 4.11, with individual cell frequencies, a, b, c, and d, giving a total of N observed events. The null hypothesis is that the *distribution* of frequency values in each row is the same, from which it follows that the *distribution* of frequency values in each column is also the same.

◢	A	B	C	D
1		Y1	Y2	Total
2	X1	a	b	a+b
3	X2	c	d	c+d
4	Total	a+c	b+d	N = a+b+c+d

Fig 4.11 2×2 contingency table

Step 1

We use the **binomial coefficient** to calculate the *number of ways* that N randomly allocated items would result in the individual numbers of items, a, b, c, and d, for the particular set of the observed *column totals*.

In the *first* row:

Number of ways of selecting exactly a items in column 1 from a total of $a + b = {}_{a+b}C_a$

and in the *second* row:

Number of ways of selecting exactly c items in column 1 from a total of $c + d = {}_{c+d}C_c$

The total *number of ways* of obtaining the specific arrangement overall $= {}_{a+b}C_a \times {}_{c+d}C_c$

Step 2

We calculate the *probability* of obtaining the distribution a, b, c, and d, by dividing the number of ways from Step 1 by the total number of ways that the particular set of *column totals* could have occurred by chance.

Number of ways of selecting exactly $a + c$ items in column 1 from a total of $N = {}_N C_{a+c}$

The probability, P, that N randomly allocated items, with the defined row and column totals, would give the specific distribution of values, a, b, c, and d is:

$$P = \frac{{}_{a+b}C_a \times {}_{c+d}C_c}{{}_N C_{a+c}} \qquad (4.6)$$

The calculations in Steps 1 and 2 are performed using Excel in Fig 4.12 for $a = 24$, $b = 19$, $c = 7$, and $d = 16$. We use the COMBIN() function to derive the binomial coefficients in F2, F3, and H2, and then Eqn 4.6 for P in H4.

Fisher's exact test: Excel analysis for Fig 4.12. Scan here to watch the video or find it via www.oxfordtext books.co.uk/ orc/currell/

◢	A	B	C	D	E	F	G	H
1		Data		Totals				
2		24	19	43	$_{a+b}C_a =$	8.005E+11	$_NC_{a+c} =$	6.406E+18
3		7	16	23	$_{c+d}C_c =$	2.452E+05		
4	Totals	31	35	66			$P =$	0.0306

Fig 4.12 Calculation of the probability of a given distribution

Step 3

The final step in calculating the *p*-value is to calculate, and sum, the probabilities that the null hypothesis could also randomly produce more *extreme* distributions, but all with the *same row and column totals*. Fig 4.13 gives the most probable of the extreme distributions by decreasing the '7' by one each time and adjusting all other values to keep row and column *totals* constant. In each case the probability, *P*, for that distribution is calculated:

25	18		26	17		27	16		28	15		etc.
6	17		5	18		4	19		3	20		
gives $P = 0.0096$			gives $P = 0.0022$			gives $P = 0.0004$			gives $P < 0.0001$			

Fig 4.13 Calculation including more 'extreme' distributions

The total probability of the null hypothesis producing the observed distribution, or one of the more extreme distributions, will be given by the sum of all probabilities, *P*:

$$p = 0.0306 + 0.0096 + 0.0022 + 0.0004 + ... = 0.0428 \ (4 \ \text{sf}).$$

This gives a *p*-value that has been *calculated* exactly. However, the scientific *interpretation* of this *p*-value is not so exact as it tends to give slightly more Type II errors than it should, due to the fact that moving from one distribution to the next (as in Fig 4.13) shows discrete jumps and is not a smooth distribution.

It is also important to note that we have calculated a one-tailed *p*-value, by only including extreme distributions in the same sense as the observed distribution–i.e. with increasing values along the diagonal from top left to bottom right. For a two-tailed test (as for the chi-squared test), we would need to include extreme distributions with increasing values along the bottom left to top right. Unfortunately, the situation is not symmetrical and we cannot just double the one-tailed value (Eqn 1.27), but if we perform the additional calculations we would find $p = 0.070$. For example, if we use the two-proportion test in Minitab (3.8.3) to compare 24/43 with 7/23 we find that $p = 0.043$ for the 'greater than' option and $p = 0.070$ for the 'not equal' option.

The use of the binomial test in SPSS is demonstrated in 6.1.7 and 6.2.9.

4.2.4 Linear by linear association

The standard contingency table is defined by *nominal* variables, and the *order* in which rows and columns are placed makes no difference to the chi-squared calculation—see for example the nominal categories, *Science*, *English,* and *History* in Fig 3.47. We now introduce the Mantel–Haenszel *linear by linear association* chi-squared test, which can be used for a contingency table when a row or column category is defined by an *ordinal* or *interval* variable.

The data in Fig 4.14 (a) and (b) both result from using two different methods, *M1* and *M2,* to lift fingerprints from difficult surfaces. The different columns represent the quality of the resultant prints recorded on an ordinal scale from 1 (poor) to 4 (excellent). The numbers of prints with the different qualities were recorded in the separate cells, giving a total of 30 prints using *M1* and 32 using *M2.*

The *linear by linear association chi-squared* is calculated, with degrees of freedom, $df = 1$, using:

$$Q_{MH} = (N-1) \times r^2 \tag{4.7}$$

where r is the Pearson correlation coefficient.

The values for Pearson's and the linear by linear association chi-squared are calculated for each of the contingency tables in Fig 4.14 (a) and (b). The results in Fig 4.14 (a) show that, although there is not enough evidence for an association just based on nominal values ($p = 0.057$), the additional information of a *progressive* change between ordinal quality levels gives $Q_{MH} = 6.33$ with a significance of association of $p = 0.012$.

Quality:	1	2	3	4
M1	9	9	7	5
M2	5	4	9	14

Pearson's chi-squared = 7.522 giving p = 0.057
Q_{HM} = 6.33 giving p = 0.012
(a) Initial column order

Quality:	1	2	3	4
M1	9	5	7	9
M2	5	14	9	4

Pearson's chi-squared = 7.522 giving p = 0.057
Q_{HM} = 0.34 giving p = 0.559
(b) Interchanging columns 2 and 4

Fig 4.14 Effect of changing the order of categories

Set (b) actually has the same values as (a), but the difference is that the results for quality '2' have been swapped with quality '4'. The Pearson's chi-squared analysis tests whether there is a significant difference in the *distribution* of values between *M1* and *M2*, but *without* any sense of progression, and because the order of the columns makes no difference, it results in the same, non-significant, *p*-value for both sets, $p = 0.057$. However, the change in the linear by linear chi-squared value from the significant $p = 0.012$ in (a) to the non-significant $p = 0.559$ demonstrates that the data for (a) has an increasing number of prints *in proportion to* the quality of the print, whereas this is not the case for (b).

It is important to note that the test here is specifically for a *linear* association, and consideration must be made of the underlying scientific processes being tested. For example, it is possible that method *M2* may have the enhanced ability to provide *medium* and *good* quality prints, but has a technical limitation in not being able to produce *excellent* quality, in which case the expected relationship with quality would not be truly linear.

Fig 4.15 shows the quality of each separate fingerprint plotted against the method (*M1* or *M2*) used. The numbers by each data point give the numbers of prints giving the same value (the same frequencies as in the contingency table).

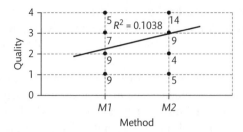

Fig 4.15 Observed qualities of fingerprint for two different lifting methods

The best-fit trendline is shown, together with the coefficient of determination, $r^2 = 0.1038$.

We can see from the slope of the best-fit line in Fig 4.15 that there is an apparent increase in the mean quality value between *M1* and *M2*. We can also test for a difference in median value by using a Mann–Whitney test for the two samples, *M1* and *M2*, from which we get $p = 0.011$ which is consistent with the Q_{HM} result of $p = 0.012$. Tests for correlation also give $p = 0.011$ and 0.010 for Pearson's and Spearman's correlation respectively.

Using the Pearson's value for $r = 0.322$ (= $\sqrt{r^2}$), for $N = 62$, we can calculate Q_{HM} using Eqn 4.7.

$$Q_{HM} = 61 \times 0.322^2 = 6.32$$

which is consistent with the value derived in SPSS.

4.3 Strength of association

In Section 3.7 we used the chi-squared statistic to perform a hypothesis test for a contingency table to test for a possible difference in the distribution of row values between different columns or vice versa. We were testing for the *existence* of an association between the factors that define the rows and columns. We now want to be able to measure the *strength* of that association.

For clarity of calculations we will use a simple 2×2 contingency table, but the concepts can be applied to larger tables.

4.3.1 Association and agreement

In describing the strength of association and agreement within a contingency table, we can consider the two variables as X and Y, each with categories described as 1 and 2.

	A	B	C	D
1		Y1	Y2	Total
2	X1	a	b	a+b
3	X2	c	d	c+d
4	Total	a+c	b+d	N = a+b+c+d

Fig 4.16 2×2 contingency table

The categories *X1*, *X2*, *Y1*, and *Y2* could be nominal values, but if they are ordinal or interval values we assume that the sense of progression would be from levels 1 to 2 in both variables.

The association between the variables could be either *symmetric*, with no specific dependence on either variable, or *directional* (asymmetric), in which the difference due to one variable *depends* on the other. A *directional* association could be in either direction (*X* dependent on *Y* or *Y* dependent on *X*), but, for consistency in our calculations, we will normally make the row variable, *X*, the *independent* variable, and the column variable, *Y*, the *dependent* variable. In this way our table is similar to an *x*–*y* scatterplot that has been rotated clockwise by 90°.

In mathematics, a table of values as in Fig 4.16 would be described as a matrix. The values *a* and *d* are said to be on the 'diagonal' of the matrix (top left to bottom right) and *c* and *b* are 'off-diagonal' elements.

Values *a* in B2 show that both *X* and *Y* record level '1', i.e. they are in *agreement*. Similarly values *d* in C3 also show agreement, both recording level '2'. However, values *b* and *c* in C2 and B3 both show *disagreement* (or negative agreement) with different values for *X* and *Y*.

An *association* (3.7.4) is demonstrated by a difference in the *distribution* of values for different rows and for different columns, and is only treated as a positive quantity.

20	18		38	3		3	38
19	22		2	37		37	2
(a)			(b)			(c)	

Fig 4.17 Examples of association and agreement

We can differentiate between association and agreement using the example tables in Fig 4.17:

(a) has an approximately even distribution of values and shows no overall association or agreement.

(b) has a strong association, e.g. with different proportions for the two rows, and shows strong positive *agreement* for ordinal data.

(c) also has strong association, but shows strong *disagreement* (or negative agreement) for ordinal data.

4.3.2 Measures of association

Tests for the *existence* of an association were given in Table 4.1, and now Table 4.2 identifies possible measures for the *strength* of an association between pairs of variables, grouped according to variable types: nominal, ordinal, and interval.

We can also divide the measured 'association' into two types: symmetrical and directional. In a *directional* (or asymmetric) association we measure how a knowledge of one

Table 4.2 Measures for the strength of an association

Variable pairs defined by type	Symmetric measures	Directional measures
Nominal / nominal	Phi, ϕ (2 × 2 tables) Cramer's V Kappa, κ	Lambda, λ
Ordinal / ordinal	Gamma, Γ Kendall's tau-b, τ Spearman's rho*, ρ Coefficient of concordance	Somers' d
Interval / interval	Pearson's coefficient*, r	Linear regression**
Nominal / interval		Eta, η

*These measures are considered more fully in Section 4.1.
**Linear regression, developed in Section 2.1, is included here as a reminder that the standard regression calculation assumes zero uncertainty in x and calculates the regression of y on x, with y being the dependent variable.

(independent) variable can be used to predict the variation of the other (dependent) variable. In a *symmetric* association, there is no sense of direction and we measure only the extent to which the two variables vary in similar ways.

When selecting a possible analysis, it is useful to remember that, due to the one-way hierarchy between variable types, it is possible to analyse an interval variable as an ordinal variable and an ordinal variable as a nominal variable, but not in the reverse order. For example, Cramer's V could be used to measure the association between categories defined by ordinal variables (although the ranking of the variables will be lost), but it would not be possible to use Kendall's τ to analyse nominal data.

4.3.3 Cramer's V and Phi

The chi-squared statistic χ^2 works well as a test to identify the *significance* of a possible association, but it is not so useful in measuring the *strength* of the association, because its value depends on sample size. In the *hypothesis test*, the sample size is automatically taken into account by including the degrees of freedom in the calculation of critical values or p-values.

Cramer's V and Phi, φ, are *symmetric measures of association* (φ applies to 2 × 2 tables) based on the chi-squared statistic χ^2 between nominal variables, but *corrected* for sample size, N:

$$V = \sqrt{\chi^2 / N(k-1)} \qquad \phi = \sqrt{\chi^2 / N} \qquad (4.8)$$

For larger than 2 × 2 tables   For 2 × 2 tables only

where N is the total of recorded trials and k is the smaller of the number of rows or columns.

An approximate qualitative evaluation of these values is given in Table 4.3.

Table 4.3 Strength of association for Cramer's *V* and Phi

Value	Strength of association
0.0 > 0.10	Weak
0.10 > 0.30	Moderate
> 0.30	Strong

4.3.4 **Goodman and Kruskal's Lambda**

Lambda is an *asymmetric* (or directional) measure between nominal variables, in which one factor is measured as being *dependent* on the other, *independent*, factor.

The statistic calculates the 'proportional *reduction* in error' in predicting the *dependent* category of the next randomly chosen answer, based on whether it is either **before**, *B*, or **after**, *A*, taking into account the information of category numbers in the *independent* factor. The value of λ is given in Eqn 4.9 by the reduction in the probability of error, $pE_B - pE_A$,

$$\lambda = \frac{pE_B - pE_A}{pE_B} \tag{4.9}$$

where pE_B and pE_A are the probabilities of error 'before' and 'after' *knowing* the values of the independent factor.

It is common practice to put the independent variable as the column variable and the dependent variable in the rows, but, for reasons of consistency with other techniques in this book, we start with an example, in Fig 4.18, where the independent variable is the row variable, *X*.

◢	A	B	C	D
1		*Y1*	*Y2*	**Total**
2	*X1*	28	12	40
3	*X2*	18	22	40
4	**Total**	46	34	80

Fig 4.18 Calculating lambda, λ

If we take *Y* as the dependent factor and *X* the independent factor, we can calculate the value of λ by trying to predict into which *Y* category the next randomly chosen value will fall. We need to calculate the probabilities that our predictions will be *wrong*, based on the *observed* cell numbers.

Step 1 (Before)

We do not know whether the *next value* will be in the *Y1* or *Y2* category.

Without any knowledge of the *X*-value, we can only use the *total* numbers for the *Y* categories as predictors,

Y1(total) = 46 and *Y2*(total) = 34

We would expect that a randomly chosen item would fall into the most probable, *Y1*, category, but the probability that it might fall into the other, *Y2*, category is given by

(80 − 46) / 80. This is the probability of error *before* knowing the value of X, and can be written in general as:

$$pE_B = \frac{N - (\text{Largest category total})}{N}. \tag{4.10}$$

In this example, $pE_B = (80 − 46) / 80 = 34 / 80 = 0.425$

The equation might seem over complicated when there are only two categories for the dependent variable, because 80 − 46 just equals the value, 34, already given in the other cell. However the equation also covers situations where there are *more than* just two categories.

Step 2 (After)

We now know the value of X and there are two elements in the calculation with X equal to *X1* or *X2*.

With X equal to *X1*, the probability of error would be (40 − 28) / 80 or,

with X equal to *X2*, the probability of error would be (40 − 22) / 80

Thus the total probability of error, given that we know the value of X, would be the sum of these two probabilities:

$$pE_A = \sum_{\text{All rows}} \left\{ \frac{\text{Row total} - \text{Largest frequency in row}}{N} \right\} \tag{4.11}$$

In this example, $pE_A = (40 − 28)/80 + (40 − 22)/80 = 0.375$

Step 3

The calculation of λ for Y dependent on X, using Eqn 4.9 gives

$$\lambda_{Y/X} = \frac{0.425 - 0.375}{0.425} = 0.118,$$

This can be interpreted by saying that a knowledge of the X values increases the predictability of Y by 11.8%.

We can also use the same data to calculate the value of lambda for X being dependent on Y:

$$pE_B = (80 - 40)/80 = 40/80 = 0.5 \text{ and}$$
$$pE_A = (46 - 28)/80 + (34 - 22)/80 = 18/80 + 12/80 = 0.375$$
$$\lambda_{X/Y} = \frac{0.500 - 0.375}{0.500} = 0.250$$

Lambda values can be used to compare the strengths of bivariate relationships, e.g. compare different pairs of answers in a questionnaire.

Note that the lambda calculation does not work when the differences in both categories of the independent variable are in the same direction, and the calculation returns a zero value even when there is an association. For example, if X is considered to be the *dependent* variable in Fig 4.19(b), Y becomes the independent variable but with differences in the same direction $24 > 7$ and $19 > 16$, giving:

$$pE_B = (66-43)/66 = 23/66 \text{ and}$$
$$pE_A = (31-24)/66 + (35-19)/66 = 7/66 + 16/66 = 23/66$$
$$\lambda_{X/Y} = \frac{23/66 - 23/66}{23/66} = 0$$

However, the calculation of Y dependent on X gives a non-zero value: $\lambda_{Y/X} = 0.161$.

4.3.5 Concordance of data pairs

A number of measures of association for *ordinal* factors are based on counting the numbers of concordant pairs, C, discordant pairs, D, and ties, T, within a whole data set.

X\Y	1	2	Total
1	a	b	$a + b$
2	c	d	$c + d$
Total	$a + c$	$b + d$	N

(a) Cell numbers

X\Y	1	2	Total
1	24	19	43
2	7	16	23
Total	31	35	66

(b) Example values

Fig 4.19 Concordance within a contingency table

We consider the 2×2 frequency table in Fig 4.19(a) with factors X and Y, in which we have the results of $N = a + b + c + d$ trials, where

a trials have resulted in values for X,Y of $(1,1)$

b trials have resulted in values for X,Y of $(1,2)$

c trials have resulted in values for X,Y of $(2,1)$

d trials have resulted in values for X,Y of $(2,2)$

We review all possible **pairs** of trials and classify them as concordant, C, or discordant, D, depending on the direction of differences in their values, using the 'rules' that:

(r,s) and (p,q) are **con**cordant if EITHER $r > p$ **and** $s > q$ OR $r < p$ and $s < q$

(r,s) and (p,q) are discordant if EITHER $r > p$ **and** $s < q$ OR $r < p$ and $s > q$

otherwise the trial 'pairs' are classified as ties, T.

Calculating the values for Fig 4.19(b):

$C = a \times d$ pairs of trials will be concordant : $(0,0)$ is concordant with $(1,1) = 24 \times 16 = 384$

$D = b \times c$ pairs of trials will be discordant : $(0,1)$ is discordant with $(1,0)$ $= 19 \times 7 = 133$

$T_X = a \times b + c \times d$ pairs will be tied with just the *same x*-values $= 24 \times 19 + 7 \times 16 = 568$

$T_Y = a \times c + b \times d$ pairs will be tied with just the *same y*-values $= 24 \times 7 + 19 \times 16 = 472$

Goodman and Kruskal's gamma

Gamma is a measure of increase of the extent to which one variable increases in step with the other:

$$G = \frac{C - D}{C + D}.$$
(4.12)

Using the values from Fig 4.19(b):

$$G = \frac{384 - 133}{384 + 133} = 0.485.$$

However, this measure does not include ties in the calculation.

Kendall's tau-b

Kendall's tau was introduced in 4.1.2 with Eqn 4.5, which had the same form as Gamma, above, and did not allow for tied pairs. However, Kendall's tau-b does allow for ties, and provides a *symmetric* measure of association:

$$\tau_b = \frac{C - D}{\sqrt{C + D + T_x} \times \sqrt{C + D + T_y}}.$$
(4.13)

Using the values from Fig 4.19(b):

$$\tau_b = \frac{384 - 133}{\sqrt{384 + 133 + 568} \times \sqrt{384 + 133 + 472}} = 0.242.$$

Somer's D

Somer's D is a *asymmetric* (directional) measure of association, which uses a similar equation to Kendall's tau-b, but only includes one set of ties depending on the measured direction. For example, Somer's D for Y dependent on X is given by:

$$D_{Y/X} = \frac{C - D}{C + D + T_y}.$$
(4.14)

Using the values from Fig 4.19(b):

$$D_{Y/X} = \frac{384 - 133}{384 + 133 + 472} = 0.254 \text{ and } D_{X/Y} = \frac{384 - 133}{384 + 133 + 568} = 0.231.$$

4.3.6 Nominal by interval association, Eta

Eta, η, is a *directional* measure of the *strength* of association between the values of an *interval* variable, v, and a *nominal* variable or factor, F. Eta has possible values ranging from 0 for no association to 1 for strong association. We can demonstrate its use with the data set A in Fig 4.20 which shows six values of interval data, v, in column B, two at each level, 1,

2, and 3, of the factor F. The same data is also presented in the contingency table format in cells E2:G7, with a '1' to indicate the data entry at each of the relevant v values and F levels. In addition, the contingency table in J2:L7 shows a *second* data set, B, which has each of the values in set A but *duplicated* to give a set with a total of 12 values.

The factor levels, F, are given *numeric* values because this is a requirement of the SPSS analysis, but they are treated as *nominal* values with no sense of progression. The Eta value does not depend on the order in which the levels of the nominal variable are presented, as with the chi-squared analysis.

	A	B	C	D	E	F	G	H	I	J	K	L
1	F	v		v\F	1	2	3		v\F	1	2	3
2	1	6.9		4.4			1		4.4			2
3	1	5.5		5.5	1				5.5	2		
4	2	6.8		5.6			1		5.6			2
5	2	7.6		6.8		1			6.8		2	
6	3	5.6		6.9	1				6.9	2		
7	3	4.4		7.6		1			7.6		2	
8		Set A				Set A					Set B	

Fig 4.20 Nominal by interval association

Using SPSS we calculate the value of Eta by selecting '☑-**Eta**', under the 'Statistics' options (8.2.4), and obtain the value 0.840.

We can also perform a one-way ANOVA or Kruskal–Wallis test using the interval data in column B to *test* whether there is any significant difference in the mean or median values of v between the nominal levels of F, and we return the values shown in Table 4.4 for the two data sets.

Table 4.4 Difference between measures for the *existence of* and *strength of* association

	ANOVA p-value	Kruskal–Wallis p-value	Eta, η, v dependent
Set A (6 values)	0.159	0.276	0.840
Set B (12 values)	0.004	0.059	0.840

The results in Table 4.4 show that, for set A with six values, there are no significant differences between the values of v for the three levels of F, but 'taking more results' with the 12 values in set B, the tests are able to detect a significant difference in mean values and almost a difference in median values. Increasing the number of data values has increased the *power* of a test for the *existence* of an association. However, the *distribution* of values has not changed (Fig 4.20) between set A and set B, and it is this characteristic that describes the *strength* of the association. This is measured by Eta, and we see in Table 4.4, that this value is the same for the two data sets.

4.4 **Agreement between variables**

A test for correlation (e.g. with a p-value) *tests* whether a relationship *exists* between two variables or factors. However, in this section we measure the *agreement* between the variables, which is the extent to which the related pairs of data have the *same values*. We approach this concept of 'agreement' through a number of different contexts.

It is important to remember that hypothesis tests for agreement are actually testing for *disagreement*, and that a null hypothesis result does not imply total agreement, only that there is no evidence of disagreement.

4.4.1 R^2 **goodness of fit**

The most familiar example of fitting experimental data to a theoretical model is through linear regression and the derivation of a best-fit straight line. We see in 2.1.1 and 5.4.6 that it is the residuals that are the underlying measure of agreement between the values predicted by the analysis and the actual experimental values. The *overall* measure of agreement is the *coefficient of determination*, r^2, which can be calculated using Eqn 2.12:

$$r^2 = 1 - \frac{SS_{RESID}}{SS_{TOT}}$$

where SS_{RESID} is the sum of squares of the residuals and SS_{TOT} is the total sum of squares in the data.

The calculated value of r^2 (or R^2) is often reported in many analytical results (e.g. ANOVAs), giving a measure of the quality of agreement between the data and the best-fit parameters of a theoretical model. However, some care has to be taken when increasing the *number* of factors to describe the experimental data, because it becomes easier to achieve a better fit purely through random chance, and it is quite possible for the value of R^2 to *increase*, suggesting a better (but spurious) agreement. An 'adjusted' value of r^2, typically written as R^2(adj), takes into account the number of variables or factors that are used to achieve the fit, and, with more factors, the *adjusted* value, R^2(adj), may begin to *decrease*, indicating that there is no real advantage in increasing the model complexity.

In addition, the graphical display of residual values is also very useful in identifying where, *within* a data set, the mathematical model deviates from the experimental data, e.g. Fig 2.6. Residuals are also very useful for assessing normality and homoscedasticity (equality of variance) in an analysis (5.4.6).

4.4.2 **Agreement between two related variables**

Two samples with *related* variables have unique links between *pairs* of data. For example, in Fig 4.2, measurements are made at the same concentration by the two assays. We have seen that the change in these values can be described by correlation and regression calculations, but in this section we consider the level of *agreement* between the values in each pair.

> **Exact agreement.** Each pair of measurements has the *same value*. Plotted on a scatterplot between the variables, this gives a straight line with slope = 1.0. The two variables must be measuring the same scientific *quantity* and with the same *units*.

Linear agreement. The change in the value of one variable is directly *proportional* to the change in the other. Plotted on a scatterplot this would give a straight line, but not necessarily with a slope equal to 1.0, or with intercept = 0.0. The two variables could be measuring different quantities.

Nonlinear agreement. The relationship between the variables follows a defined, but not linear, relationship between the values.

Rank agreement. Having ranked the variable values, the rank of each variable changes consistently with the rank of the other variable, either increasing or decreasing.

The strength of agreement for both *exact* and *linear* relationships are measured using Pearson's linear correlation coefficient, r, which is independent of the slope of the relationship (4.1.1), and the strength of *rank* agreement is measured using Spearman's correlation coefficient, ρ (rho) or Kendall's tau-b. For a curvilinear relationship it would be necessary to model the data using a nonlinear mathematical equation and calculate the goodness of fit, r^2, from the residuals. We use the following case study to develop relevant techniques.

Case study: **Toxicity assays / 3. Agreement**

—continued from 7.1.1 and 4.1.1, leading to 4.4.3

Fig 4.21 records measurements of the mortality (percentage deaths) of bacteria for increasing drug concentrations, C, (recorded as log(C)) using two assay methods, A and B, with each method repeated once. We wish to assess the *agreement between* the two methods and the *reproducibility within* each method.

The data in columns I, J, L, and M are used for the Bland–Altman plots in Fig 4.22(c) and Fig 4.23(c).

	A	B	C	D	E	F	G	H	I	J	K	L	M
1	Log(C)		A1	A2		B1	B2		(A1+A2)/2	A1-A2		(A1+B1)/2	A1-B1
2	-2.000		2.0	3.5		2.0	4.0		2.8	-1.5		2.0	0.0
3	-1.000		25.5	18.5		15.0	16.0		22.0	7.0		20.3	10.5
4	0.000		37.5	40.0		36.0	50.5		38.8	-2.5		36.8	1.5
5	0.699		57.0	62.0		33.0	54.5		59.5	-5.0		45.0	24.0
6	1.000		69.5	55.0		35.0	65.0		62.3	14.5		52.3	34.5
7	1.301		78.7	62.5		57.5	75.5		70.6	16.2		68.1	21.2

Fig 4.21 Replicate measurements of bacteria mortality using assays A and B

In this analysis, we compare *pairs* of variables separately, but in the following section (4.4.3) we develop the method of analysing multiple variables together. We first consider the *reproducibility* of assay A, by comparing the results of *A1* and *A2* in Fig 4.22.

Fig 4.22(a) plots both values of *A1* and *A2* against the *logarithm* of concentration, C. The use of the log axis presents concentrations related by multiplicative factors (e.g. dilutions) as evenly spaced points along the x-axis and avoids bunching of values at one end of the graph.

The initial options for comparing two variables include testing for a *linear* or *rank* agreement. We can then use linear regression for a *proportional* agreement and a measurement of the slope to identify *exact* agreement with slope = 1.0. Having confirmed correlation,

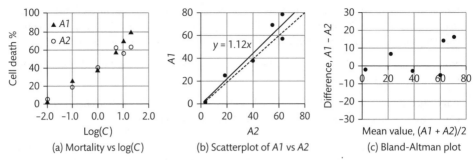

Fig 4.22 Reproducibility within A by comparing A1 and A2

we could use a Bland–Altman plot to display *disagreements* between the variables and then a paired Wilcoxon test to test whether there is a significant *net* difference between the values.

Fig 4.22(b) shows the values of *A1* plotted against the *related* values of *A2*, together with the trendline and equation. For an *exact* agreement, the values should lie along the dashed line which has a slope $m = 1.0$ and intercept $= 0$. We can calculate the confidence interval for the best-fit slope passing through the origin (Eqn 2.16) as:

$$m = 1.12 \pm 0.20.$$

As this includes the possible value of $m = 1.00$, this result does not identify any significant difference between that the two sets of measurements.

The **Bland–Altman** plot in Fig 4.22(c) provides a more *sensitive* display of the agreement between measurements of the *same variable* than the simple *x–y* scatterplot. For each *A1A2* pair, the *difference* or *disagreement* between the values, $A1_i - A2_i$, (calculated in column J of Fig 4.21) is plotted on the vertical axis against their average (mean) value, $(A1_i + A2_i)/2$ (calculated in column I). If the values are in perfect agreement then they will all lie along the '0' horizontal axis, but any disagreements will appear as vertical deviations above and below the axis. The plot is good at highlighting any drift in agreement through the range of values.

The Bland–Altman plot displays the magnitude of the differences, $x_i - y_i$, between data pairs that ideally should be giving the same value. If we believed that the *uncertainties* were normally distributed and the same across the range, then we could use a *paired t*-test to test for a *net* difference in mean values. However, given that we are often measuring over a wide range of values, this condition is unlikely to be satisfied and we can use a paired Wilcoxon test for a net difference in median values. A significant *p*-value ($p < 0.05$) would suggest that there is *disagreement* between the values, although a non-significant *p*-value would not prove that there was total *agreement*. It is also important to note that the paired test only tests for a net difference over the *whole range*, and that a Bland–Altman plot that showed a clear progression from a negative difference at one end to a positive difference at the other end may give a non-significant *net* difference for the paired test.

Fig 4.23 shows similar plots for the comparison between *A1* and *B1*. We can see the greater disagreement as the data in (b) is significantly away from the line of unit slope and the points in the Bland–Altman plot in (c) all show a positive difference between *A1* and *B1*.

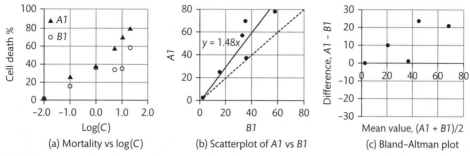

Fig 4.23 Agreement between A and B by comparing A1 and B1

We now compare each pair of variables from Fig 4.21 in Table 4.5 giving values for Spearman's correlation coefficient and the associated *p*-value, together with the result of a paired Wilcoxon test for net differences between the values and, finally, the confidence interval for the best-fit slopes of the linear relationships.

Table 4.5 Pairwise comparisons of A1, A2, B1, and B2

Variable pair	Spearman's rho	Spearman's *p*-value	Paired Wilcoxon *p*-value	Linear slope CI
A1 / A2	0.943	0.005	0.345	1.12 ± 0.20
A1 / B1	0.829	0.042	0.043*	1.48 ± 0.35**
A1 / B2	1.000	0.000	0.463	1.01 ± 0.16
A2 / B1	0.771	0.072†	0.028*	1.30 ± 0.35
A2 / B2	0.943	0.005	0.249	0.89 ± 0.15
B1 / B2	0.829	0.042	0.028*	0.67 ± 0.12**

Note: For agreement between variables, we would expect to see $p < 0.05$ for Spearman's correlation test, with $p > 0.05$ for differences using a paired Wilcoxon test, and the confidence interval, CI, for the relative slope should include 1.0.
*shows significant disagreement between variables using Wilcoxon test
**shows significant disagreement because the slope is not equal to 1.00
†shows a lack of correlation

Considering the reproducibility between A1 and A2, we see a significant correlation ($p = 0.005$), with no significant difference between the values (Wilcoxon $p = 0.345$). In addition, the linear relationship between A1 and A2 has a slope with a calculated confidence interval of 0.92 to 1.32 which includes the possible value of 1.00. These are all consistent with reproducible measurements.

Considering the agreement between A1 and B1, the correlation is significant, but the paired Wilcoxon tests indicates a significant net difference between the values ($p = 0.043$) and the confidence interval of the slope is 1.13 to 1.83 which does not include 1.00 and suggests that assay A tends to record higher values than assay B.

Comparing *A2* and *B1*, the poor correlation ($p = 0.072$) gives a large *uncertainty* in the linear slope, which results in the inability of the Wilcoxon test to detect the significant disagreement.

4.4.3 **Agreement between several variables**

We now introduce two tests that can be used to measure the *strength* of the agreement between more than two variables. Kendall's coefficient of concordance performs an analysis equivalent to *correlation* between multiple variables, and if there is good *correlation* between variables we would expect a significant result with $p < 0.05$. The Friedman test analyses differences between *linked* values across multiple variables in the same way as the paired Wilcoxon test does for paired variables, and if there is good *agreement* between the variables, we would expect the Friedman test to give a 'no significant difference' result with $p > 0.05$. The two analyses become equivalent when the data is transposed, swapping rows for columns. See 6.4.6 for performing Friedman's test in Minitab and SPSS.

Kendall's coefficient of concordance, *KCC*, is a measure of the *overall* correlation between *k* related samples of data, and is calculated as:

$$KCC: \quad W = \frac{(k-1)\bar{r}+1}{k} \tag{4.15}$$

where $\bar{r}$ is the mean of the Spearman correlations between all sample pairs, provided that all the groups have the *same direction of correlation*. For many samples (large *k*), *W* tends to equal $\bar{r}$. Compared to the *kappa* statistic (4.4.4), in which results either agree or disagree, *KCC* has the advantage of *weighting* the amount of disagreement by using the difference in ranked values.

An equivalent chi-squared value can be derived for *k* samples of size *n*:

$$\chi^2 = k(n-1)W \tag{4.16}$$

with degrees of freedom, $df = n - 1$.

Case study: **Toxicity assays / 4. Multiple comparisons**

—continued from 7.1.1 and 4.4.2

Fig 4.21 records measurements of the mortality (percentage deaths) of bacteria for increasing drug concentrations, C, (recorded as log(C)) using two assay methods, A and B, with each method repeated once. We now wish to assess the level of overall agreement between *more than two* variables.

Taking the values from Table 4.5, we can calculate the mean value of the six Spearman correlation coefficients between pairs of the four samples *A1*, *A2*, *B,1* and *B2* giving:

$$\bar{r} = (0.943 + 0.829 + 1.000 + 0.771 + 0.943 + 0.829)/6 = 0.886$$

which, using Eqn 4.15 with $k = 4$ gives

$$KCC, \; W = 0.914$$

and then with $n = 6$, Eqn 4.16 gives

$$\chi^2 = 4*(6-1)*0.914 = 18.3.$$

The p-value can be calculated from the χ^2 and df, using the Excel function

$$p = \text{CHISQ.DIST.RT}(\chi^2, df) = 0.0026.$$

If there is agreement between variables we would expect a significant result with $p < 0.05$.

The same result can be obtained in Minitab or SPSS.

Minitab > Stat > Quality Tools > Attribute Agreement Analysis...
⊙-*Multiple columns*: **Enter data columns,** e.g. *A1 A2 B1 B2*
Number of appraisers: 2–each assay is an 'appraiser'
Number of trials: 2–each assay has two trials
Known standard/attribute: $log(C)$
☑-**Categories of the attribute data are ordered**
→ Output: See Table 4.6

Table 4.6 Results from Kendall's coefficient of concordance calculations in Minitab (see 4.4.4 for Kappa)

		KCC value W	p-value for KCC	Kappa κ (using rank values)
Within assays	A	0.971	0.0837	0.6
	B	0.914	0.1035	0.4
Each assay vs standard	A	0.933	0.0003	0.8
	B	0.867	0.0009	0.7
Between assays		0.914	0.0026	0.6
All assays vs standard		0.900	<0.00005	0.75

'**Within assay**' measures the values separately between the *pairs* of results for each assay. The *KCC* values confirm that assay A is more *reproducible* than assay B. However, without the information of the standard value being measured, the reproducibility of each does not have the significance shown below.

'**Each assay vs standard**' measures the agreement between each assay pair *and* the known standard. The fact that we compare with $log(C)$ and not C makes no difference because these analyses are nonparametric, and only take rank order into account. We now see that there is highly significant agreement between *A1* and *A2* measuring C with $p = 0.0003$ and also between *B1* and *B2* measuring C with $p = 0.0009$.

'**Between assays**' measures an overall agreement between *A1, A2, B,1* and *B2*.The values, $W = 0.914$ and $p = 0.0026$, agree with the values derived above from the pairwise correlation coefficients.

'**All assays vs standard**' demonstrates the overall agreement between the two assays and the measured concentration.

To perform the same test in SPSS, it is necessary to *transpose* the data so that each of six columns in SPSS corresponds to one value of the concentration, and the separate variables are on different rows, as in Fig 4.24.

SPSS

SPSS > Analyze > Nonparametric tests > Related Samples
Fields: *C1 C2 C3 C4 C5 C6*
Settings: ◉ **Customize tests**
    ☑ **Kendall's coefficient of concordance (*k* samples)**

	C1	C2	C3	C4	C5	C6
1	2.00	25.50	37.50	57.00	69.50	78.70
2	3.50	18.50	40.00	62.00	55.00	62.50
3	2.00	15.00	36.00	33.00	35.00	57.50
4	4.00	16.00	50.50	54.50	65.00	75.50

Fig 4.24 Data entry in SPSS for Kendall's coefficient of concordance

SPSS reports the same 'between assays' results with *KCC, W* = 0.914, and *p* = 0.0026 as above.

The **Friedman test** tests for *differences* in the median values of the multiple assays. In Minitab, the Friedman test requires the data to be presented as *univariate* data with the factor being tested identified as the *Treatment* and the other factor the *Block*, and the calculation in SPSS requires the assay data to be in separate columns, as in Fig 4.21. See 6.4.6 for implementation in Minitab and SPSS, which both give *p* = 0.019 giving a significant difference between the assays *A1, A2, B,1* and *B2*.

4.4.4 Agreement within a contingency table

For *ordinal* or *nominal* data it is useful to represent the agreement, or otherwise, between two variables by recording the *frequency* of their observations in a contingency table. As an example, we can consider two variables as being the results of two tests *A* and *B*, which could be, for example, two

- assessors judging the quality of (the same) cakes in a village fete
- methods of assessing the quality of (the same) fingerprints.

Each test is performed on a number of different *subjects* (cakes, fingerprints).

Fig 4.25 shows a number, *n* = 40, of subjects being tested, or assessed, in two ways A and B, with each test recording a value on a scale of 1 to 3. In each case we can say that there are two *assessors* or *tests* each with a single *trial* of 40 *subjects*. There are no replicates in this analysis. The raw results for tests *A* and *B* are entered as related samples in columns B and C. The result of each trial pair is then *counted* into one of the nine possible combination cells

	A	B	C	D	E	F	G	H
1	Subject	Test A	Test B		A \ B	1	2	3
2	1	1	2		1	11	2	0
3	2	3	3		2	4	9	3
4	3	2	1		3	1	3	7
5	4	1	1					
41	40	3	2					

Fig 4.25 Ordinal agreement between tests *A* and *B* (Subjects 5 to 39 are in hidden rows)

shaded in F2:H4. For example, the pair of values for subject 1, '1' for *A* in B2 and '2' for *B* in C2, contribute one to the total count of two in cell G2, etc.

Agreements between *A* and *B* contribute to frequencies along the main diagonal of the table from F2 to H4. Any *disagreements* contribute to the off-diagonal frequencies. If a subject gives test *B* a larger value than test *A*, then this will appear as an entry in the top right of the table, and vice versa if the result of test *B* is less than that of *A*. Kappa, κ, is a statistic that is used to measure the amount of agreement within a contingency table.

Cohen's kappa, κ, can be understood as a symmetrical measure of the agreement between *two* assessors. It is often referred to as a measure of 'inter-rater agreement' where two assessors each rate a number of items (e.g. quality of wines), and κ is a measure of the extent to which they give the same ratings.

A similar statistic, **Fleiss's kappa**, can be used when assessing agreement between *more than two* assessors. For recording agreement between three (or more) assessors it is necessary to use a three(or more)-dimensional table of values, which is possible mathematically, but not so easy to present on a two-dimensional sheet of paper.

The κ statistic does not lead to a hypothesis test, and there is no agreement on its value in terms of an absolute qualitative scale, although Landis and Koch(Landis J R and Koch G G 'The Measurement of Observer Agreement for Categorical Data'. *Biometric* 33, pp. 159–174, 1977.) have proposed the scale in Table 4.7.

Table 4.7 Strength of inter-rater agreement

Values of κ	Amount of agreement
< 0.0	None
0.00 – 0.20	Slight
0.20 – 0.40	Fair
0.40 – 0.60	Moderate
0.60 – 0.80	Substantial
0.80 – 1.00	Almost perfect

A significant problem with *kappa* is a dependence on sample size, in that the same *percentage* of agreement will, for different *total numbers*, give different values of κ. It is nevertheless useful as a statistic to identify differences between *similar sized* sample sets.

The statistic is calculated as

$$\kappa = \frac{Pr(O) - Pr(E)}{1 - Pr(E)}$$ (4.17)

where $Pr(O)$ is the proportion of 'observed' agreements and $Pr(E)$ is the proportion of 'expected' agreements calculated from the observed data.

We will use a simple example in Fig 4.26 which represents the results of two assessors, A and B, giving binary responses, Y or N, for a total of T items.

	A	B	C	D
1	A \ B	Y	N	Row totals
2	Y	a	b	Ay
3	N	c	d	An
4	Column totals	By	Bn	T

Fig 4.26 Calculation of Kappa, κ

The total number of pairs of assessment, $T = a + b + c + d$.

Ay, An, By, and Bn are the row and column totals for A giving Y and N and for B giving Y and N respectively.

The total number of *agreements observed* in the data are:

  a items were given Y by *both* assessors and

  d items were given N by *both* assessors.

Thus, the *overall* probability of a randomly selected trial resulting in an agreement is given by:

$$Pr(O) = \frac{a+d}{a+b+c+d} = \frac{a+d}{T}$$ (4.18)

Using the data from Fig 4.19(b): $a = 24$, $b = 19$, $c = 7$, $d = 16$, $T = 66$

  $Pr(O) = 40 / 66 = 0.6061$

We also need to calculate the *expected* probability of agreement, $Pr(E)$, *given* the observed performances of *each* assessor:

Overall probability that assessor A records $Y = \dfrac{(a+b)}{(a+b+c+d)} = \dfrac{Ay}{T}$

Overall probability that assessor B records $Y = \dfrac{(a+c)}{(a+b+c+d)} = \dfrac{By}{T}$

The *random* probability that they *both* record $Y = \dfrac{Ay}{T} \times \dfrac{By}{T} = \dfrac{Ay \times By}{T^2}$

Overall probability that assessor A records $N = \dfrac{(c+d)}{(a+b+c+d)} = \dfrac{An}{T}$

Overall probability that assessor B records $N = \dfrac{(b+d)}{(a+b+c+d)} = \dfrac{Bn}{T}$

The *random* probability that they *both* record $N = \dfrac{An}{T} \times \dfrac{Bn}{T} = \dfrac{An \times Bn}{T^2}$

The random probability that A and B are in agreement is equal to the probability that they both record Y or N, which is the *sum* of the separate probabilities:

$$\Pr(E) = \dfrac{\{Ay \times By + An \times Bn\}}{T^2} \qquad (4.19)$$

Using the data from Fig 4.19(b) again: $a = 24$, $b = 19$, $c = 7$, $d = 16$, $T = 66$

$$\Pr(E) = (43 \times 31 + 23 \times 35)/66^2 = 0.4908$$

Then using Eqn 4.17: $\kappa = \dfrac{0.6061 - 0.4908}{1 - 0.4908} = 0.226$, which only rates as 'fair' agreement for this set of observed values.

A particular limitation of the kappa statistic for ordinal data with *several* levels is that kappa only assesses whether two assessors agree *exactly*, and does not assess the closeness of *disagreement*. For example, if two assessors differ by only one in a ranked answer, then this is rated the same as a much larger disagreement with a difference of five. This problem is addressed by Kendall's coefficient of concordance (4.4.3) which uses ranked values to record the magnitude of the disagreement.

4.4.5 Binary agreement

It is possible to test for agreement between *related binary* values using:

- McNemar's test for two samples
- Cochran's Q for k samples, where k may be more than two.

Case study: **Forensic questionnaire / 3. McNemar's test and Cochran's Q**

—continued from 9.2.1 and 8.2.1, leading to 6.2.1

Fig 4.27 reproduces the binary data responses given in Fig 8.10. We want to test for agreement between the responses.

McNemar's test

To understand McNemar's test, we consider the 2×2 table, H2:I3, in Fig 4.27 between the binary variables, *Q1* and *Q2*, in which

24 subjects give N to both *Q1* and *Q2*

16 subjects give Y to both *Q1* and *Q2*

◢	A	B	C	D	E	F	G	H	I
1	*Subject*	*Q1*	*Q2*	*Q3*	*Q4*		*Q1\Q2*	N	Y
2	1	N	N	N	Y		N	24	19
3	2	Y	Y	Y	Y		Y	7	16
4	3	N	N	N	N				
5	4	Y	N	Y	Y		McNemar's test:		
6	5	Y	N	N	N			$\chi^2 =$	4.65
7	6	N	N	N	N			$p =$	0.031
8	7	N	Y	Y	N				
67	66	Y	N	Y	Y				

Fig 4.27 Nine questionnaire responses from 66 subjects (rows 9 to 66 are 'hidden' in the worksheet)

19 subjects give N for *Q1* and Y for *Q2*, and

7 subjects give Y for *Q1* and N for *Q2*.

Although there is a good measure of agreement between the two questions, 40 out of 66 subjects, the essential difference between the questions lies in the way in which the 26 disagreements are distributed, and for this we can use the McNemar's test.

McNemar's test is a hypothesis test for *square* tables using the chi-squared statistic, but it specifically tests for a specific difference between the 'off-diagonal' elements of the table. The McNemar test applies to 2×2 tables whereas the McNemar–Bowker version is used for larger square tables.

It is a useful measure which detects a difference in the *changes* between samples. For example, with 'before and after' questions, it can test whether there is a difference between the numbers of people who change their answer from 'no' to 'yes' and those who change it from 'yes' to 'no'. If McNemar's test detects a difference, then the *direction* of the difference can be seen from the numbers in the table.

McNemar's test can be illustrated by using the 2×2 table in Fig 4.26, with the null hypothesis:

H_o: The probability of an item randomly appearing in '*b*' is the same as the probability for '*c*': $p_b = p_c$.

The analysis tests for a significant difference between the off-diagonal elements of the table using the statistic (with $df = 1$):

$$\chi^2 = \frac{(b-c)^2}{b+c} \qquad \text{or with a continuity} \atop \text{correction similar to Yates} \qquad \chi^2 = \frac{(|b-c|-1)^2}{b+c}. \qquad (4.20)$$

It can be seen that the statistic is dependent only on the off-diagonal elements, *b* and *c*. Using the values of $b = 19$ and $c = 7$ in Fig 4.27, we calculate $\chi^2 = 4.65$ in I6 using the continuity correction, and then calculate the *p*-value in I7 (3.7.4) using degree of freedom, $df = 1$:

$$p = \text{CHISQ.DIST.RT}(I6, 1) = 0.031.$$

Using McNemar's test identifies a significant difference in agreement which did not appear by just comparing the *proportions* of the unrelated samples. The value of McNemar's test is that it compares only the off diagonal disagreement values, whereas these differences can be masked by a larger number of diagonal agreements when using the proportions test.

In SPSS, it is possible to perform McNemar's test via contingency table statistics (8.2.4) or by using the analysis of nonparametric related samples as below.

Cochran's Q

Cochran's Q is the equivalent of the McNemar's test that can be applied to more than two samples. It is not possible to define answers to three or more variables using a 2-D contingency table, and consequently the calculation of Cochran's Q in SPSS is performed using the analysis of related nonparametric variables:

SPSS > Analyze > Nonparametric Tests > Related Samples...
Objective: ⊙-Customize analysis
Test Fields: *Q1 Q2 Q3 Q4*
Settings: ⊙-Customize tests
☑-Cochran's Q (*k* samples)
→ Output: Fig 4.28

Hypothesis test summary

	Null hypothesis	Test	Sig.	Decision
1	The distributions of Q1, Q2, Q3 and Q4 are the same.	Related-samples Cochran's Q test	.039	Reject the null hypothesis.

Asymptotic significances are displayed. The significance level is .05.

Fig 4.28 Cochran's Q for *Q1 Q2 Q3 Q4* (SPSS)

The output in Fig 4.28 reports that there is a significant difference ($p = 0.039 < 0.05$) between the four tests, but we do not immediately know where the difference might lie. We can then use McNemar's test between pairs of variables to locate possible differences, giving the results:

Q1 and Q2: $p = 0.031$ Q1 and Q3: $p = 1.000$ Q1 and Q4: $p = 0.629$
Q2 and Q3: $p = 0.052$ Q2 and Q4: $p = 0.164$ Q3 and Q4: $p = 0.648$

If we had used *only* the multiple McNemar's tests, then we would need to impose a Bonferroni correction (1.6.4) for the *significance level* by dividing it by the number of tests, giving the new significance level of $0.05/6 = 0.009$. In this case, *none* of the paired tests would show significant differences. However, the use of the single Cochran's Q test does confirm that a significant difference does exist, and we can then use the McNemar paired tests as post hoc tests to identify the principal source of the difference, i.e. between *Q1* and *Q2*.

Part II

Analysing experimental data

Part II approaches an understanding of data analysis from the 'top down' *scientific* perspective of the need to answer specific questions. It starts with the assumption that the student has a set of experimental results and wishes to know what analyses would be applicable to that data and which would address the scientific questions being asked. The book therefore provides targeted support for the student at the time when he/she is well motivated to investigate and use the techniques developed, but it also provides support for a formal course in which students are given example data to analyse. The content develops the different analytical techniques through their implementation using Excel, Minitab and SPSS, but with extensive reference made to the underlying statistics developed in Part I.

Chapter 5. Project data analysis outlines the issues that confront students, normally during their final year, when faced with their own research project or dissertation. This does *not* approach the topic from the 'research design' perspective that would be required for a 'real' research project, but reflects the exploratory nature of most student projects. It concentrates on a 'what do I do now?' approach for the student who is faced with experimental data, either within a personal project, or in an extended exercise in a taught (e.g. MSc) course.

The remaining chapters then address issues that are defined by the structure of the experimental data and the scientific questions to be answered. Each section starts with example data sets, together with possible analytical options and methods of describing and visualizing the experimental results. They then continue with the use of SPSS, Minitab and Excel to implement relevant analytical techniques.

Chapter 6. Single response variable assumes that the student wishes to test for the effect of one or more factors that might affect a single measured variable. Typical analyses include the familiar *t*-tests, ANOVAs and their nonparametric equivalents, together with general and generalized linear models.

Chapter 7. Related variables considers two or more related variables which introduces a wide range of analyses based on specific linear and nonlinear relationships as well as more general *x-y* systems. Typical analyses include linear and nonlinear regression, correlation and agreement, but extend to include convolution techniques and component spectral analysis.

Chapter 8. Frequency data relates to the counting of experimental results that fall into specific categories, either defined by the categorical nature of the measurement itself or by the use of probability density histograms in describing the distribution of interval data. This includes chi-squared calculations and associated contingency table statistics. It also addresses the specific issues raised by analysing and modelling binary data.

Chapter 9. Multiple variables provides an introduction to some of the techniques for handling data sets with many variables, including cluster and principal component analysis and multiple regression. It concludes with an example data set with multiple variables (similar to questionnaire data) and considers the different analytical 'questions' that can be asked, with reference to the techniques introduced earlier throughout the book.

5 Project data analysis

Introduction

In this chapter we are assuming that a student is conducting a 'research' project as part of a final year's study as an undergraduate, and has been asked to 'investigate' scientific behaviour within a specific topic. Having just finished collecting experimental results, he/she would, *in an ideal world*, already know what data analysis techniques to use because the experiments would have been conducted within a designed framework that fully anticipated the relevant statistical analysis. However, in common with many other students, not much thought may have been given to data analysis, or possibly the investigation has been of an exploratory nature and it was not possible to anticipate what analysis would be required.

Faced with experimental data without well-defined analytical options, the danger is that the student might grab any statistical test that will produce a p-value or even to rush off for help from a friendly statistician. However, the best approach is to sit back and carefully review the

- scientific objectives–what questions should the statistical analysis aim to answer?
- experimental results–how was the data collected, what factors were involved, replicate values, paired data, etc.?

No one can provide help until the student can explain clearly what it is he/she wants to achieve, the experimental results, and how they were collected. In fact, this actual process of reflection will often lead to a better insight of what analytical techniques can be used, and this section aims to help develop that overview.

Section 5.1 considers the initial steps in reviewing data and objectives, understanding the source of data uncertainties, and transforming the data into a format suitable for software analysis.

Section 5.2 presents examples of methods to identify experimental characteristics that are suitable for analysis.

Section 5.3 reviews the techniques and issues involved in transforming the data into a format required to match the analytical technique.

Section 5.4 discusses the relevance of normality and equality of variance for many analytical techniques and develops the approach, tests, and transformations that may be involved.

5.1 Preparing data for analysis

A student has just completed experimental work and recorded the results, but before thinking about any analysis, it is important to preserve the original data and any associated notes. It is essential to be able to check the original data, for example it may be necessary to check if a handwritten 7 has been misread as a 2. For electronic records, the original data should be saved in a secure memory store before copying the data into a new file, and all subsequent analysis should only be carried out on the duplicate data set.

5.1.1 Case studies

In this section we will meet the following case studies:

Football fantasy

Highlights the difference between *statistical* and *scientific* significances.
    5.1.4 / 1. Significance: Can a distant football fan's actions affect the team's performance?

Fingerprint quality

An investigation into methods of lifting fingerprints and the resultant quality.
    5.1.5 / 2. Organizing data entry: Considers the transfer of data from lab book to software analysis.

Ink analysis

Analyses the spectral responses of different black inks to identify a method of forensically differentiating between them.
    5.1.6 / 1. Exploratory phase: Develops an overview of the experimental investigation.

5.1.2 Identifying the variables/factors

The key to understanding data is to identify and define all the variables and factors included in the investigation. The changing values associated with the system are the measured *variables* of the experiment, and the variable that describes the condition of an experiment (e.g. the temperature of reaction) will typically be called a *factor*. The variables/factors can be categorized under the headings of Action, Type, Levels, and Variability, and this process of categorizing for a given data set can be very helpful in clarifying the *understanding* of the data, and may help in deciding on a suitable analytical technique.

Under **Action** we first identify whether we are measuring a variable that is an 'input' to the system or whether it is an 'output' variable whose value is determined by the system. However, in interrelated (e.g. correlation) measurements we can measure two outputs without an obvious input (e.g. recording height and weight of an animal without knowing its age).

Under **Type** we distinguish whether a measurement is an interval, ordinal, or nominal value (1.2.2), or a frequency (count). In some cases, a frequency can be treated as an interval variable (6.1.1).

Under **Levels** we record the range of values that the variable could take. An interval variable may be *continuous* with values within a given range, an ordinal variable may have a number of *progressing* values, and a nominal variable may have a number of specific discrete *categories*. Frequencies have *integer* values.

Under **Variability** we identify whether the value is known exactly, or how replicate values could be expected to change. They may follow a known distribution, e.g. normal, Poisson, binomial. If it is an 'input' we decide whether it is a variable/factor whose value is *fixed* in the design or operation of the experiment or whether it is a variable/factor that has been *randomly* selected.

For example, in the calibration of a spectrophotometer, the absorbance relating to the concentration of a solution would be described by the data structure in Table 5.1(a). Both variables are continuous positive interval values. We assume that the experimental variations in absorbance will follow a normal distribution and that the values of concentration are determined by the specific standard solutions prepared for the experiment. The data structure in Table 5.1(b) relates to the contingency table in Fig 8.9 which counts the numbers of people, grouped as male and female, who show none, some, good, or excellent improvement following bacteriophage treatment. The distribution of observed frequencies based on an underlying probability is expected to show a Poisson distribution.

Table 5.1 Example descriptions of experimental variables/factors

(a) Absorbance vs concentration calibration data

Variable	Action	Type	Levels	Variability
Absorbance, A	Output	Interval	Continuous 0→	Normal
Concentration, c	Input	Interval	Continuous 0→	Fixed

(b) 2×4 contingency table data

Variable	Action	Type	Levels	Variability
Frequency	Output	Frequency	Integer	Poisson
Improvement	Input	Ordinal	Category 4 levels	Fixed
Sex	Input	Nominal	Category 2 levels	Fixed

5.1.3 Understanding the uncertainty in the data

The statistical analysis of experimental data frequently calculates *significance* by comparing a possible difference in the data with the random variations in that data. For this reason, it is important to be aware of

- the sources of uncertainties in the data
- the way in which the different types of uncertainty can be differentiated, and
- the propagation of uncertainties/errors.

Section 1.4 identifies the different types of measurement, subject, and probability uncertainties and develops the techniques for their quantification.

The importance of designing the experiment to collect the relevant uncertainty data is often overlooked. For example, the identification of an *interaction* term in a multifactorial ANOVA requires *replicate* measurements at the same factor levels (3.3.2) such that the analysis can distinguish between an interaction and experimental uncertainty.

A prior knowledge of the uncertainty in a measurement process can also change the approach to analysis. For example, the *t*-test relies on the *sample* data for calculating experimental uncertainty, but prior knowledge of this uncertainty allows the use of the more precise *z*-test (3.1.5). A similar benefit occurs if the experimental uncertainty is previously known when using straight line intercept calculations (2.2.3).

5.1.4 Scientific significance

Through most of a project the student may have been immersed in the science of the experiment, but it is easy to forget this context as soon as one enters the alien world of statistics to analyse the results. For example, *statistical* significance is a measure of the statistical probability with which an observed set of results could have occurred by chance, but, before we can conclude a *scientific* significance, we must be aware of the wider scientific context of the problem, as illustrated by the following case study.

Case study: Football fantasy / Significance

A football fan regularly watches his team play on television. He typically leaves the room for about 15 minutes during each match to collect refreshments, and, over several matches, gets the impression that his team is more likely to concede a goal if he is *out* of the room. Will his team do better if he stays in the room? He decides to conduct a hypothesis test which has the null hypothesis:

H_0: The probability of his team conceding a goal is not affected by whether he is in, or out, of the living room in his own house.

During each of the next six matches, he leaves the room for a randomly selected period of 15 minutes out of the 90 minutes of play. He would then expect, assuming that the null hypothesis is correct, that the average *proportion* of goals conceded during his absence would equal 15/90 = 0.167.

In fact, his team concede a total of 12 goals, five of which were while he was out of the room, which is greater than the randomly expected average of 12 × 0.167 = 2 goals, confirming his original impression. Should he now publish his findings?

The statistical approach to the results uses Fisher's exact test (3.8.2) to test whether the measured proportion of 5/12 is greater than 0.167 for a sample size of 12, and returns a statistically significant *p*-value = 0.037. Based on this statistical analysis he rejects the null hypothesis and concludes that leaving his living room significantly increases the probability of his football team conceding a goal!

His conclusion is clearly wrong—there is no *scientific* mechanism by which his presence in his own living room can affect how his team play on the football pitch many miles away. A *statistically* significant result does not necessarily mean that there must be a significant *scientific* effect.

5.1.5 Data entry into software

Initially, experimental data is likely to be organized in a 'collection' format, defined by the order in which the experiments were carried out, and is likely to consist of a number of sets of tabulated data recorded in different conditions and times.

Case study: Fingerprint quality / 2.Organizing data entry

—continued from 6.4.1

As a simple example, the columns A to D in Fig 5.1 record the assessed quality of fingerprints as *Quality* values (on a 0 to 5 ordinal scale), obtained using three different lifting methods, *Method* (*A, B, C*) at three different temperatures, *Temp* (10, 20, 30). It is also possible that some 'initial' analyses have been performed, as represented by the calculations of mean and standard deviations (SD) for replicate data values, but we now need to organize the data for entry into analytical software.

	A	B	C	D	E	F	G	H	I	J	K
1	Collection format:							Analysis format:			
2											
3	Method A:	Quality measurements			Mean	SD		*Rec*	*Temp*	*Quality*	*Method*
4	Temp = 10	2	1	2	1.67	0.58		1	10	2	A
5	Temp = 20	3	3	2	2.67	0.58		2	20	3	A
6	Temp = 30	2	1	1	1.33	0.58		3	30	2	A
7								4	10	2	B
8	Method B:	Quality measurements			Mean	SD		5	20	3	B
9	Temp = 10	2	3	2	2.33	0.58		6	30	3	B
10	Temp = 20	3	4	2	3.00	1.00		7	10	3	C
11	Temp = 30	3	4	4	3.67	0.58		8	20	4	C
12								9	30	4	C
13	Method C:	Quality measurements			Mean	SD		10	10	1	A
14	Temp = 10	3	2	3	2.67	0.58		11	20	3	A
15	Temp = 20	4	3	5	4.00	1.00		12	30	1	A
16	Temp = 30	4	3	3	3.33	0.58		13	10	3	B
17								14	20	4	B

Fig 5.1 Data in both 'collection' and 'analysis' formats (data in columns H to K extend to other rows)

Excel is very accommodating in the possible layouts for data analysis, generally accepting data for analysis in either rows or columns within the two-dimensional spreadsheet. This can be very useful for presenting data and laying out complex calculations, and is the ideal medium for collecting and organizing experimental results. It is also possible to carry out a range of basic statistical analyses using Excel, e.g. initial calculations for the means and standard deviations as in Fig 5.1. However, for more advanced analysis it is probably necessary to use other dedicated software packages. These include a number of possible Excel 'add-ins' that can expand Excel's capability, but in this book we will mainly consider the separate packages: Minitab and SPSS.

Both Minitab and SPSS accept data in a strictly column format, and are *not* two-dimensional spreadsheets. This has the advantage of establishing discipline in organizing the data and requires careful thought about how the data is structured. Hence, the next task may be transcribing data from a 'collecting' format in Excel to an 'analysing' format for Minitab or SPSS. The columns I to K in Fig 5.1 record the same data as in columns A to D, but reorganized into the required column (stacked) format with each column identifying a separate variable or factor. When moving data from *rows* to *columns* in Excel, it is often useful to use the 'paste special' method and check the 'transpose' option which will automatically switch the copied data from a row to column format (or vice versa).

It is important to remember that most software analyses expect to use the *individual raw data values* and *not* calculated means and standard deviations. The software calculations use the variation between *individual* values as the means of estimating the inherent experimental uncertainty for comparison with any differences that may be due to factor effects.

Each row in the column format will record a 'unit' of measurement, which is often described as a 'record' or as a 'subject' (from the analysis of questionnaires). Each *row* will have either:

- **Univariate response variable.** The same *single* variable (e.g. *pH*, fish weight, response on a Likert scale) is recorded in a single column (column J in Fig 5.1), and the conditions under which it was recorded for each subject are identified by values in the other columns.

- **Related response variables**. Related measurements have two, or more, response variables associated with the same record or subject being measured. These may be either:

  - **bivariate or multivariate data** in which *different* variables measure aspects of the same experimental system, e.g. *Quality, Temp, Method* recorded as columns C1, C2, and C3-T in Fig 5.2(a),

    or

  - **repeated measures** in which the *same* variable is measured under different conditions for the *same* subject, e.g.*%T* for different inks in Fig 3.40(a) at the same wavelength. If just two values are measured, e.g. the measurements of bacterial contamination of the same surface *before* and *after* cleaning, then this may lead to the familiar *paired t*-test.

Data in **Minitab** is entered into a worksheet (Fig 5.2(a)) in which each *numeric* column has a label, C1, C2 etc. and a space to add a column name. If *text* is entered into a column it adds a '−T' to the column number, but it is possible to change between *text* and *numeric* columns using:

Minitab > Data > Change Data Type

↓	C1	C2	C3-T	C4
	Entry Direction▷	Quality	Method	
1	10	2	A	
2	20	3	A	
3	30	2	A	
4	10	2	B	
5	20	3	B	
6	30	3	B	

	Temp	Quality	Method	MethodN
1	10.00	2.00	A	1.00
2	20.00	3.00	A	1.00
3	30.00	2.00	A	1.00
4	10.00	2.00	B	2.00
5	20.00	3.00	B	2.00
6	30.00	3.00	B	2.00

(a) Minitab worksheet   (b) SPSS data editor: Data view

Fig 5.2 Extracts from the data files for Minitab and SPSS for data from Fig 5.1

The printed outputs in Minitab appear in the session window and graphs appear in separate pop-up windows. A *project* file with the '.mpj' extension consists of a worksheet(s) to hold the data and a session window to record the results, although it is possible to save the worksheets separately as '.mtw' files.

SPSS holds data in data editor files with the extension '.sav' and separately saves results in viewer files with the extension '.spv'. The SPSS data editor has two 'views': Data view as in Fig 5.2(b) for the data values, and variable view as in Fig 5.3 for defining the qualities for each variable. In Fig 5.2(b) we have added a variable *MethodN* which is the same as *Method*, but coded 1 to 3 so that it could be used as a scale variable instead of a nominal variable.

	Name	Type	Width	Decimals	Label	Values	Missing	Columns	Align	Measure	Role
1	Temp	Numeric	8	2		None	None	6	Center	Scale	Input
2	Quality	Numeric	8	2		None	None	5	Center	Scale	Input
3	Method	String	1	0		None	None	5	Center	Nominal	Input
4	MethodN	Numeric	8	2		None	None	6	Center	Scale	Input

Fig 5.3 SPSS data editor: Variable view

The data values appear in the data view, and their characteristics are defined in the variable view, where the key values to define are:

Name which cannot include spaces or punctuation marks (except for a full stop in the body of the name)

Type which describes the data entry, e.g. numeric, string, date, etc.

Measure which can be scale, ordinal, or nominal. In general it is useful to define an ordinal *variable* as having a scale *measure*, but this can depend on the analysis being carried out.

Label provides an additional and more flexible description of the variable.

The SPSS output, including printouts and graphs, appear in the output viewer (.spv) file.

5.1.6 Reviewing data and objectives

When completing a final year project, a student's motives for using statistical analysis can easily become confused. The student may feel (probably correctly) that the final grade will depend on demonstrating familiarity with statistical analysis. In which case, he/she may start looking for *any* analysis that might accept the data rather than an analysis that matches the *objectives* of the research. It is important to first carefully review the experimental results and then consider possible statistical analyses in the *context* of the original scientific objectives.

Describing data

A useful first step that is often overlooked is to *describe* raw data values. Not only will it be necessary to describe the experimental results for the project report, but the process of visualizing the data will help immensely in seeing what is required for any further analysis. We use the following case study as an example, but we also put 'describing the data' as a first step in identifying analytical methods in most sections of Chapters 6, 7, and 8.

Case study: **Ink analysis / 1. Exploratory phase (overview)**
—leading to 5.2.3, 2.2.2, 3.6.2, 3.3.3, and 6.4.8

As part of a forensic investigation to characterize and distinguish between four black inks, the percentage transmission, *%T*, of light (Fig 5.4) is measured using a visual spectral comparator over wavelengths from 500 nm to 800 nm. For the purposes of forensic classification, we need to identify a common characteristic that will generate a simple measureable parameter that *differentiates* between the inks. We first develop an overview of the experimental investigation.

  5.2.3 / 2. Analytical characteristics: Identifies test statistics from raw data.

  2.2.2 / 3. Exact *y*-intercept: Calculates the confidence intervals for intercepts at 50%T.

  3.6.2 / 4. Repeated measures: Tests for a difference in %T at 'linked' wavelengths.

  3.3.3 / 5. ANCOVA analysis 1: Differentiates between inks with wavelength as a covariate using Excel and Minitab.

  6.4.8 / 6. ANCOVA analysis 2: Differentiates between inks with wavelength as a covariate using SPSS.

By plotting their spectral responses on the graph in Fig 5.4 we see that all inks show a near zero transmittance over the main visual range up to about 650 nm—hence they all appear 'black'. However, the plot does allow us to identify a possible method of differentiation, by recording the long wavelength 'cut-off' points at which their spectra cross the 50%T level. In the next step in this case study example (5.2.3), we focus on different *statistical* methods that can be used to test for these possible differences.

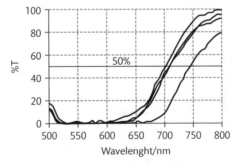

Fig 5.4 Transmission spectra for black inks

Reviewing objectives

With a good understanding of what the experimental data looks like, it is useful to sit back and review the ultimate aims of the research and see how they may help to identify any specific analysis. Table 5.2. uses the 'Ink analysis' case study to establish a hierarchy of objectives, starting with the title of the project.

Table 5.2 Hierarchy of objectives in the 'Ink analysis' case study

Overall aim (title)	Ink analysis. An investigation into the forensic analysis of black ink
Objectives	Identify methods of differentiating between black inks
Experimental data	Univariate values of %T as a function of wavelength provide a spectral analysis of the different inks, Fig 5.4
Analytical characteristics	Select long wavelength cut-off values at 50%T in the spectrum
Types of analysis	Test for significant differences at 50%T
Specific analyses	– Confidence intervals for 50%T intercepts (2.2.2) – Repeated measures in %T with linked wavelengths (3.6.2) – ANCOVA for differences in %T with wavelength a covariate (3.3.3)

Having reviewed both the experimental data and objectives, it may be necessary to identify a specific data characteristic for testing (Section 5.2), transform the data (Section 5.3), or consider the normality and variance conditions for analysis (Section 5.4). Otherwise, the various analytical options in Chapters 6, 7, 8, and 9 should be investigated, according to the data type and objectives for analysis.

5.2 Deriving test characteristics

If the research was somewhat exploratory (typical for a student project), it may have raised questions and possible analyses that were not expected. It would be fortunate if it were possible to take the raw results from the exploratory data and enter them directly into a statistical test, but in many cases it will be necessary to manipulate the original data in some way. However, the possibilities for data rationalization are endless, and the best we can do here is to provide some of the major techniques together with some examples which may provide some basis of experience for future problems.

For example, it may be necessary to

- select a sub-set of the data relevant to the focus of the investigation (5.2.3).
- combine data to arrive at new variables that represent relevant specific scientific values (5.2.4).
- transform the response variable according to a theoretical relationship (5.2.5).

Even after identifying analytical data it may still be necessary to

- transform data to provide a near-normal distribution of experimental data (5.4.7).
- use linearization to transform one, or both, of two interrelated variables such that their theoretical relationship can be described by a straight line (Section 2.3).

5.2.1 **Case studies**

In this section we meet the following case studies:

Bacterial growth

Investigates the effect of different antibacterial cleaning agents on the growth of bacteria.

> 5.2.2 / 1. Exploratory phase: Identifies characteristics capable of analysis.

Ink analysis

Analyses the spectral responses of different black inks to identify a method of forensically differentiating between them.

> 5.2.3 / 2. Analytical characteristics: Identifies test statistics from raw data.

Chemotaxis index

The chemotaxis experiment initially treats nematodes with different concentrations of a therapeutic drug, and then measures their capacity for migration towards a food supply in a segmented agar plate.

> 5.2.4 / 2. Deriving analytical statistic: Starting with *raw data values*, it calculates the *chemotaxis index* that is then used to analyse the data.

Mean kinetic temperature

Models the effect of raised temperature on the deterioration of stored pharmaceutical products.

> 5.2.5 / Modelling the analytical variable: Uses a theoretical model to *weight* the output data.

5.2.2 **Beyond the exploratory phase**

The following case study illustrates how an exploratory study into the growth of bacteria can be used to identify aspects for further research. In principle, the *same* data should not be used to *both* identify possible test characteristics *and* then to carry out the analysis, because the data for the analysis should be *independently* obtained. However, in the student project there is usually no time to collect new data, but it is acceptable to use the existing data to *illustrate* the statistical techniques that could be used, provided that reservations are stated about using the same data twice for both identification and analysis.

Case study: **Bacterial growth / 1. Exploratory phase (overview)**
—linking to: 3.1.1, 3.4.3, 7.3.5

The general aim of this exploratory project was to assess the effect of using different antibacterial cleaning agents on the growth of bacteria. The investigation records the effect on the quantity of genetically modified bioluminescent E-coli as a function of time. The graphs in Fig 5.5 plot luminesce, L, for three increasing concentrations, C1, C2, and C3, of the cleaning agent. In this exploratory phase we first identify possible test statistics from the raw experimental data.

3.1.1 / 2. Difference in slopes using t-test: Demonstrates the essential statistics of a t-test.

3.4.3 / 3. Difference in slopes as an interaction: Demonstrate interaction in an ANOVA.

7.3.5 / 4. Using smoothing convolutes: Develops a best-fit curve to raw data as drawn in Fig 5.5, and calculates maximum and minimum values.

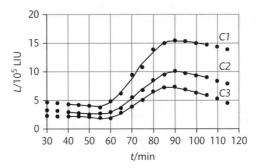

Fig 5.5 Growth of bacterial populations in presence of antibacterial agents. The process of drawing these best-fit *curved* lines is developed in 7.3.5

Following the exploratory phase, represented by the curves in Fig 5.5, it may become evident that there is some effect of particular interest that could be tested more carefully, e.g. defining the difference between the curves.

There is an initial lag before the bacteria enter a growth phase, and the student wishes to compare the different curves just within this growth phase plus find characteristics that define the *scientific* performance and which can be tested *statistically*.

The two most obvious statistical choices are:

- The slope of the line between 60 min and 85 min, which would be the rate of bacterial growth. Statistical tests for difference in slopes are developed in 3.1.1 and 3.4.3.
- The difference between the minimum and maximum values, either as a difference or a ratio. This calculation is considered in 7.3.5 by using smoothing convolutes to calculate the best estimates for difference between maximum and minimum.

The actual choice will depend on the science of the problem.

5.2.3 **Selecting analyses**

Exploratory data typically covers a wide range of values hoping to identify regions of analytical interest. We need then to pick out a reduced range of values suitable for specific analysis. In 2.1.5 we see how a 'linear' calibration graph may have a high end curvature, but, by selecting a useful working range, the calibration may still be linear, and in 5.2.2 we see the identification of the growth phase of bacteria for analysis.

In the following case study, the full exploratory data from the spectrophotometric analysis of three inks gives an overall description, but if we wish to differentiate between the spectra, we need to concentrate on a specific range of values.

Case study: Ink analysis / 2. Analytical characteristics

—continued from 5.1.6, leading to 2.2.2, 3.6.2, 3.3.3, and 6.4.8

We saw in Fig 5.4 the spectral responses for four black inks, but we now need to identify simple measureable parameters that differentiate between the inks.

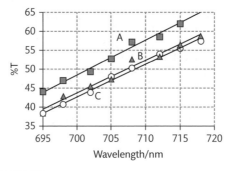

Fig 5.6 Black ink transmission spectra

The only visual difference between the ink spectra given in Fig 5.4 is that they have different cut-off points at the long wavelength end of their spectra. We can define this mathematically by recording the wavelength, λ_{50}, where the transmittance passes the 50 %T point. One ink is clearly different, with a cut-off wavelength, λ_{50}, at about 745 nm. We can then compare the other three inks, A, B, and C, by analysing the specific section of the graph in Fig 5.6, between 40 %T and 60%T.

The most direct method is to calculate the wavelengths with which they each cross the 50%T line and test for any significant *horizontal* difference (2.2.2).

An alternative method is to test for a *vertical* difference between the lines, but the problem here is that the change in %T with wavelength (a covariate) is an additional vertical variation which confounds a simple analysis such as a t-test. The first method uses the fact that measurements for each line are made at the *same* wavelengths, and this is a unique factor that links the corresponding measurement in each sample. This allows us to use a *paired* t-test between each pair of lines or a *repeated measures* test between all three lines (3.6.2). For the second method, we consider a situation where the wavelength measurements for each line are not at the same values, and we *demonstrate* the use of an ANCOVA analysis (3.3.3) by testing for a vertical difference while taking into account the wavelength covariate.

5.2.4 Combining data

Depending on the science of the system, it may be necessary to take more than one response variable and mathematically combine them with a defined formula to obtain a *derived* analytical characteristic.

Case study: Chemotaxis / 2. Deriving analytical characteristics
—continued from 6.3.1, leading to 5.4.6

The aim of the investigation of is to see whether the pre-treatment with a drug affects the mobility of nematodes. The experimental procedure of chemotaxis uses an agar plate which is divided into three zones with an attractive food supply (NaCl) in section A and a control in section B. The nematodes are introduced into section C and might then be expected to migrate preferentially to the food in section A. Sections A and B also contain spots of sodium azide to paralyse the nematodes once they arrive in either of the sections.

The numbers of worms in each section A, B, and C are recorded in Fig 5.7, where each row represents one measurement. The experiment is repeated at three levels of pre-treatment, *Treat*, using three concentrations of a chemotherapeutic drug, with eight replicates at treatment level 0, seven at level 1 and nine at level 2. A number of rows have been hidden in the worksheet (shown by the horizontal lines) to save space.

	A	B	C	D	E	F	G	H	I	J
1	RecNo	Treat	A	B	C	A-B	A+B+C	CI	Mean	St. Dev
2	1	0	35	6	54	29	95	0.31		
3	2	0	40	8	62	32	110	0.29		
8	7	0	37	14	66	23	117	0.20		
9	8	0	34	8	61	26	103	0.25	0.27	0.08
10										
11	9	1	14	20	52	-6	86	-0.07		
12	10	1	30	10	57	20	97	0.21		
16	14	1	40	19	65	21	124	0.17		
17	15	1	29	17	48	12	94	0.13	0.15	0.11
18										
19	16	2	19	32	55	-13	106	-0.12		
20	17	2	27	11	50	16	88	0.18		
26	23	2	38	29	41	9	108	0.08		
27	24	2	23	26	59	-3	108	-0.03	0.01	0.09

Fig 5.7 Results data for chemotaxis measurements

Note that some rows have been hidden at the horizontal lines to save space

The chemotaxis index is calculated as the ratio

$$CI = (A-B)/(A+B+C)$$

where A, B, and C are the numbers of nematodes in the various sections after one hour.

The aim is to test whether different pre-treatment affects the value of the index. There are then two *apparent* options for calculating values for *CI*:

1. (Incorrect option) For each of the three *treatment levels*, calculate total values of *A*, *B*, and *C* separately, and then calculate *one* value of *CI*.

2. For each *measurement* (row) calculate the value of *CI* for that row as in column H, and then calculate the mean and standard deviations of the *CI* values for each treatment as in columns I and J.

This is similar to the problem in the 'DIY dice' case study (1.4.6) in that case option 1 loses any information about the *experimental variation* in replicate measurements of *CI*. It is only in option 2 that we can calculate the standard deviation variation. The software analysis must calculate the experimental uncertainty to be able to test whether any observed differences could have occurred by random chance.

The calculated mean and standard deviation, *CI*, values suggest that there is an apparent difference between the treatments, but we now want to perform a hypothesis test to test whether this difference is statistically significant. The next step in the analysis is to copy the *individual* calculated values of *CI* for each measurement into a data column in Minitab or SPSS, together with a second column which identifies the value of *Treat* related to that measurement, and then use an ANOVA analysis (5.4.6 and 6.3.1).

5.2.5 Modelling response variables

Perhaps the most common use of a theoretical transformation is in the transformation of variables to obtain a straight line relationship (Section 2.3). For example, in the decay of a bacterial colony the population will fall according to an exponential relationship with time, but by taking the logarithm of the population, *N*, it is possible to see a linear relationship of $\ln(N)$ against time, *t*. However, we now consider the use of a theoretical transformation to develop a *weighting* factor for the experimental data.

Case study: Mean kinetic temperature / Modelling the analytical variable

In this case study, a pharmaceutical product is intended to be kept refrigerated at a temperature which will fluctuate around 5°C. However, due possibly to refrigerator failure or delayed transport between refrigerators, the product is exposed to a relatively high temperature surge for a short period, as shown in Fig 5.8.

The rate of decay, *v*, of the product increases with temperature according to the Arrhenius's equation:

$$v = A \times \exp\left\{-\frac{\Delta H}{R \times T_K}\right\}$$

where T_K is the temperature in degrees Kelvin, ΔH is the activation energy, *R* is the gas constant = 8.31 J mol^{-1}K^{-1}, and *A* is a constant which will cancel out in our calculations.

We wish to calculate the mean kinetic temperature (*MKT*) which is the *constant* storage temperature that would result in the same overall decay as the varying temperature profile shown in the graph. We estimate that ΔH = 85 kJ mol^{-1} for this product.

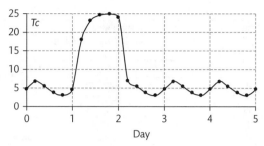

Fig 5.8 Temperature fluctuation with time

The exponential relationship means that the rate of decay is not linear with temperature, and a high temperature will have a proportionally greater effect than a lower temperature. This means that simply taking the *arithmetic* average of all temperatures will underestimate the amount of decay. We will need to provide the higher temperatures with an increased 'weighting' of importance.

The exponential term in the equation becomes the 'weighting' factor for each data point. Hence to calculate an average for the rate of decay for all n data points we need to calculate the average of the 'weights' from each point, i, which will be given by:

$$\bar{v} = \frac{A \times \sum_i \exp\left\{-\dfrac{\Delta H}{R \times T_K}\right\}}{n}$$

Defining the MKT_K (in degrees K) as that constant temperature that would produce the same overall ageing as observed, the average rate will then be given by

$$\bar{v} = A \times \exp\left\{-\dfrac{\Delta H}{R \times (MKT)_K}\right\}$$

By combining the above two equations, and with some rearrangement, we get

$$(MKT)_K = \frac{-\left(\Delta H / R\right)}{\ln\left\{\dfrac{\sum_i \exp\left\{-\dfrac{\Delta H}{R \times T_K}\right\}}{n}\right\}}$$

We can implement this calculation easily in Excel.

◢	A	B	C	D	E	F	G
1	Time	Data			Weighting		
2		*T /C*			exp{-(ΔH/R)/(T+273.15)}		
3							
4	0	4.9	Average *T*_C =	8.31	1.08E-16	*MKT*_C =	12.58
5	0.2	6.8			1.38E-16		
6	0.4	5.5			1.17E-16		
7	0.6	3.82			9.33E-17		
8	0.8	3.1			8.47E-17		
9	1	4.7			1.05E-16	ΔH =	85000
10	1.2	18.1	Temp surge		5.70E-16	R =	8.31
11	1.4	23.2			1.04E-15	ΔH/R =	10223.23
12	1.6	24.6			1.23E-15		
13	1.8	24.9			1.27E-15	Sum / *n* =	2.89E-16
14	2	24			1.14E-15	ln(Sum/*n*) =	-35.78
15	2.2	7.1			1.44E-16	*MKT*_K =	285.73
16	2.4	5.5			1.17E-16		
28	4.8	3.1			8.47E-17		
29	5	4.8			1.06E-16		

Fig 5.9 MKT calculation (rows 17 to 27 are hidden)

The values for ΔH and R are entered in cells G9 and G10 and the ratio calculated in [G11] = G9/G10.

The calculation of weighted values is performed in column E, with, for example,

$$[E4] = EXP(-G\$11/(B4+273.15))$$

with the addition of 273.15 to convert °C to °K.

The above equation for MKT_K is then evaluated, starting with the exponential term:

$$\sum_i \exp\left\{-\frac{\Delta H}{R \times T_K}\right\}: \qquad [G13] = SUM(E4: E29)/COUNT(E4: E29) = 2.89 \times 10^{-16}$$

(we could have just used)   [G13] = AVERAGE(E4:E29)

We then take the natural log: [G14] = LN(G13) = −35.78

MKT in °K is then calculated: [G15] = G11/G14 = 285.73

Finally, the MKT in °C is calculated in G4 by subtracting 273.15 from G15:

$$MKT_C = 12.58$$

For comparison, we also calculate in D4 the simple *arithmetic* mean of all the temperatures

Average temperature $= 8.31$.

We see how the weighted average gives a much higher effective 'mean' temperature.

5.3 Transforming and weighting data

There are a number of reasons why the data might not be suitable for use directly in analysis, and it may be necessary to apply a *mathematical transformation* to all the values or to *weight the importance* of individual values.

The most common reasons to *transform* data include:

- improving graphical presentation
- linearization for regression analysis
- meeting normality and equality of variance conditions.

Data transformation also occurs as a *link function* within the GsdLM analysis (3.4.7).

Weighting is used to accommodate the fact that different data, perhaps measured with greater accuracy, should be given greater effect, or leverage, in the calculations (5.2.5, 5.3.4).

We start in 5.3.2 with the practicalities of data transformation in Excel, SPSS, and Minitab, and then 5.3.3 reviews common transformations and refers to other sections of the book where these transformations are also addressed. Finally 5.3.4 considers the situation in which the uncertainty varies for different measured values of *y*, and introduces the method of *weighting* to give a measure of importance to each data value in the analysis that matches the known uncertainty in the value.

5.3.1 Case studies

In this section we will meet the following case studies:

Species abundance

The investigation aims to identify whether different management has affected the species abundance on different sites.

5.3.2 / 2. Transforming data: Demonstrates the transformation processes in Minitab and SPSS.

Experimental uncertainties

This case study links together related issues on managing the errors and uncertainties in experimental data.

5.3.4 / 5. Weighting: Demonstrates the use of 'weighting' to combine values with different uncertainties.

5.3.2 Software transformation

In transforming data, we use a mathematical function (e.g. log) to generate a new value, y', for each original data value, y: $y' = \mathrm{f}(y)$.

In *x*–*y* analyses, we may wish to transform either or both variables, for example:

A *loglog* transformation would take the logs of both variables:
$y' = \ln(y)$ and $x' = \ln(x)$.

A *loglinear* transformation would only take the logs of the *y*-variable:
$y' = \ln(y)$ and $x' = x$.

Transforming data can be performed easily within Excel, SPSS, and Minitab. Starting with the *y*-data in one column, the process creates a new column of the transformed *y'*-values.

Transforming data: Minitab and SPSS transformations in Fig 5.10. Scan here to watch the video or find it via www.oxfordtextbooks.co.uk/orc/currell

Case study: **Species abundance / 2. Transforming data**

—continued from 5.4.7, returning to 5.4.7

The data values in Fig 5.10 (a) show the first six random rows in Minitab of the number, *Species*, of plant species observed per quadrat in ten quadrats in four sites, *S1*, *S2*, *S3*, and *S4*, taken at the same time of year in each of two years. We use this as example data for demonstrating the mathematical transformations processes in Minitab and SPSS, and we look at the effect of transformations on the normality of the data in 5.4.7.

(a) Original data   (b) Minitab using 'calculator'   (c) SPSS using 'compute variable'

Fig 5.10 Transforming a variable in Minitab and SPSS

In Excel we can use the relevant *functions*, e.g. SQRT(), LN(), but in Minitab and SPSS we use the processes illustrated in Fig 5.10 (b) and (c). We label a new data column, e.g. *SqRtSpecies*, and then direct the software to generate the transformed values within that column. The relevant expressions can be typed in directly or they can be selected from the options list within dialogue boxes:

Minitab > Calc > Calculator...
Store result in variable: *SqRtSpecies*
Expression: SQRT(*Species*)

SPSS > Transform > Compute Variable...
Target variable: *SqRtSpecies*
Numeric Expression: SQRT(*Species*)

The *arcsine* transformation (5.4.7) requires the entries ASIN(SQRT()) in Excel and Minitab, and ARSIN(SQRT()) in SPSS.

5.3.3 Common transformations

Graphical presentation

Data transformation can be used to improve the *visualization* of data. If, when the data is plotted on an *x–y* graph, it is too spread out or too tightly packed towards one end of a scale, then it may be useful to transform the *scale* variable. The most common remedy is to use a log transformation for one or both variables. The logarithm to base 10 can be convenient for *display* because the unit divisions on the scale now represent *multiplicative factors* of ten, giving a very clear visual interpretation of scale. However, if already using natural logarithms in the data analysis, it may be better to continue to use these logs to avoid confusions.

Linearization

The process of linearization can be used if it is expected that the *x–y* data follows a specific nonlinear mathematical relationship (e.g. exponential decay or power equation). After linearizing the data, it is then possible to use the powerful technique of linear regression to produce and analyse a best-fit straight line.

These techniques include:

- changing the variable, 2.3.1.
- linearizing an exponential relationship, 2.3.4.
- using logarithms for power relationships, 2.3.5.

Normality and equality of variance transformations

In Section 5.4, we consider the situations in which the data distribution and the variances of samples do not meet the requirements for an analytical procedure (e.g. ANOVA), but can be transformed into an acceptable form.

5.3.4 Weighting data

The technique of 'weighting' data is frequently used, both in pure statistics and in a scientific context, to represent the varying importance, or effect, of different values in the data.

In the 'Mean kinetic temperature' case study (5.2.5), we used a theoretical equation to derive weightings that reflected the increased importance of higher temperatures in the decay of a pharmaceutical product.

The following case study demonstrates how the results from samples of different sizes and uncertainties can be combined to give one overall result.

Case study: **Experimental uncertainties / 5. Weighting values**

−continued from 1.4.6

We can demonstrate the use of weighting by considering a situation where five students are asked to measure blood alcohol level, in units of mg of alcohol per 100 ml of blood, using a method which has a known standard deviation uncertainty of 2.0. Without receiving specific instructions, they all used different numbers of replicate measurements (5, 1, 8, 10, and 2) to calculate their best estimate values as shown in column B of Fig 5.11. Each student has also calculated, in column E, the uncertainty, u, in their best estimate using the standard error from Eqn 1.21 based on the known standard deviation of 2.0 and their individual sample size.

The problem is 'how to combine their results into one best estimate value?'

	A	B	C	D	E	F	G
1	Student	Mean	No. of	Approx.	Measurement	Weighting	Weighted
2		value, v	replicates, n	sum = $v \times n$	uncertainty, u	$w = 1/u^2$	value = $v \times w$
3	1	65.2	5	326	0.89	1.25	81.5
4	2	64.3	1	64.3	2.00	0.25	16.075
5	3	62.9	8	503.2	0.71	2	125.8
6	4	63.2	10	632	0.63	2.5	158
7	5	64.8	2	129.6	1.41	0.5	32.4
8							
9	Totals	320.4	26	1655.1		6.5	413.775
10	Mean	64.08		63.66			63.66

Fig 5.11 Weighting results according to their uncertainties

The simple mean value of their results is calculated by taking the total of values in B9 and dividing by five to get the simple average in B10 of 64.08. However, this calculation has given the same *importance* to every value, even though the separate values clearly have different associated uncertainties, and it is likely to be unduly biased by the less accurate values.

We can correct this bias because we are given the numbers of replicates for each measurement, and we can work out, in column D, the sum of the values recorded by each student. For example, for student 1, the mean value of 65.2 from five measurements tells us that the sum of his/her five measurements was $65.2 \times 5 = 326.0$. In this way, we can calculate backwards to the overall sum of all 26 measurements made by all students, obtaining the value of 1655.1 in D9. We can then calculate the true mean value of *all* measurements in D10 as 1655.1 / 26 = 63.66.

We now consider the situation where *we do not know* the number of replicates and ignore the data in columns C and D, so that the only information we have is the uncertainties in column E quoted by each student. We choose to work *backwards* from these uncertainties, u, to infer relative sample sizes, n. We start with Eqn 1.21 which gives the uncertainty, u, as being inversely proportional to the square root of the sample size, and then rearrange this equation such that the number of replicates, n, is *inversely proportional to the square of uncertainty*:

$$u \propto \frac{1}{\sqrt{n}} \quad \text{giving} \quad n \propto \frac{1}{u^2} \quad \text{and hence} \quad w = \frac{1}{u^2} \tag{5.1}$$

We can therefore calculate a *weighting factor*, *w*, proportional to the sample size, *n*, which is the reciprocal of the square of the uncertainty. The individual weighting factor values are calculated in column F, e.g.

[F3] = 1/E3^2 = 1.25

For each datum, the value is multiplied by its associated weighting factor to derive the weighted value in column G, e.g.

[G3] = B3*F3 = 81.5

The mean value in G10 is then calculated by adding all the weighted values in G9 and dividing by the sum of the weighting factors in F9, giving 413.775 / 6.5 = 63.66, the same result as previously calculated above using replicate numbers.

In this example we see how a weighting factor which is the *inverse* of the *square* of the uncertainty provides the correct importance weighting for data values with different uncertainties. When used in software packages, it is only necessary to give the weighting factors as *relative* values as we do in this example (i.e. they do not add up to 1.00), because the calculation automatically normalizes the proportion by dividing by the sum of the factors (in F9).

5.4 Normality and homoscedasticity

Many analytical techniques make the assumptions that:

- sample data is derived from populations with *normal distributions*.
- different sample groups are homoscedastic, i.e. have the *same variance*.

In this section we review the methods for *testing* these assumptions, and consider data *transformations* that convert a data set which does not meet these requirements into one which does (or nearly does).

However it is important to emphasize that the robust *t*-tests and GLM/ANOVAs remain viable analytical techniques unless the assumptions are *grossly* violated, and, for *exploratory* investigations, these analyses continue to give useful output information, provided that reservations about the validity of any *p*-values are made explicit by a discussion of the statistical limitations. In particular, in a *student project*, a full discussion of possible tests and their limitations is an appropriate way to demonstrate a sound understanding of all the issues involved. The exploratory results can then be used to develop new investigations which can be designed to satisfy the statistical criteria.

5.4.1 Case studies

In this section we meet the following case studies:

Chemotaxis index

The chemotaxis experiment initially treats nematodes with different concentrations of a therapeutic drug, and then measures their capacity for migration towards a food supply in a segmented agar plate.

> 5.4.6 / 3. Normality and homoscedasticity: Demonstrates the use of *residuals* to assess the normality and equal variance conditions necessary for the ANOVA.

Species abundance

The investigation aims to identify whether different management techniques have affected the species abundance at different sites.

5.4.7 / 1. Normality transformation: Uses transformations to improve data normality.

5.4.2 **Analytical approach**

The first step in assessing the normality and homoscedasticity of experimental data is to consider the *science* involved (5.4.3), and only use *statistical* analysis as a subsequent check of the assumptions. The first question is then: 'Is there any evidence that the data will *not* have a normal distribution?'

If there is *no prior information* (5.4.3) suggesting that the data is *not* normal, proceed with the analysis and then perform the necessary tests to confirm the *validity* of the analysis. There are then two approaches, depending on the data structure and sample sizes:

- For simple tests, such as a two sample *t*-test, there may be enough data values in *each* sample to make reasonable assessments of the normality (5.4.5) of each sample separately and test for a difference in variance (5.4.4). It is then possible to decide, beforehand, whether to use the parametric or nonparametric alternative.

- For more complex analyses with multiple levels of a number of factors, there may be only a few (or even just one) values in each combination of conditions, and the most direct method of assessing normality and homoscedasticity is to analyse the *residuals* (5.4.6) that remain *after* performing the parametric analysis (e.g. ANOVA). This is effectively a 'post hoc' method of checking the validity of the assumptions for using the parametric test in the first place

If the data is *probably* not normal, due to some well-defined reasons for non-normality, then there are established transformations that can be used (5.4.7). In other cases it may be appropriate to try a series of different transformations or go directly to nonparametric tests.

5.4.3 Anticipating normality

Fortunately, we can treat very many routine experimental measurements as normal. For relatively small random uncertainties, with no specific bias in either direction, the variations will usually follow a normal distribution. In addition, the central limit theorem confirms that when we take the mean values of several measurements, the distribution of the mean values will tend to be normal, even if the distribution of individual measurements is not normal.

The best approach to assessing the normality of experimental data is therefore to consider the reasons that it might *not* be normal. There are situations in which data is clearly *not* normal, including:

- binary and ordinal data in which only specific values can occur.
- systems where there are additional factors occurring *within* the 'sample' measurements, e.g. the exam results of a student cohort might show a bimodal (two peaks) distribution. corresponding to two sub groups entering the course with different previous knowledge.

- frequency data, in which the observed value is dependent on an underlying *low value* of *probability*, will follow a Poisson distribution (1.3.3), e.g. radioactive decay.

There are also situations, for replicate interval data, in which the normality condition *might* be violated because it is *skewed* in one direction or the other:

- Experimental values that are close to a *limiting value*, e.g. recording values of 4.2 with an 'uncertainty' of 0.5 when it is known that the maximum possible value is 5.0. The distribution is likely to be *skewed* away from the limiting value.
- Proportions, particularly with values 0 to 0.3 or 0.7 to 1.0, or percentages with values 0 to 30% or 70% to 100%. The distribution can be skewed away from either 0 (0%) or 1.0 (100%). However, proportions in the middle range of 0.3 to 0.7 are likely to be symmetrical and thus more likely to be normal.
- Experimental variations that are greater than about 5% to 10% of the measured value (assuming 0 is a limiting value).

Finally, it is useful to ask if there is any *previous knowledge* of the distribution of similar types of measurements. The approach reported by others (e.g. from published papers) can also be a useful starting point, but it is important to check whether it also works for one's own data.

5.4.4 Differences in variance

Differences in variance are not often a problem when using a *t*-test for small differences in mean values between two *similar* data samples. However, the variance, σ^2, of a measurement often increases with the mean value, μ, being measured, so that ANOVA analyses involving a wide range of measured values may also find significant differences in variance.

For data whose uncertainty is determined by the Poisson distribution (e.g. types of frequency data), the variance equals the mean value, $\sigma^2 = \mu$. This specific relationship can then lead to the successful use of a square root transformation (5.4.7) to counteract the problem. The GsdLM also has the Poisson distribution as a standard distribution option (3.4.7).

However, the relationship between variance and measured value is often not well defined, and may be dependent on characteristics of the different samples, e.g. any characteristic that differentiates between well and ill patients is likely to have a greater range of possible values for the 'ill' condition than for the 'well' condition. In such cases it may be possible to try different transformation options.

The various tests to compare variances are given in 6.2.4 and 6.3.4, and their analysis through residuals is introduced in 5.4.6.

5.4.5 Testing normality

The key parameters that indicate deviations from normality are:

- **skewness** in which the data which does not have a symmetrical probability, with one 'tail' extending further than the other.
- **kurtosis** in which, compared to the ideal bell-shaped curve, the data may be either flatter or more sharply peaked at the central value.

Although the most frequently used analyses(e.g. *t*-tests, ANOVAs) are known to be robust for deviations from normality, they can become unreliable when the distribution is asymmetric with a long tailed, i.e. with significant, *skewness*.

The main hypothesis tests for normality are

Anderson–Darling (Minitab)

Shapiro–Wilk (SPSS)

which have the null hypothesis

H_0: Data source population has a normal distribution.

Consequently, if $p > 0.05$, then there is no significant evidence that the distribution is *not* normal and, unless we have other reasons to suspect non-normality, we would usually treat the data as normal. The use of Minitab and SPSS to test for normality in a single data set is given in 6.1.6 and 8.1.6.

The normality criterion requires that the sample values have been randomly chosen from a source *population* which has the normal probability distribution, but, unless we have a very *large* sample of replicate measurements, it is unlikely that our *sample* will also appear with the neat bell shape. This is illustrated in Fig 5.12 which shows the frequency distributions of four samples, each of ten data values, all selected at random from the same source population which has a normal distribution with a mean of 9.5 with a standard deviation of 1.5. Each sample is shown with the results of the Shapiro–Wilk (SPSS) and the Anderson–Darling (Minitab) normality tests.

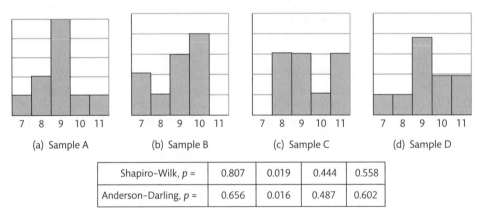

Shapiro-Wilk, $p =$	0.807	0.019	0.444	0.558
Anderson-Darling, $p =$	0.656	0.016	0.487	0.602

Fig 5.12 Random samples from a normally distributed population

Just by *looking* at the sample distributions it would be impossible to say whether they were drawn from a normal distribution or not, but the *p*-values suggest (with $p > 0.05$) that samples, A, C, and D, can be accepted as normal data. However, for sample B the normality tests with $p < 0.05$ suggest, *incorrectly*, that the data is not normally distributed. These examples illustrate the difficulties in assessing normality based on small samples of data. In fact, for experimental data, the first approach should be a consideration (5.4.3) of the *scientific* system from which the data was collected.

In addition to the hypothesis tests, it is useful to use normality plots that provide a useful graphical presentation.

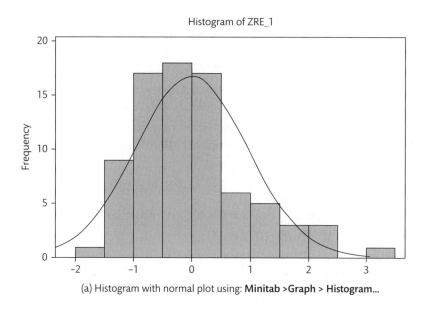

(a) Histogram with normal plot using: **Minitab >Graph > Histogram...**

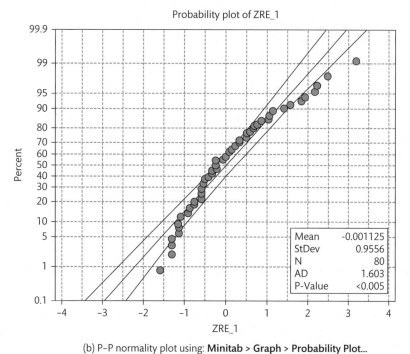

(b) P–P normality plot using: **Minitab > Graph > Probability Plot...**

Fig 5.13 Graphical assessment of normality using Minitab

Fig 5.13(a) shows the histogram for the residuals, ZRE_1, from Fig 5.18(b), together with a best-fit normal curve. The data is skewed to the right with skewness = 1.064 (from Table 5.3).

Fig 5.13(b) shows the same data plotted on a P–P normality curve which plots the individual data values (on the x-axis) against the expected proportional values of a normal data set. A data set with a near normal distribution would have values that all lie along the central diagonal line and within the curved confidence intervals on either side.

> If the data is *skewed* then the observed values will extend further towards one end of the curve. Fig 5.13(b) shows the *positively* skewed data with values spread out to the right.

> If it has significant *kurtosis*, then it will be curved away from the line symmetrically in both directions, e.g. Fig 5.13(b) shows positive kurtosi.

The output also gives the p-value (<0.005) for the Anderson–Darling normality test, confirming the deviation from normality.

The Q–Q plot (Fig 5.19(a)) provides a similar presentation to the P–P plot, but using *quartile* values instead of proportions.

Analysing residuals (Minitab): Analysis for Fig 5.14 data. See also 6.4.5. Scan here to watch the video or find it via www.oxfordtextbooks.co.uk/orc/currell

5.4.6 Using residuals

Residuals are the differences between the experimentally measured values and the values predicted by the analytical model (2.1.1). The normality and homoscedasticity requirements for the data *also* apply to residuals in the analysis, and we can use the distribution of residual values as a 'post hoc' test for the overall quality of fit of the analysis (4.4.1).

If our experimental design is such that we have a large number of *replicates* in each sample then it would be possible to assess the normality of *each sample* separately, but we usually only have a few replicate measurements at each of the possible conditions of the experiment. For example, in linear regression there is often only one measurement at each x-value, or in a factorial investigation there may be only a very few replicates at each possible combination of factor levels. By using the residuals we can group all values into just one data set for analysis.

Analysing residuals (SPSS): Analysis for Fig 5.14 data. See also 6.4.5. Scan here to watch the video or find it via www.oxfordtextbooks.co.uk/orc/currell

The residual value from each data point is calculated, and then 'standardized' by dividing by the standard deviation of all the residuals. The relevance of standardization is that we have a reasonable idea of how we expect normally distributed data to be spread out. For example, the probability of observing a value more than two standard deviations from the mean is about 1 in 20.

Case study: **Chemotaxis index / 3. Normality and homoscedasticity**

—continued from 6.3.1 and 5.2.4, returning to 6.3.1

Using the data derived from Fig 5.7, we have values for the chemotaxis index, *CIndex*, in Fig 5.14 with 24 values distributed over three treatment levels. We wish to test whether the data meets the normality and homoscedasticity (equality of variance) conditions for a parametric ANOVA analysis.

	A	B	C	D	E	F
1	Treat	Cindex	Treat	Cindex	Treat	Cindex
2	0	0.31	1	-0.07	2	-0.12
3	0	0.29	1	0.21	2	0.18
4	0	0.17	1	0.18	2	-0.09
5	0	0.26	1	0.30	2	0.01
6	0	0.44	1	0.17	2	0.06
7	0	0.26	1	0.17	2	0.02
8	0	0.20	1	0.13	2	-0.06
9	0	0.25			2	0.08
10					2	-0.03

Fig 5.14 Chemotaxis index values (shown unstacked in separate treatment groups)

The one-way GLM/ANOVA analysis of this data is described more extensively in Section 6.3, but we will concentrate here on testing that the data meets the necessary parametric criteria. The data must be first stacked with all the *CIndex* values in one column.

Minitab

In Minitab we can produce a direct analysis of residuals and/or choose to save the residual values as *SRES1* in an empty column for later analysis.

> **Minitab > Stat > ANOVA > General Linear Model > Fit General Linear Model ...**
> **Reponses:** *CIndex*   *Factors: Treat*
> **> Graphs ...** It is useful to select ▼ *Standardized* for **Residuals for plots**
>   Select ⊙ *Individual plots,* e.g.
>   ☑ *Normal plot of residuals* to show normality
>   ☑ *Residuals versus fits* show equality of variance
>   or
>   Select ⊙ *Four in one*
> **> Storage ...** Use ⊙ *Standardized residuals* to save residuals into the next empty column.
> → Output in Fig 5.15

The top two plots in Fig 5.15 are similar to those using SPSS and are discussed below. In addition, this 'four in one' plot also gives a histogram of residual values which is not inconsistent with a normal distribution. It also shows the residuals for each data value stepping through the data, which does not show any significant non-random patterns of behaviour, with only one residual more than two standard deviations from the mean.

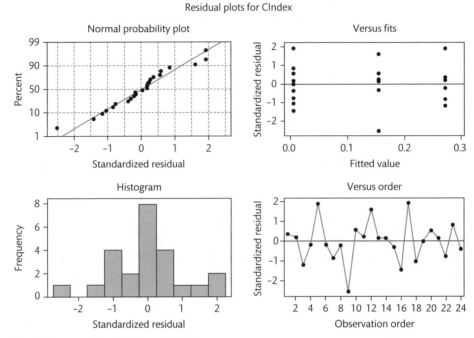

Fig 5.15 'Four in one' residual plots (Minitab)

The saved residuals, *SRES1*, can be analysed as a single data sample (8.1), using

Minitab >Stat > Basic Statistics > Graphical Summary ...

which will perform the Anderson–Darling normality tests and give values for skewness and kurtosis.

SPSS

In SPSS, it is easier to run the GLM/ANOVA analysis first, requesting that the residuals for each data value are saved into a new column in the data editor:

SPSS > Analyze > General Linear Model > Univariate ...
Dependent variable: *CIndex*   **Fixed factor(s):** *Treat*
> Save: ☑ **Standardized residuals.** which saves the residuals in a new column with the name *ZRE_1* and label ***Standardized Residual***
> Options: Display ☑ ***Homogeneity tests***
→ The homogeneity test reports Levene's test (Fig 5.16 (a)), which, in this case, shows no significant differences ($p = 0.845$) in variances between the levels of *Treat*.

Levene's Test of Equality of Error Variances[a]

Dependent Variable: CIndex

F	df1	df2	Sig.
.170	2	21	.845

Tests the null hypothesis that the error variance of the dependent variable is equal across groups.

a. Design: Intercept + Treat

Tests of Normality

	Kolmogorov-Smirnov[a]			Shapiro-Wilk		
	Statistic	df	Sig.	Statistic	df	Sig.
Standardized Residual for CIndex	.109	24	.200	.965	24	.541

*. This is a lower bound of the true significance.

a. Lilliefors Significance Correction

(a) Test for homoscedasticity   (b) Tests for normality

Fig 5.16 Tests for homoscedasticity and normality (SPSS)

The saved residuals can then be analysed using:

SPSS > Analyze > Descriptive Statistics > Explore...
Dependent List: *Standardized residual*
> Plots: ☑ *Normality plot with test.*

→ The output will include the standard statistics with skewness and kurtosis, tests for normality, Fig 5.16 (b), and the Q–Q normality plot in Fig 5.17 (a).

It is also possible to plot the variation of residuals for the different levels of *Treat*:

SPSS > Graphs > Legacy Dialogs > Scatter/Dot...
Simple scatter: Define
Y axis: *Standardized residual*
X axis: *Treat*
→ Output Fig 5.17(b)

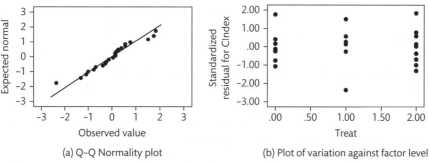

(a) Q–Q Normality plot   (b) Plot of variation against factor level

Fig 5.17 Normality and homoscedasticity plots (SPSS)

The normality plots in both SPSS and Minitab show that the data follows a reasonably normal distribution. The standardized residual plots show a similar spread of values at each of the three treatment levels, hence we can also assume that we have a homogeneity of variance.

5.4.7 **Data transformations**

Any data transformation will affect both normality and the homogeneity of variance, but generally both conditions react in a similar way and a suitable transformation for one often also addresses problems with the other. In practice, it is probably best to focus on developing the equality of variance, as deviations from normality are well tolerated by the main analytical techniques.

If there is no guidance from previous knowledge in choosing a suitable transformation (e.g. arcsine for proportions), then try the simplest transformations first: square root, logarithmic, before more complex options.

We now consider the most common types of transformation for general data values.

Square root transformation

takes the square root of all the data values:

$$y' = \sqrt{(y)} \text{ or } y' = \sqrt{(y+0.5)} \tag{5.2}$$

(the '+0.5' is often included for data with low values of y).
This equalizes variance for samples where $\sigma^2 = \mu$ and reduces *positive* skewness.

The square root transformation is a mild transformation that is particular useful for positively skewed data and when the sample variances increase in proportion to their mean values. The Poisson distribution (1.3.3) is a particular example which occurs if there is a relatively low probability of observing a specific event, e.g. in counting appearances of rare plants. This distribution is positively skewed, and has the specific characteristic that the expected variance in the data will be equal to its mean value. See the 'Species abundance' case study later in this section.

Logarithmic transformation

takes the logarithm of all the data values:

$$y' = \log(y) \text{ or } y' = \log(y+1) \tag{5.3}$$

(the '+1' is often included for data including 0 values, because log (0) becomes 'minus infinity').
This also reduces *positive* skewness, and reduces variances for samples with larger mean values.

The log to base 10 transformation has the advantage that it produces a data axis that is conveniently defined by 'powers of ten'. See the 'Species abundance' case study later in this section.

Box–Cox transformations

The Box–Cox approach uses a family of transformations which are defined by a parameter, λ, whose value can be adjusted to suit the experimental distribution. The transformation can only be applied to non-zero data values.

$$y' = y^\lambda \text{ if } \lambda \neq 1 \quad \text{or} \quad y' = \ln(y) \text{ if } \lambda = 1 \tag{5.4}$$

If $\lambda = 1$, the transformation becomes a simple logarithmic transformation, and if $\lambda = 0.5$, the transformation becomes a square root transformation. When selecting a Box–Cox transformation, the software analyses the data before recommending the optimum value of λ for the transformation. See the general regression example in 7.2.5.

Arcsine transformation

This is used for the transformation of proportions, P, or percentages, $P\%$.

$$P' = \arcsin(\sqrt{P}) \quad \text{or} \quad P' = \arcsin(\sqrt{P\%/100}) \tag{5.5}$$

A measured proportion has values that are limited at both ends, 0.0 and 1.0, of its possible range. The data tends to show a *positive* skewness for proportions between 0 and 0.3 which is corrected by the *square root* factor, and *negative* skewness between 0.7 and 1.0 which is corrected by the *arcsine* function. For proportions within the central region, 0.3 to 0.7, skewness is less likely to be a problem and transformation may not be necessary.

Case study: **Species abundance/ 1. Normality transformation (overview)**
—leading to 5.3.2

The data values in Fig 5.18(a) show the first six random rows in SPSS of the number, *Species*, of plant species observed per quadrat in ten quadrats in four sites, *S1*, *S2*, *S3*, and *S4* taken at the same time of year in each of two years. Two of the sites received different management compared to the other two, and the aim of the experiment is to test whether the different management has had any effect on the numbers of observed species. We start by trying different transformations to improve data normality.

5.3.2 / 2. Transforming variables: Demonstrates the transformation process in Minitab and SPSS

Transforming for normality: SPSS analysis leading to Fig 5.18 and Table 5.3. See also 5.3.2. Scan here to watch the video or find it via www. oxfordtextbooks. co.uk/orc/currell

	Year	Site	Species	var
1	2012	S2	7	
2	2012	S4	7	
3	2013	S3	8	
4	2013	S1	7	
5	2012	S1	7	
6	2012	S3	6	

	Year	Site	Species	ZRE_1	SqRtSpecies	ZRE_2	LnSpecies	ZRE_3
1	2012	S2	7	.70	2.65	.78	1.95	.85
2	2012	S4	7	-.60	2.65	-.52	1.95	-.44
3	2013	S3	8	.32	2.83	.37	2.08	.40
4	2013	S1	7	.76	2.65	.86	1.95	.95
5	2012	S1	7	.32	2.65	.41	1.95	.49
6	2012	S3	6	-1.08	2.45	-1.05	1.79	-.99

(a) Initial data entry   (b) Transformed data and saved residuals

Fig 5.18 Numbers of species at four sites (Six rows shown randomly out of a total of 80 rows)

Table 5.3 shows the results of performing the ANOVA/GLM analysis (as in 6.4.4), testing for the main effects of the *Year* and *Site* and also the interaction *Year*Site*. A significant result for *Year*Site* would indicate that the *different* management of *some* sites resulted in a difference in the numbers of species, but $p = 0.312$ in Table 5.3 suggests that this interaction is not significant. However our main interest in this data is in the *normality* and *homoscedasticity* of the values.

Within the operation of the ANOVA/GLM, the option of saving the standardized residuals as ZRE_1 in an available column allows an analysis of these residuals using *Explore*, giving the values in Table 5.3. The results for *Species* show that, although there is no significant difference in variance between samples ($p = 0.888$), the data is *not* normally distributed, with $p = 0.000$ for the Shapiro–Wilk test, and both skewness and kurtosis are significant:

Skewness = 1.064, which is more than twice the standard error of 0.269

Kurtosis = 1.163, which is more than twice the standard error of 0.532.

The resultant P–P normality plot using Minitab is given in Fig 5.13(b) and Q–Q plot in Fig 5.19(a) using SPSS, in which the positive skewness is shown by the points extending further to the top right of the plot, and the positive kurtosis is shown by the curvature of the data.

Table 5.3 Results of GLM/ANOVA analysis of data in Fig 5.18(b)

ANOVA results:	Species	SqRtSpecies	LnSpecies
Year	$p = 0.057$	$p = 0.047$	$p = 0.041$
Site	$p = 0.004$	$p = 0.003$	$p = 0.002$
Year*Site	$p = 0.312$	$p = 0.311$	$p = 0.321$
Levene's test for homogeneity of variance	$p = 0.888$	$p = 0.874$	$p = 0.669$
Residuals analysis:	ZRE_1	ZRE_2	ZRE_3
Skewness (St error = 0.269)	1.064	0.651	0.257
Kurtosis (St error = 0.532)	1.163	0.215	− 0.114
Normality test, Shapiro-Wilk	$p = 0.000$	$p = 0.022$	$p = 0.241$
Normality plots	Fig 5.13 (a)/(b) Fig 5.19 (a)		Fig 5.19 (b)

Two transformations of the data are made using

SPSS > Transform > Compute Variable...

producing respectively the square root of the values, *SqRtSpecies*, and the log of the values, *LnSpecies*. The ANOVA/GLM analysis is repeated for each new data set, producing new sets of residuals ZRE_2 and ZRE_3 and giving the results in Table 5.3, and the final SPSS data set in Fig 5.18(b).

Although the gentle square root transformation reduces both skewness and kurtosis, the data is still significantly non-normal ($p = 0.022$), but using the log transformation the data can now be considered to be sufficiently near normal ($p = 0.241$). The transformation has reduced skewness and kurtosis to acceptable levels, with the result that the Q–Q normality plot in Fig 5.19(b) now has the data values lying close to the 'normal' line.

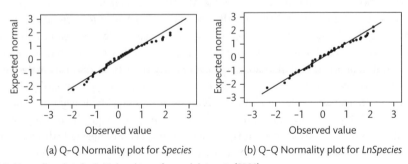

(a) Q–Q Normality plot for *Species*   (b) Q–Q Normality plot for *LnSpecies*

Fig 5.19 Normality plots for initial and transformed data sets (SPSS)

Single response variable

Introduction

This chapter presents the techniques used to analyse the effect that one or more factors may have on a single response variable. The data can be described as *univariate* and may be either *interval* or *ordinal* data. Some *frequency* values can be treated as interval values, provided that the values are sufficiently large (6.1.1), but, for the analysis of frequencies or probabilities, also refer to Chapter 8.

Section 6.1 presents analyses relevant to *single samples* of replicate data (e.g. data descriptions, one sample *t*-test, Wilcoxon test).

Section 6.2 assumes *two data samples*, which is essentially a special case of a one factor analysis with just two levels (e.g. independent samples *t*-test, Mann–Whitney test).

Section 6.3 relates to investigations that test, or measure, the effect of just *one factor*, but with more than two levels (e.g. one-way ANOVA, Kruskal–Wallis test, repeated measures).

Section 6.4 develops investigations of more than one factor together with possible interactions. The GLM is used as a flexible approach for conducting ANOVA calculations, and the effect of a covariate term is addressed through the ANCOVA.

6.1 One sample

The simplest set of experimental results is a set of replicate measurements of a *single* value, for which we will normally be aiming to

- *derive the best-estimate values* (1.4.1) for the mean, median, standard deviation, etc. of the population from which these experimental results have been drawn, or to
- *test for a difference* between the experimental data and a specified test value.

If the data leads to a 'counting of values' or *frequency* values (e.g. data set *D9* in Fig 6.1), then you should also refer to Section 8.1, although for larger frequency values, you can treat the values as *continuous* data (e.g. data set *D10* in Fig 6.1).

6.1.1 Example data

Single data samples can be of different data *types* (5.1.2) and expressed using different *level* formats, as illustrated in Fig 6.1.

	A	B	C	D	E	F	G	H	I	J	K	L	M	N	O
1	Data set:	D1		D2	D3	D4		D5	D6		D7	D8		D9	D10
2	Type:	Interval		Ordinal	Ordinal	Ordinal		Binary	Binary		Nominal	Nominal		Frequency	Frequency
3	Levels:	continuous		text	integer	rank		text	integer		text	coded		integer	'continuous'
4		7.57		SA	2	1		F	1		AB	1		6	52
5		7.15		A	1	2		F	1		AB	1		2	55
6		7.33		N	0	3		M	0		Ab	2		9	49
7		6.43		D	-1	4		F	1		ab	4		4	46
8		6.68		SD	-2	5		F	1		AB	1		3	57
9		6.97		D	-1	4		M	0		aB	3		6	53
10		7.39		N	0	3		M	0		aB	3		4	49
11		7.75		A	1	2		F	1		AB	1		7	53
12		6.75		N	0	3		F	1		AB	1		3	39
13		7.18		N	0	3		F	1		AB	1		3	52
14		7.18		SA	2	1		F	1		aB	3		8	52
15		6.22		A	1	2		M	0		ab	4		8	41

Fig 6.1 The first 12 values from each of 10 data samples (50 values in each)

Each data sample in Fig 6.1 represents a set of 50 *replicate* measurements made under the *same experimental conditions*, and with the important assumption that every value is assumed to be *independent* of any other value, i.e. the recorded value of one measurement is not affected by any other measurement in the sample.

- *D1* shows 12 values of a set of 50 continuous *interval* values randomly drawn from a normal population with mean, $\mu = 7.00$ and standard deviation, $\sigma = 0.5$.

- *D2*, *D3*, and *D4* all relate to the same set of *ordinal* values of Likert questionnaire responses, but expressed using different level values. The text form (*D2*) gives the abbreviations *SA*– strongly agree, *N*– neutral, *D*–disagree, etc., and these are coded (*D3*) by an integer value from +2 to −2. The same data is also given a ranked value (*D4*).

- *D5* and *D6* record *binary* data in text and numeric forms respectively. Binary data can be considered as either ordinal or nominal data with only two possible states. This data records the observations of 50 frogs as being either male, *M*, or female, *F*, leading to a measurement of proportions (6.1.7).

- *D7* and *D8* record *nominal* values in text and *coded* forms respectively. See Section 8.1 for the analysis of this type of data using categorical frequencies.

- *D9* and *D10* record observed *frequency* values which must be integer. Frequency data is considered in Chapter 8. However, if the frequency values are sufficiently high, as for data set *D10*, it is possible to treat the data as *continuous interval* data.

6.1.2 Analytical options

We list below *some* of the most common analyses used for one sample data, together with links for further information. The issues associated with choosing whether to use *parametric* or *nonparametric* tests are developed in Section 5.4.

Describing data (6.1.3)

- Graphical plots: Boxplot, bar chart, histogram, stem and leaf plot
- Numerical statistics: Mean, median, standard deviation, confidence interval, etc.

Tests / Measurements:

- Is the **mean** value different from a *specified value*? One sample *t*-test for normal data (6.1.4, 3.1.2)
- Is the **median** value different from a *specified value*? Wilcoxon test (nonparametric) (6.1.5, 3.5.2)
- Is the data distribution **normal**? Anderson–Darling, Kolmogorov–Smirnov, Ryan–Joiner (similar to Shapiro–Wilk) tests (6.1.6, 8.1.6)
- Is the **distribution** different from a *specified distribution*? Kolmogorov–Smirnov test (6.1.6, 8.1.6)
- Do the values occur **randomly** above and below a *specified value*? Runs test (nonparametric) (6.1.6)
- Use a **goodness of fit** test to compare with a *specified distribution* of frequencies: Chi-squared goodness of fit test (8.1.5, 3.7.2)
- Is the **proportion** of numerical values above and below a defined value different from a *specified proportion*? Binomial test (6.1.6)
- Is the **proportion** of binary values different from a *specified proportion*? Proportion tests (6.1.7)

Describing sample data (Excel): Descriptives and graphs. Scan here to watch the video or find it via www.oxfordtextbooks.co.uk/orc/currell/

6.1.3 Describing the data

It is useful to summarize the data values using numerical and/or categorical statistics and graphical plots.

Numerical statistics (Section 1.5) can be calculated for any data with *numerical* values, e.g. integer, ordinal, ranked, or coded data. The sample statistics (e.g. mean, standard deviation) are best estimates that describe the population from which the sample was drawn. The *relevance* of particular statistics will depend on the data type (for example a measure of skewness has no meaning for ordinal data), and the *reliability* of the distribution statistics (e.g. skewness, kurtosis, Fig 6.2) will be low for small data sets. An important derived statistic is the *confidence interval* (1.5.2) which gives the best estimate range for the true *mean* of the data.

Categorical statistics (Section 8.1) can be applied to all data types and counts the frequencies with which specific values occur in the data set. The distribution of values can be represented using a stem and leaf plot (Fig 6.3), bar graphs, or histograms (Figs 6.4, 6.5).

The simplest **graphical plot** that gives an immediate overview of a data set is the **boxplot** introduced in 1.1.2. In addition, various software options, **Excel (> Insert > Charts)**, **Minitab (> Graph)** and **SPSS (> Graphs > Legacy Dialogs)**, have the facilities for drawing a wide range of individual graphs and data plots—see also 8.1.3. The following options for Excel, SPSS, and Minitab give examples of methods for summarizing and presenting experimental data.

Describing sample data (Minitab): Descriptives and graphs. Scan here to watch the video or find it via www.oxfordtextbooks.co.uk/orc/currell/

Describing sample data (SPSS): Descriptives and graphs. Scan here to watch the video or find it via www.oxfordtextbooks.co.uk/orc/currell/

Excel

Excel, with its two-dimensional worksheet, is a very useful format for tabulating, presenting, and calculating results. It also provides a range of graphical outputs, which can be easily edited for printing.

Excel uses a variety of *dynamic* functions, *fx*, to calculate different statistics individually, e.g. AVERAGE(), STDEV.S(), etc., and a set of summary statistics is also available through the Data Analysis Add-In:

Excel > Data > Data analysis > Descriptive statistics
– calculates a range of statistics.

SPSS

Descriptive statistics can be accessed through three main menu choices as below, with 'Descriptives' giving the main numerical statistics, 'Frequencies' adding data plots and 'Explore' providing greater flexibility with the ability to split the data on the basis of factors in other columns.

SPSS > Analyze > Descriptive statistics >Descriptives...
Variable(s): e.g. *D1*
> **Options:** e.g. mean, standard deviation, standard error of the mean, skewness, kurtosis
→ Fig 6.2

An example of the output in Fig 6.2 gives statistic values and their standard errors. When using these statistics, we can make a rough estimate of the *confidence deviation* (Eqn 1.22) as being *twice* the standard error. For example, the skewness in Fig 6.2, would have a *confidence interval* of $0.130 \pm 2 \times 0.337$ giving a range from -0.54 to 0.80, which, because it includes 0.0, suggests that the distribution could be normal.

Descriptive Statistics

	N	Mean		Std. Deviation	Skewness		Kurtosis	
	Statistic	Statistic	Std. Error	Statistic	Statistic	Std. Error	Statistic	Std. Error
D1	50	7.0262	.06995	.49465	.130	.337	-.287	.662
Valid N (listwise)	50							

Fig 6.2 SPSS: Example of output from the 'Descriptives' option

SPSS > Analyze > Descriptive statistics > Frequency...
Variable(s): e.g. *D2 D9*
> **Statistics:** mean, mode, median, standard deviation, skewness, kurtosis, quartiles, etc.
> **Charts:** Bar charts (Fig 6.4 for *D2*), histograms (similar to Fig 6.6 for *D9*) or pie charts (similar to Fig 6.7 for *D7*)

SPSS > Analyze > Descriptive statistics > Explore...
Dependent List: e.g. *D1*
> **Statistics:** Option to select numerical parameters
> **Plots:** Boxplot and outliers, histogram, stem and leaf plot (Fig 6.3), normality plot and test (Fig 5.19)

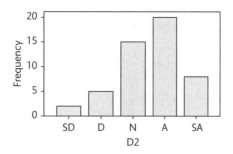

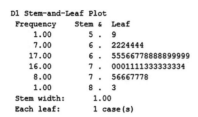

```
D1 Stem-and-Leaf Plot
  Frequency    Stem &  Leaf
      1.00       5 .  9
      7.00       6 .  2224444
     17.00       6 .  55566778888899999
     16.00       7 .  0001111333333334
      8.00       7 .  56667778
      1.00       8 .  3
  Stem width:     1.00
  Each leaf:      1 case(s)
```

Fig 6.3 SPSS stem and leaf plot for *D1*

Fig 6.4 SPSS bar chart for *D2*

It is often necessary to edit the graph produced in the SPSS output window. For example, in Fig 6.4, it was necessary to change the horizontal axis labels from a default *alphabetical* order into the *ranked* order. This was achieved by double-clicking on the chart to open the chart editor, then using a right click on the data bars to open the relevant properties window and changing the order under 'Categories'.

Descriptive Statistics: D9

Variable	Mean	SE Mean	StDev	Q1	Median	Q3	Skewness	Kurtosis
D9	6.120	0.347	2.455	4.000	6.000	8.000	0.56	-0.12

Fig 6.5 Minitab: Example of descriptive statistics for data set D9

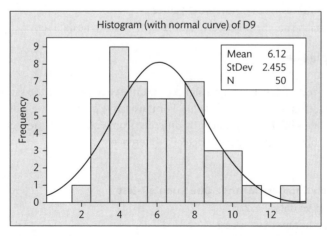

Fig 6.6 Minitab histogram with normal curve for *D9*

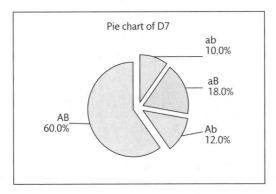

Fig 6.7 Minitab pie chart of *D7*

Minitab

In addition to the individual graph options under > **Graph** (e.g. pie chart in Fig 6.7), Minitab has a range of options for displaying data:

Minitab > Stat > Basic Statistics > Display Descriptive Statistics...
Variable(s): e.g. *D9*
By variable(s): Used to identify subgroups within the column of data.
> **Statistics:** mean, mode, median, standard deviation, skewness, kurtosis, quartiles, etc.
> **Graphs:** ☑-**Histogram of data, with normal curve**
→ Output: Fig 6.5 displays the sample statistics and Fig 6.6 gives the histogram

Minitab > Stat > Basic Statistics > Graphical Summary...
Variable(s): e.g. *D9*
By variable(s): Used to identify subgroups within the column of data.
→ Output: Produces a graphical output which includes the histogram in Fig 6.6, the Anderson–Darling test for normality, the sample statistics in Fig 6.5, and confidence intervals for the mean, median, and standard deviation.

Minitab > Stat > Tables > Tally Individual Values...
→ Output: Counts the number of occurrences of each value in the data set.

6.1.4 **One sample *t*-test**

The **one sample *t*-test** (3.1.2) tests for a significant difference between the sample *mean* of *normally* distributed data and a *specific value*. For illustrative purposes, we will test whether the mean of the data set, *D1*, is significantly different from 7.2, and, as the true population mean is actually 7.0, we might expect to detect a difference.

SPSS

SPSS > Analyze > Compare Means > One Sample T-test
Test variable: Select data, e.g. *D1*
Test value: Enter test mean value, e.g. 7.2
→ Output: Fig 6.8

One sample tests (Minitab): Analysis for *t*-test and Wilcoxon test. Scan here to watch the video or find it via www.oxfordtextbooks.co.uk/orc/currell/

One sample tests (SPSS): Analysis for *t*-test and Wilcoxon test. Scan here to watch the video or find it via www.oxfordtextbooks.co.uk/orc/currell/

One-Sample Test

	Test Value = 7.2					
	t	df	Sig. (2-tailed)	Mean Difference	95% Confidence Interval of the Difference	
					Lower	Upper
D1	-2.484	49	.016	-.17380	-.3144	-.0332

Fig 6.8 SPSS output for *t*-test

With $p = 0.016$ it is possible to state that there is a significant difference between the mean and the test value of 7.2.

Minitab

Minitab > Stat > Basic Statistics > 1-Sample t ...

▼ One or more samples, each in a column

Enter sample: e.g. *D1*

☑-**Perform hypothesis test**

   Hypothesized mean: e.g. 7.2

Minitab gives the same results in Fig 6.9 as SPSS, except that the confidence interval is expressed differently, e.g. the upper *CI limit* of 7.1668 in Minitab is equal to the *difference*, -0.0332, in SPSS from the test mean of 7.2 with $7.2 - 0.0332 = 7.1668$.

```
Variable   N   Mean    StDev   SE Mean     95% CI        T      P
D1        50  7.0262  0.4946   0.0700  (6.8856, 7.1668) -2.48  0.016
```

Fig 6.9 Minitab output for *t*-test

6.1.5 **Wilcoxon test**

The Wilcoxon test is the *nonparametric* equivalent of the *t*-test, testing for a significant difference between the sample *median* and a specific value.

SPSS

The Wilcoxon test is performed in SPSS through '**Analyze > Nonparametric Tests > One Sample...**' and is given in 6.1.6.

Minitab

Minitab > Stat > Nonparametrics > 1-Sample Wilcoxon...

Variables: e.g. *D1*

◉-**Test median:** e.g. 7.2

This gives results in Fig 6.10, which show a significant difference ($p = 0.020$) from the specified value of 7.2.

```
Test of median = 7.200 versus median not = 7.200

         N for  Wilcoxon          Estimated
      N  Test  Statistic    P     Median
D1   50    50      396.0  0.020     7.020
```

Fig 6.10 Minitab output for Wilcoxon test

Nonparametric tests (SPSS): Nonparametric tests. See also 6.4.6. Scan here to watch the video or find it via www. oxfordtextbooks. co.uk/orc/currell/

6.1.6 **SPSS nonparametric tests**

The 'nonparametric tests' option in SPSS provides a range of tests (including the above Wilcoxon test) that do not assume a normal distribution, including the option of testing for normality itself:

SPSS > Analyze > Nonparametric Tests > One Sample...

Fields: Select data, e.g. *D1*

Settings: ⊙-Customize tests

☑-**Compare observed binary probability to hypothesized (binomial test)** (3.8.2)

 Options: For continuous field, enter *cutpoint*, e.g. 7.2

☑-**Test observed distribution against hypothesized (Kolmogorov–Smirnov** test) (8.1.6)

 Options: Select a distribution(s) for comparison: e.g. normal

☑-**Compare median to hypothesized (Wilcoxon** signed-rank test) (3.5.2)

 Hypothesized median: Enter defined median value, e.g. 7.2

☑-**Runs test**

 Options: For continuous field, leave the *cutpoint* as the default sample median

The output for the example data, *D1*, is given in Fig 6.11. For reference, the values in the data set *D1* were actually derived randomly from a normal population with a mean (and median) value of 7.00 and a standard deviation of 0.50, giving a sample mean of 7.03 and a sample median of 7.00.

1. The runs test records no significant evidence ($p = 0.568$) for any *non-random ordering* of the data values above and below the sample median. This supports the expectation that the data values have been obtained *independently* of each other.

Hypothesis Test Summary

	Null Hypothesis	Test	Sig.	Decision
1	The sequence of values defined by D1<=7.00 and >7.00 is random.	One-Sample Runs Test	.568	Retain the null hypothesis.
2	The categories defined by D1 <=7.20 and >7.20 occur with probabilities 0.5 and 0.5.	One-Sample Binomial Test	.066	Retain the null hypothesis.
3	The median of D1 equals 7.20.	One-Sample Wilcoxon Signed Rank Test	.020	Reject the null hypothesis.
4	The distribution of D1 is normal with mean 7.03 and standard deviation 0.49.	One-Sample Kolmogorov-Smirnov Test	.967	Retain the null hypothesis.

Asymptotic significances are displayed. The significance level is .05.

Fig 6.11 SPSS: Nonparametric analyses of randomly selected normal data with mean of 7.0

2. Given that the value of 7.20 is not equal to the true median value, we would expect that the proportions of randomly selected values would *not* be the same above and below this value. However the binomial test is not quite *powerful* (1.6.3) enough ($p = 0.066$) to find evidence for a difference for this particular data set.

3. The Wilcoxon signed rank test identifies a significant difference ($p = 0.020$) between the sample median of 7.00 and the test value of 7.20, agreeing with the Minitab result in Fig 6.10.

4. Test for normality. The Kolmogorov–Smirnov test accepts ($p = 0.967$) that the distribution could be normal with calculated values of mean and standard deviation close to the true values.

6.1.7 **Proportions**

One proportion (SPSS): Analysis for Fig 6.1 data. See also 3.8.2. Scan here to watch the video or find it via www. oxfordtextbooks. co.uk/orc/ currell/

> ### Case study: **Frogs / 1. Introduction (overview)**
> — leading to 3.8.2 and 3.8.3
>
> Randomly selecting a sample of 50 frogs in column H in Fig 6.1 from a large lake, it was found that 37 were female. We wish to test whether the observed proportion of 37/50 is significantly *greater than* an expected test proportion of 0.6, or significantly *different from* an observed proportion of 30 females out of 50 randomly selected from a *second* lake.
>
> > 3.8.2 /2. One proportion: Develops the binomial and other statistics for the exact one sample test.
> >
> > 3.8.3 /3. Two proportions: Develops the statistics for the two sample test.

The 'Frogs' case study sets two problems:

One proportion test, observing 37 female frogs in a sample of 50, has the null hypothesis:

H_0: The probability that each randomly selected frog will be female is 0.6.

Two proportion test, observing 37 female frogs in a sample of 50 from one lake and 30 out of 50 from a second lake, has the null hypothesis:

H_0: The probability that a randomly selected frog will be female is the same in both lakes.

The statistics of possible analytical techniques are developed in 3.8.2 and 3.8.3 together with the use of **Minitab** to perform tests based on the normal approximation, exact binomial tests, and a chi-squared test for the two proportion problem.

If the data is presented as in Fig 6.1 column H, with *individual* observations, then we use SPSS to perform the binomial test as in 6.1.6 using:

SPSS > Analyze > Nonparametric Tests > One Sample...

with

☑-**Compare observed binary probability to hypothesized (binomial test)**
> Options...: Hypothesized proportion: 0.6

which reports a one-sided p-value $= 0.030$ for a significant difference.

However, if we have just the two frequency *totals* then we can still perform the binomial test as above by first weighing the *F* and *M* categories with the frequencies 37 and 13 respectively as in Fig 8.4.

6.2 Two samples

This section considers two samples of measurements of the *same property*, but measured under different conditions. It uses the example of *pH* measurements from two rivers, in which case, the specific rivers can be considered to be two *levels* of one river *factor*. The analysis here also differentiates between *unrelated* and *related* measurements of the same property, but for *interrelated* data between two samples measuring *different* properties see Chapter 7.

6.2.1 **Example data**

A	B	C	D	E	F	G	H	I	J	K	L	M	N	O	P	Q	R	S
1 D1				D2				D3			D4							
2 pH	River		Meter	A	B	A-B		Grp	Q5		Q1	Q2	Q3	Q4		Q1\Q2	N	Y
3 6.56	A		M1	6.56	6.47	0.09		G2	1		N	N	Y	Y		N	24	19
4 6.52	A		M2	6.52	6.4	0.12		G1	0		Y	Y	Y	Y		Y	7	16
5 6.7	A		M3	6.7	6.49	0.21		G2	-2		N	N	N	N				
6 6.61	A		M4	6.61	6.55	0.06		G2	0		Y	N	Y	Y		McNemar's test:		
7 6.47	B							G2	0		Y	N	N	N			$\chi^2 =$	4.65
8 6.4	B							G1	0		N	N	N	N			$p =$	0.031
9 6.49	B							G1	2		N	Y	Y	N				
10 6.55	B							G1	0		N	Y	N	N				
11								G1	0		N	N	Y	N				
68								G2	-1		Y	N	Y	Y				
69																		
70								Sum of "Y"s =		23	35	23	26					
71								Proportion of "Y"s =		23/66	35/66	23/66	26/66					

Fig 6.12 Example data from the River *pH* and Forensic Questionnaire case studies (rows 12 to 67 are 'hidden' in the worksheet)

Case study: **River pH / 1. Overview**

In Fig 6.12, data set *D1* in column A shows four *replicate pH* measurements recorded in *each of two* rivers, A and B. The data source is identified in column B by the *text* variable, *River*, either A or B. In this format, there is no link or relationship between a specific value measured in one river and a specific value in the other river, and consequently the data samples are described as *independent* or *unrelated*.

Data set, *D2*, in columns E and F, shows the same *pH* measurements, but in this case, four different *pH* meters, *M1*, *M2*, *M3*, and *M4* were used, each making one of the measurements from each river. There are now specific links between pairs of measurements from each river. For example, the value 6.56 in A is *uniquely* linked to 6.47 in B because they were both measured using *M1* and no other value was measured using *M1*. The two *related* samples *A* and *B* are also described as *paired*. In this section we test for differences as both *unrelated* and *related* samples.

3.1.3 / 2. Two sample *t*-test: Develops the statistics of the *independent samples t-test*.

3.2.2 / 3. ANOVA calculations: Develops the statistics underpinning the *analysis of variance* calculations.

3.4.2/ 4. GLM, ANOVA, and the *t*-test: Demonstrates the *equivalence* of the ANOVA and *t*-test, leading to the GLM.

3.5.1 / 5. Mann–Whitney test: Develops the statistics of the Mann–Whitney test as an example of *nonparametric* testing.

3.6.1 / 6. Paired *t*-test: Introduces the use of four separate *pH* meters as *unique links* between pairs of data to develop the statistics of the paired *t*-test.

3.9.2 / 7. Resampling technique: Demonstrates the use of resampling to *estimate* the same results as the direct tests.

Case study: **Forensic questionnaire / 4. Ordinal and binary responses**
—continued from 9.2.1, 8.2.1, and 4.4.5

Data sets *D3* in columns I and J and *D4* in columns L to O in Fig 6.12 are extracts of the responses to a forensic questionnaire on the interpretation of evidence. The *same* questions are answered by two groups of people, *G1* and *G2* (possibly different age groups or sexes), with the answers *Q5* on an ordinal scale from −2 to + 2 and *Q1* to *Q4* as binary responses.
In 6.2.6 we use nonparametric tests for differences in the median values and distributions of Q5 responses between the different groups and in 6.2.9 we test for differences in the proportions of binary answers.

The variables in Fig 6.12 are summarized in Table 6.1.

Table 6.1 Variable characteristics (5.1.2)

Variable	Action	Type	Levels	Variability
Sets: D1 and D2				
pH	Output	Interval	Continuous	Normal ?
River	Input	Nominal	*A / B*	Fixed
Meter	Input	Nominal	*M1/ M2* etc.	Fixed
Set: D3				
Grp	Input	Nominal	*G1 / G2*	Fixed
Q5	Output	Ordinal	− 2 to + 2	
Set: D4				
Q1 to Q4	Output	Binary	*Y / N*	

In Table 6.1, the data marked 'normal ?' is expected to have a normal distribution, but this can be checked using *residuals* within the analysis (5.4.6). The nominal variables, *River*, *Meter,* and *Grp*, are identified as 'Fixed' because the subjects have not been selected randomly as representative of *all* rivers, meters, and groups.

6.2.2 **Analytical options**

When dealing with two samples, it is essential to distinguish between *independent* (*unrelated*) samples and *paired* (*related*) samples. Related data with three or more samples is called 'repeated measures' (Section 3.6). *Independent* sample data is normally entered in a *single* column, as data set *D1* in Fig 6.12, and it would make no difference if we changed the order of the values in sample *A* (A3:A6) without changing the order of values in sample *B* (A7:A10). *Related* data is normally entered in *separate* columns and, as for the data set *D2*, there are specific links between paired values in the same rows, and it would not be possible to change the order in E2:E6 without changing the order in F2:F6.

The issues associated with choosing whether to use *parametric* or *nonparametric* tests are developed in Section 5.4.

Describing data (6.2.3)

● Graphical plots: Combined boxplots, differences vs value for paired data

Tests / measurements:

● Are the sample distributions **normal**?: Section 5.4
● Is there a difference in **variance** (and standard deviation)?: *F*-test (normal distribution), Levine's test (any continuous distribution) (6.2.4, 3.2.1)

Unrelated samples:

● Is there a difference between **mean** values (for normally distributed data)?: Independent samples *t*-test for equal variances, Welch's modified test for unequal variances (6.2.5, 3.1.3)
● Is there a difference between **median** values (nonparametric test)?: Mann–Whitney test (6.2.6, 3.5.1)
● Comparing **distributions** of values: Kolmogorov–Smirnov test (not suitable for small samples) (6.2.6), Chi-squared test for association (suitable for a few categories of nominal or ordinal data) (3.7.4)

Related samples:

● Is there a difference between **mean** values (for normally distributed data)?: Paired samples *t*-test (6.2.7, 3.6.1)
● Is there a difference between **median** values (nonparametric test)?: Paired Wilcoxon test (6.2.8)
● Comparing **distributions** of nominal or ordinal values: Cross-tabulation and chi-squared contingency table (8.2.4)

6.2.3 **Describing the data**

The characteristics of *individual* data samples can be described using the methods in 6.1.3. For *unrelated* samples, the 'Explore' analysis in SPSS provides both numerical and graphical data descriptions, e.g.

SPSS > Analyze > Descriptive statistics > Explore...
Dependent List: e.g. *Q5*   **Factor List:** *Grp*
> Statistics: Option to select numerical parameters
> Plots: Boxplots and outliers (Fig 6.13), histogram, stem and leaf plot, normality plot and test.

Specific graphs can also be produced under the graph menu in both Minitab and SPSS:

Minitab	**SPSS**
Minitab > Graph > Boxplots...	**SPSS > Graphs > Legacy dialogs > Boxplot...**
Select *One Y* and *With Groups*	**Select** *Simple* - **Define**
Graph variable: *Q5*	**Variable:** *Q5*
Categorical variables: e.g. *Grp*	**Category Axis:** *Grp*
→ Output: Similar to Fig 6.13	→ Output: Fig 6.13

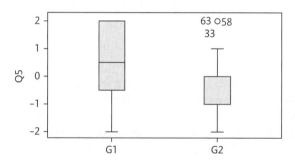

Fig 6.13 SPSS boxplots for *Q5* responses for groups *G1* and *G2*

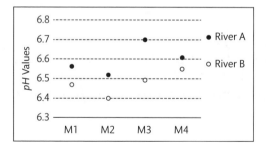

Fig 6.14 Related *pH* data values plotted against meter

The *Q5* boxplots for *G1* and *G2* in Fig 6.13 both record values over the whole range −2 to + 2, but with most *G1* values being '+ 2' and most *G2* values being '0' with just three '+ 2' values identified as outliers (record numbers 33, 58, and 63). It is possible to view the numbers in each category by using separate stem and leaf plots (Fig 6.3) for each variable.

For the *related pH* samples in Fig 6.14 it can be useful to plot both sets of data on a 'line' graph in Excel with each row (meter) being the categories on the *x*-axis. This can show whether there is any pattern in the differences between the pair values, and, for example, we can see that *M2* appears to be biased, giving the lowest readings for both rivers.

6.2.4 Comparing variances

We can use the parametric *F*-test and nonparametric **Levene's test** for a difference in **variance**:

Excel > Function > F.TEST (*array1, array2*)
For Fig 6.12 data, enter A3:A6, A7:A10 for *array1, array2*
→ Returns the two-tail *p*-value.

Minitab > Stat > Basic Statistics > 2 Variances...
→ Returns *p*-values for *F*-test and Levene's test

SPSS > Analyze > Compare Means > Independent Samples T test
Levene's test is performed *within* the *t*-test analysis, as in 6.2.5 below

For the *D1 pH* data in Fig 6.12, the two-tailed *F*-test gives $p = 0.718$ and Levene's test $p = 0.618$. There is therefore no evidence of a difference in variance, and we then assume 'equal variance' for further analyses.

6.2.5 Two sample *t*-test

Two sample tests (Minitab): Analysis of variance, means, and medians. Scan here to watch the video or find it via www.oxfordtextbooks.co.uk/orc/currell/

See 3.1.3 for the development of the two sample *t*-test for unrelated data, which assumes that the two samples are normally distributed and have equal variance. If the variances are not equal, it is possible to use the Welch's modified *t*-test, but, as the standard *t*-test is a robust test and the variance test can be unreliable for small samples, it is not necessary to routinely test for a difference in variance unless there is reason to believe that the variances are likely to be different. The tests can be performed using:

Excel

Excel > Function > T.TEST (*array1,array2, tails, type*)
For Fig 6.12 data, enterA3:A6,A7:A10 for *array1,array2*. Enter 1 or 2 for the number of *tails*, and, for *type* enter 1 for related/paired data, 2 for equal variance, or 3 for unequal variance.
→ Output: Equal variance gives, *p* (2-tailed) = 0.0520 and for unequal variance, *p* (2-tailed) = 0.0540

SPSS

Two sample tests (SPSS): Analysis of variance, means, and medians. Scan here to watch the video or find it via www.oxfordtextbooks.co.uk/orc/currell/

SPSS > Analyze > Compare Means > Independent samples T-test...
Test variable(s): *pH*
Grouping variable: *River*
Define groups...: Enter *A* and *B* to identify the two samples
→ Output: Fig 6.15

Independent Samples Test

		Levene's Test for Equality of Variances		t-test for Equality of Means							
										95% Confidence Interval of the Difference	
		F	Sig.	t	df	Sig. (2-tailed)	Mean Difference	Std. Error Difference		Lower	Upper
pH	Equal variances assumed	.297	.606	2.418	6	.052	.12000	.04962		-.00142	.24142
	Equal variances not assumed			2.418	5.715	.054	.12000	.04962		-.00291	.24291

Fig 6.15 SPSS output from two sample t-test

In Fig 6.15, Levene's test gives $p = 0.606$, indicating that there is no evidence for a difference in variance, and hence it would be safe to use the 'not significant' result, $p = 0.052$, of the standard t-test, with 'equal variances assumed'.

Minitab

Minitab > Stat > Basic Statistics > 2-Sample t...

▼ *Both samples are in one column*

Samples: e.g. pH

Sample IDs: e.g. *River* (in Minitab 16 these are called *subscripts*)

> Options...

Define expected difference between means and select either 1- or 2-tailed test

☑-**Assume equal variances** - choose equal or unequal variances

→ Output: Fig 6.16

```
Difference = mu (A) - mu (B)
Estimate for difference:  0.1200
95% CI for difference:  (-0.0014, 0.2414)
T-Test of difference = 0 (vs not =): T-Value = 2.42  P-Value = 0.052  DF = 6
Both use Pooled StDev = 0.0702
```

Fig 6.16 Minitab output from two sample t-test assuming equal variance

When calculating with *equal* variance, Minitab reports the value of the pooled standard deviation (0.0702) calculated using Eqn. 3.6. For calculations with *unequal* variance, Minitab would give p (2-tailed) = 0.054, and with no 'pooled standard deviation'. Both results fail to find any significant difference between the two rivers.

It is useful to compare the data in the above results for SPSS and Minitab with those obtained in Fig 3.3.

6.2.6 **Nonparametric tests**

The Mann–Whitney test is the nonparametric equivalent to the two sample t-test for unrelated data, testing for a difference in *median* values.

The Kolmogorov–Smirnov test tests for a difference in the rank *distribution* of values in the two samples.

Using **SPSS** for the **Q5 data** in respect of the different groups, *G1* and *G2*:

SPSS > Analyze > Nonparametric Tests... > Independent Samples...

Objective: ⊙ **Customize analysis**

Test Fields: *Q5*   **Groups:** *Grp*

(this option will not accept the groups in *scale* format)

Settings: ⊙ **Customize tests**

☑-**Mann–Whitney U test (2 samples)**

☑-**Kolmogorov–Smirnov test (2 samples)**

→ Output: Fig 6.17

Hypothesis Test Summary

	Null Hypothesis	Test	Sig.	Decision
1	The distribution of Q5 is the same across categories of Grp.	Independent-Samples Mann-Whitney U Test	.031	Reject the null hypothesis.
2	The distribution of Q5 is the same across categories of Grp.	Independent-Samples Kolmogorov-Smirnov Test	.195	Retain the null hypothesis.

Asymptotic significances are displayed. The significance level is .05.

Fig 6.17 SPSS output from independent samples testing

The differences between the two results in Fig 6.17 is that the Mann–Whitney U test identifies a difference ($p = 0.031$) just between the *median* values of the two samples, whereas the Kolmogorov–Smirnov test is unable to identify a significant difference ($p = 0.195$) in the *relative shapes of the distribution* of values within each sample.

It is also possible to perform the Mann–Whitney test in SPSS using the 'Legacy' option:

SPSS > Analyze > Nonparametric Tests... > Legacy dialogs > 2 Independent Samples

Using **Minitab** for the *pH* **data** for a nonparametric test between rivers *A* and *B*:

Minitab > Stat > Nonparametrics > Mann-Whitney

(The two samples must be in different columns)

First Sample: *A*

Second Sample: *B*

→ Output: Fig 6.18

```
Point estimate for ETA1-ETA2 is 0.1200
97.0 Percent CI for ETA1-ETA2 is (-0.0300,0.3000)
W = 25.0
Test of ETA1 = ETA2 vs ETA1 not = ETA2 is significant at 0.0606
```

Fig 6.18 Mann-Whitney test using Minitab

The nonparametric Mann–Whitney test for a difference between the two rivers returns a 'no significance' result ($p = 0.061$) in agreement with that of the independent samples *t*-test above.

6.2.7 **Paired *t*-test**

For related samples, the parametric paired *t*-test can be performed for normally distributed data, using:

Paired tests (Minitab): Analysis for paired *t*-test and Wilcoxon test. Scan here to watch the video or find it via www. oxfordtextbooks. co.uk/orc/currell/

Excel

Excel > Function > T.TEST (*array1,array2, tails, type*)

For Fig 6.12 data, enterE3:E6,F3:F6 for *array1,array2*. Enter 1 or 2 for the number of *tails*, and, for *type*, enter 1 for related data.

→ Output gives the two-tailed *p*-value: $p = 0.0342$

SPSS

SPSS > Analyze > Compare Means > Paired-Samples T-test

Paired variables: Enter *A* and *B* as the two variables for **Pair 1**

→ Output: Fig 6.19

Paired tests (SPSS): Analysis for paired *t*-test and Wilcoxon test. Scan here to watch the video or find it via www. oxfordtextbooks. co.uk/orc/currell/

Paired Samples Test

		Paired Differences					t	df	Sig. (2-tailed)
		Mean	Std. Deviation	Std. Error Mean	95% Confidence Interval of the Difference				
					Lower	Upper			
Pair 1	A - B	.12000	.06481	.03240	.01688	.22312	3.703	3	.034

Fig 6.19 SPSS output for paired samples *t*-test

Minitab

Minitab > Stat > Basic Statistics > Paired t...

▼ *Each sample is in a column*

Sample 1: *A*

Sample 2: *B*

(Minitab calculates the first sample minus the second sample values)

→ Output: Fig 6.20

```
95% CI for mean difference: (0.0169, 0.2231)
T-Test of mean difference = 0 (vs not = 0): T-Value = 3.70   P-Value = 0.034
```

Fig 6.20 Minitab output for paired samples *t*-test

We can now see that, by taking into account the variations between the meters, the paired *t*-tests are able to detect a significant difference ($p = 0.034 < 0.05$) between the mean *pH* values of the two rivers.

6.2.8 **Paired Wilcoxon test**

The paired Wilcoxon test is the nonparametric equivalent to the paired *t*-test. This is not performed in the standard version of Excel.

SPSS

> **SPSS > Analyze > Nonparametric Tests > Related Samples...**
> **Objective: ⊙-Customize analysis**
> **Test Fields:** *A B*
> **Settings: ⊙-Customize tests**
> ☑-**Wilcoxon matched-pair signed-rank (2 samples)**
> → Output: Fig 6.21 giving a 'no significance' result with $p = 0.068$

Hypothesis Test Summary

	Null Hypothesis	Test	Sig.	Decision
1	The median of differences between A and B equals 0.	Related-Samples Wilcoxon Signed Rank Test	.068	Retain the null hypothesis.

Asymptotic significances are displayed. The significance level is .05.

Fig 6.21 SPSS output for paired Wilcoxon signed rank test

Minitab

To perform the Wilcoxon paired test in Minitab, we first calculate the differences between the values of *A* and *B* for each data pair, and record the values in column G in Fig 6.12. We then carry out a one sample Wilcoxon test to test whether the *A–B* data has a median value that is significantly different from 0.

> **Minitab > Stat > Nonparametrics > 1-Sample Wilcoxon...**
> **Variables:** *A–B*
> ⊙-**Test median:** 0
> → Output: $p = 0.100$, which does not show a significant difference.

The nonparametric test is not as *powerful* as its parametric equivalent, failing to find a difference in median values between the two rivers.

6.2.9 Unrelated binary data

Two proportions (SPSS): Analysis for Fig 6.22. See also 3.8.3. Scan here to watch the video or find it via www.oxfordtextbooks.co.uk/orc/currell/

Considering the Forensic Questionnaire binary data in Fig 6.12, we *could* choose to treat the four data samples, *Q1, Q2, Q3* and *Q4* as *unrelated*. In this case the best that we could do is to calculate the proportions of 'Y' outcomes in each sample as shown in row 71, and test for any significant differences.

To test whether there is a significant difference between the two extreme *proportion* values 23/66 for *Q1* and 35/66 for *Q2* we use the two-proportion test in Minitab (see 3.8.3), and find a non-significant $p = 0.053$ for Fisher's exact test, which suggests that there is insufficient evidence to find a difference between the samples. For SPSS we need to weight (8.1.3) the frequency values as in Fig 6.22(a) and then use crosstabs and contingency table statistics (8.2.4) to give the results in Fig 6.22(b). The two proportion test produces a 2×2 contingency table which means that we should use either the Yates continuity correction or the Fisher's exact test, which give $p = 0.054$ and 0.053 respectively, agreeing with the Minitab result.

	Answer	Question	Freq
1	Y	Q1	23
2	N	Q1	43
3	Y	Q2	35
4	N	Q2	31

(a) Data entry for 'weighting'

Chi-Square Tests

	Value	df	Asymp. Sig. (2-sided)	Exact Sig. (2-sided)	Exact Sig. (1-sided)
Pearson Chi-Square	4.429[a]	1	.035		
Continuity Correction[b]	3.721	1	.054		
Likelihood Ratio	4.455	1	.035		
Fisher's Exact Test				.053	.027
N of Valid Cases	132				

a. 0 cells (0.0%) have expected count less than 5. The minimum expected count is 29.00.

(b) Calculated chi-squared values

Fig 6.22 Crosstabs and 2×2 contingency table analysis (SPSS)

However, we can use the full information available by treating the data as *related*, and test for *agreement* between samples (4.4.5) by either:

- comparing two *variables* at a time, using *cross-tabulation* to derive a 2×2 contingency table (8.2.4) whose analysis can include McNemar's test for agreement between the values or

- comparing more than two variables together, using Cochran's Q, which is equivalent to the McNemar test for multiple samples.

6.3 One factor

In Section 6.2 we analysed differences between two data samples, which could be considered as *two levels* of one factor. In this section we extend the analysis to *more than two levels* of the one factor.

6.3.1 Example data

	A	B	C	D	E	F	G	H	I	J
1	Chemotaxis data:							Forensic questionnaire data:		
2	*Treat*	*Cindex*	*Treat*	*Cindex*	*Treat*	*Cindex*		*Year*	*T1*	*T2*
3	0	0.31	1	-0.07	2	-0.12		3	88	87
4	0	0.29	1	0.21	2	0.18		3	57	56
5	0	0.17	1	0.18	2	-0.09		3	68	67
6	0	0.26	1	0.30	2	0.01		2	36	33
7	0	0.44	1	0.17	2	0.06		3	81	78
8	0	0.26	1	0.17	2	0.02		2	42	31
9	0	0.20	1	0.13	2	-0.06		3	60	63
10	0	0.25			2	0.08		3	71	77
11					2	-0.03		2	41	34
12								2	53	51
68								2	43	48

Fig 6.23 Data from the Chemotaxis Index (unstacked) and Forensic Questionnaire case studies (rows 13 to 67 are 'hidden' in the worksheet)

Case study: **Chemotaxis index / 1. One factor analysis (overview)**

The chemotaxis experiment initially treats nematodes with different concentrations of a therapeutic drug, and then measures their capacity for migration towards a food supply in a segmented agar plate. The effect of the pre-treatment factor, *Treat*, on the derived chemotaxis index, *CIndex*, is tested using a one-way ANOVA.

The data in Fig 6.23, columns A to F, gives 24 calculated (5.2.4) chemotaxis index values, *CIndex*, for three levels of the pre-treatment factor, *Treat*. Note that it is *not* necessary that the samples are all the same size. For convenience of display, the data in Fig 6.23 has been shown 'unstacked', in that the *unrelated* replicate samples for the different levels are given in separate columns. For entry into software, we must put all response values, *CIndex*, into the *same* column (stacked) with the levels of *Treat* for each entry given in a separate column.

In this section we demonstrate the one-way ANOVA analysis (parametric and nonparametric):

6.3.3 Clustered boxplots used for *describing* data

6.3.4 Test for normality and homoscedasticity

6.3.5 One-way ANOVA

6.3.6 post hoc tests

6.3.7 Kruskal–Wallis test (nonparametric one-way ANOVA)

5.2.4 / 2. Deriving the analytical statistics: Starting with *raw data values* in a lab notebook, we consider the calculation of the *chemotaxis index* that is then used to analyse the data.

5.4.6 / 3. Normality and homoscedasticity: We use *residuals* to assess the normality and equal variance conditions necessary for the ANOVA analysis.

Case study: **Forensic questionnaire / 5. Repeated measures**

The data in columns H to J in Fig 6.23 is an extract from the full forensic questionnaire data set in Fig 9.12, and gives the results *T1* and *T2* of two tests performed by 66 subjects who are identified as falling into two *year* groups. The tests were conducted *before* and *after* some additional tuition classes, and in 6.3.8 we are interested in comparing the effect of the classes on the performances of the two groups. We use this data as an example of *repeated measures* (Section 3.6) in which the same measurement (i.e. the test) is performed more than once on the same subject.

The variables in Fig 6.23 are summarized in Table 6.2. Note that it is not known initially whether *CIndex* is normally distributed within each *treat* group, although there is no prior information that it is *not* normal. We see in Fig 9.13 that the distribution of *T1* (and probably *T2*) is actually bimodal due to the two *year* groups, but we will initially treat the distribution *within* each *year* as being normal.

Table 6.2 Description of variables in Fig 6.23

Variable	Action	Type	Levels	Variability
Chemotaxis				
CIndex	Output	Interval	Continuous	Normal ?
Treat	Input	Ordinal	0, 1, and 2	Fixed
Forensic				
T1 and *T2*	Output	Interval	Continuous	Normal ?
Year	Input	Ordinal	2 and 3	Fixed

6.3.2 Analytical options

The issues association with choosing whether to use *parametric* or *nonparametric* tests are developed in Section 5.4.

Describing data (6.3.3)

- Graphical plots: Boxplots, factor plots

Tests / Measurements:

- Test for **normality** and **homoscedasticity** (homogeneity of variance): 6.3.4 and 5.4.6
- Is there a difference between the **mean/median** values for different levels of the factor?: GLM/ANOVA (normal data) (6.3.5), Kruskal–Wallis test (nonparametric) (6.3.7), GsdLM (6.4.7)
- Use **post hoc** tests to locate significant differences: 6.3.6, 3.2.4

   Repeated measures data:

- Is there a difference between **repeated** values (within-subject)?: 3.6.2, 6.3.8
- Is there a difference between the **mean** values for different levels of the factor (between-subject)?: 6.3.8

6.3.3 Describing the data

The methods for describing data here are similar to those used for unrelated samples in 6.2.3 and for individual data in 6.1.3. In particular, the 'Explore' option in SPSS provides a comprehensive range of numerical and graphical outputs for data that can be separated by factor levels.

One of the most effective plots for the raw data is the comparative boxplot in Fig 6.24(a) which can be produced through the graph menu in both Minitab and SPSS:

Minitab > Graph > Boxplot... > One Y, With Groups
Graph variables: *CIndex*   **Categorical variables:** *Treat*

SPSS > Graphs > Legacy dialogs> Boxplot...>Simple
> Define: Variable *CIndex*   **Category axis:** *Treat*

The fairly symmetrical boxplots suggest that the data for each treatment will be normally distributed (5.4.6). There are just three outliers identified, with one, marked by a star, considered to be an extreme outlier.

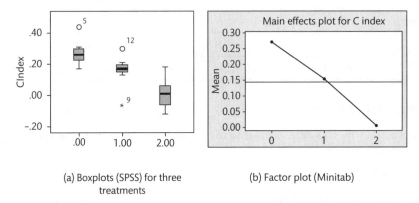

(a) Boxplots (SPSS) for three treatments   (b) Factor plot (Minitab)

Fig 6.24 Displaying one factor variations

Factor plots, as in Fig 6.24(b) also show the variation of the *mean* values for the different samples. They can be requested from within the ANOVA analyses (6.3.5).

6.3.4 Normality and equality of variance (homoscedasticity)

The assessment for both the normality and homoscedasticity is developed in Section 5.4 by analysing the distribution of the *residuals* after fitting a GLM to the data. It is often convenient to perform these checks *within the analysis itself* as a *confirmation* that the analysis is valid. The ANOVA analysis of residuals for the Chemotaxis data is performed in detail using **Minitab and SPSS** in 5.4.6.

Minitab also has a separate test for the equality of variance.

Minitab >Stat > ANOVA > Tests for Equal Variances...
▼ *Response data are in a column for all factor levels*
Reponse: *CIndex*   **Factors:** *Treat*
> Options...: Opting to base the test on the normal distribution will use Bartlett's test, otherwise the analysis uses Levene's test
→ Output : Fig 6.25

```
Bartlett's Test (Normal Distribution)
Test statistic = 0.55, p-value = 0.758

Levene's Test (Any Continuous Distribution)
Test statistic = 0.16, p-value = 0.853
```

Fig 6.25 Minitab tests for differences in variance between samples

SPSS performs Levene's test for variance within the ANOVA calculation (6.3.5), with the results given in Fig 6.26.

Levene's Test of Equality of Error Variances[a]

Dependent Variable: CIndex

F	df1	df2	Sig.
.170	2	21	.845

Tests the null hypothesis that the error variance of the dependent variable is equal across groups.

a. Design: Intercept + Treat

Fig 6.26 Levene's test for homoscedasticity from within GLM/ANOVA (SPSS)

GLM/ANOVA (Minitab): Analysis leading to Figs 6.27(a), 6.28, and 6.30. Scan here to watch the video or find it via www.oxfordtextbooks.co.uk/orc/currell/

The results in Figs 6.25 and 6.26 indicate that there are no significant differences in the variances of the samples for the three *Treat* levels.

If the data does not meet the normality and homoscedasticity criteria then we cannot rely on the significance of the *p*-values from the analysis, and we would need to use some other analysis or transform the data (5.4.7) to meet these conditions.

6.3.5 GLM / ANOVA

An analysis of variance (ANOVA) calculation tests for a significant difference between the mean values of three or more samples. The basic calculation and interpretation of ANOVAs are developed in Section 3.2, but, in practice, it is often more convenient to use the ANOVA analysis within the GLM analysis (3.4.2).

The criteria for an ANOVA is that the random distribution of values has a normal distribution and the variances of all samples are equal (homoscedasticity). However, the GLM/ANOVA techniques are very robust and still produce reliable results for limited deviations from these conditions.

GLM/ANOVA (SPSS): Analysis leading to Figs 6.27(b), 6.29, and 6.31. Scan here to watch the video or find it via www.oxfordtextbooks.co.uk/orc/currell/

Minitab

Minitab > Stat > ANOVA > General Linear Model > Fit General Linear Model...
Reponses: *CIndex*
Factors: *Treat*
→ Output: Fig 6.27(a)

and then for the *post hoc* tests and data plots:

Minitab > Stat > ANOVA > General Linear Model > Comparisons...
Response: *CIndex*   **Type of comparison: ▼ Pairwise**
Select a *post hoc* test (3.2.4), e.g. ☑-*Tukey*
Choose terms for comparison: Double click on *Treat* to see: C Treat
> Graphs...   ☑-**Interval plot for difference in means**
> Results...
    ☑-**Grouping information**
    ☑-**Tests and confidence intervals**
→ Output: Same information as in Fig 6.28

and factor plots can be obtained through:

Minitab > Stat > ANOVA > General Linear Model > Factorial plots...

Tests of Between-Subjects Effects

Dependent Variable: CIndex

Source	Type III Sum of Squares	df	Mean Square	F	Sig.
Corrected Model	.305[a]	2	.152	16.638	.000
Intercept	.496	1	.496	54.207	.000
Treat	.305	2	.152	16.638	.000
Error	.192	21	.009		
Total	.956	24			
Corrected Total	.497	23			

a. R Squared = .613 (Adjusted R Squared = .576)

```
Analysis of Variance for CIndex, using Adjusted SS for Tests

Source  DF   Seq SS    Adj SS    Adj MS      F      P
Treat    2  0.30098   0.30098   0.15049   16.13  0.000
Error   21  0.19593   0.19593   0.00933
Total   23  0.49691

S = 0.0965919   R-Sq = 60.57%   R-Sq(adj) = 56.82%
```

(a) Minitab   (b) SPSS

Fig 6.27 ANOVA outputs from Minitab and SPSS

SPSS

SPSS > Analyze > General Linear Model > Univariate
Dependent variable: *CIndex*   **Fixed factor(s):** *Treat*
Random factor(s): Identify any random factors
Covariate(s): Include any covariates
> Model... Allows choice of factors to be included in the test–not relevant for one-way ANOVA
> Post Hoc...Post HocTests for: *Treat* and select tests e.g. ☑-**Bonferroni**, ☑-**Sidak**
> Options... Allows additional analyses, e.g.
    ☑-**Tests for homogeneity of variance** and
    ☑-**Residual plots**
→ Output : Fig 6.27(b), Fig 6.26, and Fig 6.29

Fig 6.27 shows the presentation of the standard ANOVA table of results (3.2.3) from both Minitab and SPSS, with $p = 0.000$ indicating difference(s) between the levels with a highly significant p-value that is less than 0.0005. However, we need to perform post hoc tests (3.2.4) to decide which *pairs* of samples are significantly different.

6.3.6 Post hoc comparison tests

The post hoc comparison tests are requested within GLM/ANOVA for both Minitab and SPSS (6.3.5). The interpretation and the variety of possible tests is discussed in 3.2.4.

The results given in Fig 6.28 are obtained for the 'Chemotaxis' case study.

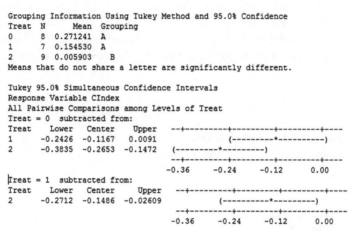

```
Grouping Information Using Tukey Method and 95.0% Confidence
Treat   N      Mean  Grouping
0       8   0.271241  A
1       7   0.154530  A
2       9   0.005903     B
Means that do not share a letter are significantly different.

Tukey 95.0% Simultaneous Confidence Intervals
Response Variable CIndex
All Pairwise Comparisons among Levels of Treat
Treat = 0  subtracted from:
Treat    Lower    Center    Upper    --+---------+---------+---------+----
1       -0.2426  -0.1167   0.0091                    (---------*---------)
2       -0.3835  -0.2653  -0.1472    (---------*---------)
                                     --+---------+---------+---------+----
                                     -0.36    -0.24    -0.12     0.00

Treat = 1  subtracted from:
Treat    Lower    Center    Upper    --+---------+---------+---------+----
2       -0.2712  -0.1486  -0.02609                (----------*---------)
                                     --+---------+---------+---------+----
                                     -0.36    -0.24    -0.12     0.00
```

Fig 6.28 Minitab 16 output for Tukey post hoc test
(Minitab 17 provides the same information in a rearranged format)

In Fig 6.28, the *grouping information* shows that Treatments 0 and 1 both contain the letter 'A' and Treatment 2 only contains the letter 'B'. The interpretation is that there is a significant difference between Treatments 2 and both 0 and 1, and that there is no significant difference between 0 and 1. This is confirmed by reference to the confidence intervals shown as bracketed ranges, because the difference between 0 and 1 has a confidence interval from −0.24 to + 0.009, which includes *zero* and therefore gives the possibility that there is *no* difference. The confidence intervals between 2 and both 0 and 1 do not overlap zero showing that their differences are significant.

Multiple Comparisons

Dependent Variable: CIndex
Bonferroni

(I) Treat	(J) Treat	Mean Difference (I-J)	Std. Error	Sig.	95% Confidence Interval	
					Lower Bound	Upper Bound
.00	1.00	.1168	.04953	.084	-.0121	.2456
	2.00	.2669*	.04650	.000	.1460	.3879
1.00	.00	-.1168	.04953	.084	-.2456	.0121
	2.00	.1502*	.04823	.016	.0247	.2756
2.00	.00	-.2669*	.04650	.000	-.3879	-.1460
	1.00	-.1502*	.04823	.016	-.2756	-.0247

Based on observed means.
The error term is Mean Square(Error) = .009.
*. The mean difference is significant at the .05 level.

Fig 6.29 Bonferroni post hoc test with SPSS

The Bonferroni post hoc tests in Fig 6.29 report both confidence intervals of differences and p-values. For example, the difference between Treatments 0 and 1 has a confidence interval (reversed in signs) between -0.01 to $+0.25$, similar to the Tukey results in Fig 6.28. The p-values confirm the significant differences between 2 and 0 ($p < 0.0005$) and 1 ($p = 0.016$), in addition to the 'no significant' difference between 0 and 1 ($p = 0.084$).

6.3.7 Kruskal–Wallis test

The Kruskal–Wallis test is the nonparametric equivalent of a one-way ANOVA, and is based on the distribution probabilities of the data when expressed as ranked values (3.5.2). It is necessary to use this test for ordinal or ranked data, or for interval data that does not satisfy the normality and homoscedasticity conditions required by the ANOVA. However, for the purposes of illustration we use the Chemotaxis data here as a comparative example.

Minitab

Stat > Nonparametrics > Kruskal–Wallis...

Response: *CIndex*   **Factor:** *Treat*

→ Output: Fig 6.30

```
Kruskal-Wallis Test on CIndex
Treat    N    Median  Ave Rank      Z
0        8  0.256819      19.0   3.18
1        7  0.170732      13.0   0.22
2        9  0.008065       6.3  -3.31
Overall 24                12.5
H = 13.64  DF = 2  P = 0.001
```

Fig 6.30 Kruskal-Wallis test output for Minitab

SPSS

Analyze > Non Parametric Tests > Independent Samples...

Test Fields: *CIndex*

Groups: *Treat* (cannot be a scale variable—if necessary change description to ordinal)

Settings: ⊙-Customize tests

☑-**Kruskal-Wallis 1-way ANOVA (*k* samples)**

→ Output : Fig 6.31

Using the Legacy Dialogs in SPSS, it is possible to enter the factor (group) as a *scale* variable, and it is possible (and necessary) to define the range of factor levels to be tested.

Analyze > Nonparametric Tests > Legacy Dialogs > *K* Independent Samples

Test variable: *CIndex*

Grouping variable: *Treat* (0,2) (define the ends of the chosen range of values)

☑-**Kruskal-Wallis H**

→ Output : Same as Fig 6.31

Hypothesis Test Summary

	Null Hypothesis	Test	Sig.	Decision
1	The distribution of CIndex is the same across categories of Treat.	Independent-Samples Kruskal-Wallis Test	.001	Reject the null hypothesis.

Asymptotic significances are displayed. The significance level is .05.

Fig 6.31 Kruskal-Wallis test output for SPSS

The results of $p = 0.001$ for both Minitab and SPSS shows a highly significant difference between at least one pair of the three samples, consistent with the ANOVA results.

6.3.8 Repeated measures

We have met the power of related measurements in the paired t-test and Wilcoxon test in 6.2.7 and 6.2.8, and, extending this to more than two variables, we have already used a *repeated measures* analysis in 3.6.2 to identify a significant difference between three variables.

In this analysis, the data in columns H to J in Fig 6.23 has repeated measures between the tests *T1* and *T2* taken *before* and *after* additional classes. Applying the terminology, the 'within-subject' variation is the *horizontal* difference between *T1* and *T2* for each subject row, and the 'between-subject' variation describes how the *T1-T2* difference might vary *vertically* between the two year groups.

Using **SPSS**

Repeated measures 2: SPSS analysis leading to 6.32. See also 3.6.2. Scan here to watch the video or find it via www.oxfordtextbooks.co.uk/orc/currell/

SPSS > Analyze > General Linear Model >Repeated Measures...

The first step is to describe the 'within subjects' variation by giving it a name, and, if you wish, a name for the 'between subjects' measure.

Within Subject Factor Name: e.g. *TestDiff*

and entering the number of related variables

Number of Levels: 2   **> Add**

Measure Name: e.g. *YearGroup*   **> Add**

> Define

Within Subjects Variables: Click across *T1* and *T2* to give T1(1.Year Group) and T2(2.Year Group)

Between Subjects Factors: Click across *Year*

For more than two factor groups it is also possible to include post hoc tests

→ Output : Fig 6.32

Tests of Between-Subjects Effects

Measure: YearGroup
Transformed Variable: Average

Source	Type III Sum of Squares	df	Mean Square	F	Sig.
Intercept	369453.986	1	369453.986	2532.365	.000
Year	34168.168	1	34168.168	234.200	.000
Error	9337.142	64	145.893		

(a) Between subjects

Tests of Within-Subjects Contrasts

Measure: YearGroup

Source	TestDiff	Type III Sum of Squares	df	Mean Square	F	Sig.
TestDiff	Linear	.073	1	.073	.008	.927
TestDiff * Year	Linear	53.103	1	53.103	6.172	.016
Error(TestDiff)	Linear	550.692	64	8.605		

(b) Within subjects

Fig 6.32 Repeated measures output (SPSS)

The primary result in Fig 6.32(a) is for the test for a possible difference in overall mean values *between* the two *year* groups, and concludes ($p = 0.000$) that there is a significant difference in mean values between the two groups. This is confirmed by the clearly bimodal distribution of scores for *T1* in Fig 9.13.

The result of particular interest here is in Fig 6.32(b), which concentrates on the differences *within* individual subjects and concludes that the average difference between *T1* and *T2* is not significantly different ($p = 0.927$) from 0, i.e. the additional classes do not appear to have made any *overall* difference. However, the interaction (3.3.2) term, *TestDiff* * *Year*, between the *TestDiff* and the *Year* group, is significant with $p = 0.016$, which suggests that the effect of the additional classes is different between the two *Year* groups. This can be demonstrated by analysing the differences, *T2* − *T1*, separately for the two *Year* group samples. Using Explore in SPSS, we get:

Mean difference for Year 3 = +1.35

Mean difference for Year 2 = −1.25

It appears that the extra classes have helped Year 3 giving them an increased score but have confused Year 2 and actually reduced their score.

Refer to 3.6.2 for the interpretation of Mauchly's test for *sphericity* which tests for the equality of variance when there are three or more repeated measures.

6.4 Multiple factors and interactions

This section analyses a single interval or ordinal response variable which may be dependent on a *number* of input factors or variables. It extends the use of the ANOVA for multiple levels of one factor to analyse multiple *factors* and possible *interactions* between those factors, and in 6.4.8 we also introduce the analysis of covariance (ANCOVA) which includes the effect of a continuous input *variable* that is correlated with the response variable.

6.4.1 Example data

We use two main case studies to introduce various aspects of multiple factor analysis, one with *interval* response data and one with *ordinal* data. We also include, in 6.4.8, data from the 'Ink analysis' case study to demonstrate the use of an ANCOVA analysis.

Case study: Boxing performance / 1. Multifactorial analysis (overview)

In an experiment to test for the effect of dehydration on boxing performance, six amateur boxers (a,b,c,d,e,f) each carried out two simulated boxing bouts (of three rounds each) in each of two states of hydration: euhydration (*E*) (normal state) and dehydration (*D*). The measured performance variable is the number of punches in each round, and a random extract from the results is given in Fig 6.33.
In this section:

6.4.3 Clustered boxplots describe the data and interaction plots identify key behaviour patterns.

6.4.4 The main analysis using GLM/ANOVA identifies significant factors and interaction.

6.4.5 The tests for normality and homoscedasticity support the validity of the analysis.

9.1.6 / 2. Multiple regression: Uses stepwise regression and general regression to identify the significant factors in a best-fit model.

	A	B	C	D	E	F	G	H	I
1	RecNo	Punches	Subject	Hydrat	HydratN	Round	Bout	H*R	H*B
2	11	131	e	E	1	2	1	2	1
3	3	129	c	E	1	1	1	1	1
4	62	117	b	D	2	5	2	10	4
5	42	127	f	D	2	1	1	2	2
6	24	135	f	E	1	4	2	4	2
7	16	150	d	E	1	3	1	3	1
8	55	147	a	D	2	4	2	8	4
9	51	132	c	D	2	3	1	6	2
10	40	132	d	D	2	1	1	2	2
11	59	122	e	D	2	4	2	8	4
12	31	159	a	E	1	6	2	6	2

Fig 6.33 Extract of data from the 'Boxing' case study

When entering data into Excel, it is useful to give all the data values a unique record number, as in column A of Fig 6.33. This allows the data to be sorted under various headings when editing the data, but then to use the record number as a key to re-sort the data back into its original order.

The *Punches* data is an integer frequency, but, given the high values, we can treat it as an *interval* variable. The level of hydration has been recorded as *E* or *D* under *Hydrat*, but we have also added a numeric code under *HydratN*, because SPSS will only recognize the numeric value for certain analyses. There are six possible rounds, with one to three occurring within *Bout* = 1 and four to six occurring within *Bout* = 2. The values of the variables *H*R* and *H*B* are calculated by multiplying *HydratN* by *Round* and *Bout* respectively, and their use is introduced for multiple regression in 9.1.6. The aim is to investigate whether the level of hydration affects performance over different rounds and bouts.

The data structure (5.1.2) is summarized in Table 6.3.

Table 6.3 Data structure for the variables in Fig 6.33

Variable/factor	Action	Type	Levels	Variability
Punches	Output	Frequency / interval	Continuous	Normal
Hydrat	Input	Nominal	E and D	Fixed
HydratN	Input	Coded	1 and 2	Fixed
Round	Input	Ordinal	1 to 6	Fixed
Bout	Input	Ordinal	1 and 2	Fixed
*H*R*	Input	Nominal	Derived, 1 to 12	Fixed
*H*B*	Input	Nominal	Derived, 1 to 4	Fixed

	A	B	C	D	E	F	G	H	I
1	Rec	Temp	Quality	Method		Temp\Method	A	B	C
2	1	10	2	A		10	2, 1, 2	2, 3 ,2	3, 2, 3
3	2	20	3	A		20	3, 3 ,2	3, 4, 2	4, 3, 5
4	3	30	2	A		30	2, 1, 1	3, 4, 4	4, 3, 3
5	4	10	2	B					
6	5	20	3	B		Temp	A	B	C
7	6	30	3	B		10	2	2	3
8	7	10	3	C		10	1	3	2
9	8	20	4	C		10	2	2	3
10	9	30	4	C		20	3	3	4
11	10	10	1	A		20	3	4	3
12	11	20	3	A		20	2	2	5
13	12	30	1	A		30	2	3	4
14	13	10	3	B		30	1	4	3
15	14	20	4	B		30	1	4	3
16	15	30	4	B					

Fig 6.34 Three formats for presenting *quality* as function of two factors *temp* and *method* (data in columns A, B, C, and D extend to further rows)

Case study: Fingerprint quality / 1. Multifactorial analysis (overview)

The data in Fig 6.34 records the assessed quality of fingerprints as *Quality* (on a 0 to 5 ordinal scale), obtained using three different lifting methods, *Method* (A, B, C) at three different temperatures, *Temp* (10, 20, 30). There are three replicate measurements at each of the nine possible combinations of factor levels, and the data is presented in three ways:

- *stacked* as in columns A, B, C, and D
- *unstacked* in cells F6:I15 (used for Friedman two-way ANOVA)
- replicate measurements *tabulated* in cells F1:I4.

The case study investigates different methods and conditions for collecting fingerprints for forensic analysis.
In this section:

6.4.3 Identify key behaviour patterns using clustered boxplots and interaction plots.

6.4.6 The main analysis uses nonparametric Friedman test and Kendall's coefficient of concordance to identify factor significance.

6.4.7 Uses the generalized linear model with a logit transformation for ordinal data.

5.1.5 / 2. Organizing data entry: We consider the transfer of data from lab book to software analysis.

The data structure (5.1.2) is summarized in Table 6.4.

Table 6.4 Data structure for the variables in Fig 6.34

Variable/factor	Action	Type	Levels	Variability
Quality	Output	Ordinal	0 to 5	Ordinal
Temp	Input	Interval	10, 20, 30	Fixed
Method	Input	Nominal	A, B, C	Fixed

The data would normally be entered into software analysis using the column format with the individual response values, *quality*, in column C, and *identified* by the relevant levels of the two factors, *temp* and *method*, in columns B and D respectively. Only the first 15 records are illustrated in the figure.

It is also possible to enter the data in an 'unstacked' format in F7:I15 with three samples representing the three *method* levels, *blocked* by the different temperatures. This is used for the Friedman nonparametric two-way ANOVA in 6.4.6 for a difference between methods, but it is also possible to test for a *temp* effect by re-entering and 'rotating' the data, to give between three columns for the three *temp* levels blocked by *method*.

6.4.2 Analytical options

The issues association with choosing whether to use *parametric* or *nonparametric* tests are developed in Section 5.4.

Factor plots (Minitab): Analysis leading to Figs 6.35(a), 6.36(b). Scan here to watch the video or find it via www. oxfordtextbooks. co.uk/orc/ currell/

Describing data

- Graphical plots: Clustered boxplots, Factor and interaction plots: 6.4.3

Tests / Measurements:

- Testing for **normality** and **homoscedasticity** (homogeneity of variance): 5.4 and 6.4.5
- Is there a difference between the **mean/median** values due to factors?: GLM/ANOVA (normal data) (6.4.4), Friedman test (nonparametric) (6.4.6), GsdLM (6.4.7)
- Is there an **interaction** between factors?:GLM/ANOVA (6.4.4 and 3.3.2)
- Use **post hoc** tests to locate significant differences: 6.3.6, 3.2.4
- Including **covariant** factors: 6.4.8

Factor plots (SPSS): Analysis leading to Figs 6.35(b), 6.36(a). Scan here to watch the video or find it via www. oxfordtextbooks. co.uk/orc/ currell/

  Repeated measures data:

- Is there a difference between **repeated** values (within-subject)?: 3.6.2, 6.3.8
- Is there a difference between the **mean** values for different levels of the factor (between-subject)?: 6.3.8

6.4.3 Describing the data

Graphical plots are particularly useful for multifactorial data as they can begin to show some of the structure that is not immediately obvious in tables of numbers. The boxplots in Fig 6.35 are quick and effective ways of displaying the *raw data* in your results, and are produced through the graph menu in both Minitab and SPSS, using

    Minitab > Graph > Boxplot... > One Y, With Groups
    Graph variables: *Punches*   **Categorical variables:** *Hydrat Bout*
    (entering the factors in the order in which the boxplots are to be grouped)

SPSS > Graphs > Legacy dialogs ... > Boxplot... > Clustered
> Define: Variable: *Quality*   **Category Axis:** *Method*   **Define Clusters by:** *Temp*

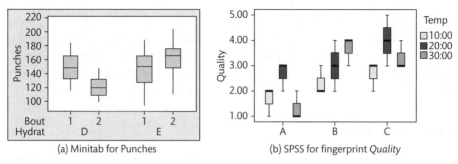

(a) Minitab for Punches   (b) SPSS for fingerprint *Quality*

Fig 6.35 Clustered boxplots in Minitab and SPSS

The clustered (grouped) boxplots in Fig 6.35(a) relate to the 'Boxing' case study in which the numbers of *Punches* are recorded for two factors, *Bout* (1 or 2), and *Hydrat* (E or D). The *Bout* boxplots can be grouped hierarchically within *Hydrat* as shown, or vice versa. The boxplots show no major differences between the different samples, except that it is useful to note that going from *Bout* 1 to 2 for dehydration, *D*, the number of punches falls, but going from *Bout* 1 to 2 for euhydration, *E*, the number rises. We will see that this is the effect of a significant *interaction* between bout and hydration.

The boxplots in Fig 6.35(b) relate to the 'Fingerprints' case study with *Quality* recorded for *Temp* and *Method*. The only clear variation is a general increase in quality from method *A* to *C*.

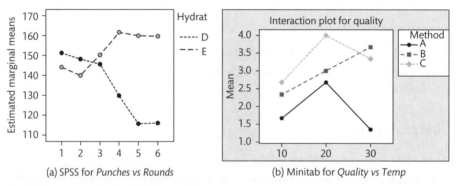

(a) SPSS for *Punches vs Rounds*   (b) Minitab for *Quality vs Temp*

Fig 6.36 Interaction plots in Minitab and SPSS

Interaction and factor plots, as in Fig 6.36, calculate the *mean values* of any replicate data values and plot these against the different combinations of factor levels. The plots use the GLM/ANOVA analyses to calculate the sample mean values, and are a useful *visualization* of the data variations *even for data that fails to meet the normality and homoscedasticity require-ments* for the full ANOVA analysis.

In Minitab, the plots are available under the ANOVA submenu:

Minitab > Stat > ANOVA > Main Effects Plot... or Interactions Plot...

In SPSS, the plots are available under the Plot option within the GLM analysis

SPSS > Analyze > General Linear Model > Univariate ...> Plots...

Fig 6.36(a) is from the 'Boxing' case study and shows how the numbers of *Punches* varies for six rounds for different levels of *Hydrat*, D and E. Again we have evidence of an *interaction* in that the number of punches drops in the later rounds, but *only* for those subjects who are dehydrated.

Fig 6.36(b) relates to the Fingerprint case study and shows how the mean value of *Quality* changes with temperature for the three methods, *A, B,* and *C*. The main consistent characteristic is that, overall, method *C* has a greater average quality than *B* or *A*.

6.4.4 GLM/ANOVA

The ANOVA calculation (3.2, 3.3) can test simultaneously for the significance of multiple factors and their interactions, and is easily performed using the GLM (Section 3.4). The key data requirements for performing an ANOVA are normality and homoscedasticity (equal variance at all factor levels), and checks to confirm that the data meets these requirements are performed *within* the ANOVA group of analyses (5.4.5 and 6.4.5). These results can be reported as the 'quality of the model', but if the data fails to meet the requirements, it is necessary to move to nonparametric analyses.

With two or more factors, we may choose to test for the significance of *individual* factors (e.g. *Hydrat* and *Bout*) and also for a possible *interaction* between factors (*Hydrat*Bout*). The identification of interactions is developed in Section 3.3.2. However, we can only test for an interaction if there are *replicate* measurements, i.e. if there are at least two measurements made at each combination of factor levels.

Comparison (post hoc) tests (3.2.4 and 6.3.6) are requested within the analysis dialogue windows, and it is also possible to include the effect of *covariates* (6.4.8) on the analysis.

The data for Minitab and SPSS must be in the form of *unrelated* samples entered as a single column of univariate data (B in Fig 6.33 and C in Fig 6.34), but with the factor levels defined by the values in separate columns.

SPSS

SPSS > Analyze > General Linear Model > Univariate...
Dependent variable: *Punches*
Fixed factor(s): *Subject Hydrat Bout*
Random factors(s): Include any random factors (3.4.4)
Covariate(s): Include any covariates (6.3.8)
> Model... *Subject Hydrat Bout Hydrat*Bout* (we identify just one of the possible interactions)
> Plots... Select any factor plots

Multifactorial GLM/ANOVA (Minitab): Analysis leading to Fig 6.37. Scan here to watch the video or find it via www.oxfordtextbooks.co.uk/orc/currell/

Multifactorial GLM/ANOVA (SPSS): Analysis leading to Fig 6.37. Scan here to watch the video or find it via www.oxfordtextbooks.co.uk/orc/currell/

> **Post Hoc...** Select any post hoc tests (3.2.4 and 6.3.6)
> **Options...** e.g. Homogeneity test and spread vs level plot

The results in Fig 6.37 show that there is a significant difference between subjects ($p < 0.0005$) as would be expected, with hydration a significant factor with *Hydrat* $p < 0.0005$ and the interaction between hydration and the bout also significant with *Hydrat*Bout* $p < 0.0005$. Although the *p*-value for the bout is greater than 0.05, we would still consider it to be a significant factor because of its involvement in the interaction term. The use of post hoc tests to identify the location of the differences between factor levels is introduced in 3.2.4 and demonstrated in 6.3.6.

Tests of Between-Subjects Effects

Dependent Variable: Punches

Source	Type III Sum of Squares	df	Mean Square	F	Sig.
Corrected Model	28228.944[a]	8	3528.618	11.421	.000
Intercept	1485226.125	1	1485226.125	4807.315	.000
Subject	12991.458	5	2598.292	8.410	.000
Hydrat	6068.347	1	6068.347	19.642	.000
Bout	654.014	1	654.014	2.117	.151
Hydrat * Bout	8515.125	1	8515.125	27.561	.000
Error	19463.931	63	308.951		
Total	1532919.000	72			
Corrected Total	47692.875	71			

a. R Squared = .592 (Adjusted R Squared = .540)

Fig 6.37 Two-way ANOVAs with interaction for the 'Boxing' case study data (SPSS)

The analysis in **Minitab** follows a very similar structure, with the interaction term explicitly defined as a model factor:

Minitab > Stat > ANOVA > General Linear Model > Fit General Linear Model...
Reponses: *Punches*   **Factors:** *Subject, Hydrat, Bout*
> **Model:** Highlight *Hydrat* and *Bout* in **Terms in the model**
   Cross predictors and terms in the model -Add
The terms in the model should now include *Subject, Hydrat, Bout, Hydrat*Bout*
→ Output: Standard ANOVA table with results as in Fig 6.37

The inclusion of the product *Hydrat*Bout* under Model requests that the ANOVA tests for an interaction between these factors, and the output gives the standard ANOVA table of results with the same values as in Fig 6.37. Post hoc test and factorial plots can be obtained by now using:

Minitab > Stat > ANOVA > General Linear Model > Comparisons...
Minitab > Stat > ANOVA > General Linear Model > Factorial plots...

Excel provides a two factor ANOVA test through the add-in 'Data Analysis'. The data for Excel must be held in a table of values with the two factors defining the rows and columns (as in F6 to I15 in Fig 6.34), *together* with the row and column labels.

Excel provides separate tests depending on replication:

Excel > Data > Data analysis > ANOVA: Two-Factor *with* Replication
(replicate values must be in successive *rows* as in G7:I15)
Input range: *F6:I15* (this range must include row and column labels)
Rows per sample: *3*
Output range: Select a new worksheet or a cell for the top left hand cell of data output area.
(Excel automatically tests for an interaction between the factors.)

Note that, if there are no replicates, it is not possible to test for an interaction:

Excel > Data > Data analysis > ANOVA: Two-Factor *without* Replication
Input range: Define the data range from top left cell to bottom right cell.
☑**-Labels:** - Check if the defined range includes the row and column labels.
Output range: Select top left hand cell of data output or new worksheet.

Excel provides the ANOVA output table in the standard format.

6.4.5 Checking for normality and homoscedasticity

The ANOVA calculation assumes that the random data variations are described by a normal distribution and with an equal variance for all measurement conditions. The consideration of these criteria is developed in more detail in Section 5.4, in which we see that the first step is to consider whether there is any evidence in the raw data to suggest that it does *not* comply with the requirements.

In respect of the 'Boxing' case study, the relevant variable is *Punches,* which is actually an *integer* variable, but, because the variation is considerably greater than just one unit and does not approach any limiting value (e.g. '0'), we can treat it as a *continuous* variable. We can also see from the boxplots in Fig 6.35(a) that the random variations due to different subjects are small compared to the median values, and we can conclude that the variations are likely to be *symmetrical* about mean values. In addition, the *spreads* of the boxplots are similar for all the experiment conditions so that we do not expect to see any significant differences in variance.

In respect of the 'Fingerprint' case study, we know immediately that the response data, *Quality*, is an *ordinal* variable, and, with just six possible levels limited by 0 and 5, we would not expect a parametric ANOVA calculation to give reliable estimations of significance (*p*-values).

The main approach to *testing* these criteria is from within the ANOVA analysis itself, by analysing the *residuals* between the best-fit model and the experimental data (5.4.5), either directly and/or by saving the individual residual values and testing them afterwards. Within the **Minitab** dialogue we can include:

> **Graphs ... It is useful to select ▼ *Standardized* for Residuals for plots**
Useful plots are:
    ☑**-Normal plot of residuals** to show normality
    ☑**-Residuals versus fits** show equality of variance or
    select ◉**-Four in one** (Fig 5.15)
> **Storage ...** Use ☑**-Standardized residuals** to save residuals an empty column as *SRES1*

Normality and homoscedasticity (Minitab): Analysis of the 'Boxing' case study. See also 5.4.6. Scan here to watch the video or find it via www.oxfordtextbooks.co.uk/orc/currell/

Normality and homoscedasticity (SPSS): Analysis of the 'Boxing' case study. See also 5.4.6 and 6.3.4. Scan here to watch the video or find it via www.oxfordtextbooks.co.uk/orc/currell/

The saved residuals can then be tested for normality:

Minitab>Stat>Basic Statistics>Normality Test...
Variables: *SRES1*
Select a specific test, e.g. ◉**-Anderson–Darling**

Within the **SPSS** dialogue we can include:

> **Save:** ☑**-*Standardized residuals,*** which saves the residuals in a new column with the
Name *ZRE_1* and Label *Standardized Residual*
> **Options: Display** ☑**-Homogeneity tests**

The saved residuals can then be analysed using:

SPSS > Analyze > Descriptive Statistics > Explore...
Dependent List: *Standardized residual*
> **Plots:** ☑**-Normality plot with test**

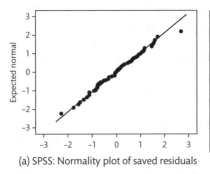

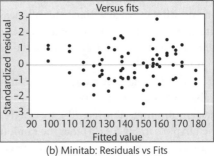

(a) SPSS: Normality plot of saved residuals   (b) Minitab: Residuals vs Fits

Fig 6.38 Testing for normality and homoscedasticity in the 'Boxing' case study

Examples of the results for the 'Boxing' case study are given in Fig 6.38. The Q–Q plot
(5.4.5) of residuals in Fig 6.38(a) shows the *residuals* fitting closely to the diagonal normality
line, and the normality tests give Shapiro–Wilk, $p = 0.941$ (SPSS) and Anderson–Darling,
$p = 0.214$ (Minitab), both confirming that these is no significant deviation from normality.

The plot of residuals 'Versus Fits' in Fig 6.38(b) shows the spread of experimental data
about the calculated values (Fits), and there does not appear to be any significant change in
the spread (variance) for the different results along the *x*-axis.

Minitab also provides a separate test for the homogeneity of variance requirement:

Minitab > Stat > ANOVA > Test for Equal Variances...
▼ ***Response data are in a column for all factor levels***
Response: *Punches*   **Factors:** *Subject Hydrat Bout*

and gives the following results:

Levene's test (for any continuous distribution), $p = 0.999$

or by choosing the normal distribution under **Options…**

Bartlett's test (assuming a normal distribution), $p = 0.974$

both of which confirm that we can assume homoscedasticity.

6.4.6 Nonparametric ANOVAs

Nonparametric ANOVA (Minitab): Analysis leading to Fig 6.39(b). Scan here to watch the video or find it via www.oxfordtextbooks.co.uk/orc/currell/

The Friedman test is the *nonparametric* equivalent of the two-way ANOVA, testing for a difference in the median values of one factor while *blocking* the other factor. For the data in Fig 6.34 in cells G7:I15, it tests for a significant difference between the ranked values of methods A, B, and C, by treating each row as measurements made under the *same* conditions of the factor, *Temp*. The test provides no information about any significance in the row factor, and, if we wish to test for the significance of *Temp*, we would need to *transpose* the rows and columns such that there was a column of data for each value of *Temp*.

The statistics of Kendall's coefficient of concordance are introduced in 4.4.3 where it is developed to test for agreement between samples. It provides a similar analysis to the Friedman test in that it tests for a difference between the column values while blocking the rows.

We also see in the next section that the GsdLM can be used to perform the same analysis using a logit transformation for the ordinal data.

SPSS

Nonparametric ANOVA (SPSS): Analysis leading to Fig 6.39(a). See also 6.1.6. Scan here to watch the video or find it via www.oxfordtextbooks.co.uk/orc/currell/

> **SPSS > Analyze > Nonparametric Tests > Related Samples…**
> **Fields:** $A B C$
> **Settings:** ⊙-**Customize tests**
> ☑-**Kendall's coefficient of concordance (k samples)**
> ☑-**Friedman's 2-way ANOVA by ranks (k samples)**

Hypothesis Test Summary

	Null Hypothesis	Test	Sig.	Decision
1	The distributions of A, B and C are the same.	Related-Samples Friedman's Two-Way Analysis of Variance by Ranks	.014	Reject the null hypothesis.
2	The distributions of A, B and C are the same.	Related-Samples Kendall's Coefficient of Concordance	.014	Reject the null hypothesis.

Asymptotic significances are displayed. The significance level is .05.

(a) SPSS summary

```
Kendall's Coefficient of Concordance

For Method being blocked by Temp:
    Coef  Chi - Sq  DF       P
0.476703  8.58065    2   0.0137

For Temp being blocked by Method:
    Coef  Chi - Sq  DF       P
0.456790  8.22222    2   0.0164
```

(b) Minitab

Fig 6.39 Nonparametric two-way ANOVAs

In these analyses, each *row* of data is blocked separately, effectively giving nine levels to the factor *Temp*, and does not take into account the *replicates* at each level. Fig 6.39 (a) gives the same p-value $= 0.014$ for both tests which identifies a difference between the distribution of values in *A*, *B*, and *C*, showing a significant effect for *Method*. To identify where this difference lies it is possible to view the plot in Fig 6.35(b), but it would also be possible to perform paired Wilcoxon tests separately between pairs of *A*, *B*, and *C*, using a Bonferroni correction for their significance (1.6.4).

In Minitab, the **Friedman** test requires that the data is presented as *univariate* data with the test factor identified as the Treatment and the other factor the Block. However, Minitab requires that there are *no replicate* Block (i.e. *Temp*) values, and cannot directly analyse the data in Fig 6.34. This can be overcome by replacing *Temp* with a *dummy* variable with values from 0 to 9 in column F, in which case the test gives the same p-value, $p = 0.014$.

Minitab

> **Minitab > Stat > Nonparametrics > Friedman...**
> **Response:** Enter column holding data
> **Treatment:** Enter column with factor being tested
> **Blocks:** Enter column with factor to be blocked

To perform the **Kendall's concordance** test, Minitab will accept the data either in univariate form or as related data in the same format as SPSS. Using the univariate input:

Minitab

> **Minitab > Stat > Quality Tools > Attribute Agreement Analysis...**
> **Attribute:** *Quality* (response variable)
> **Samples:** *Method* (factor being *tested*)
> **Appraisers:** *Temp* (factor being *blocked*)
> **Known standard/attribute:** Include any 'correct answer' data if known
> ☑-**Categories...** ordered (check when response variable is *ordinal*)

The result in Fig 6.39(b) again gives $p = 0.014$ for the significance of a *Method* effect, and, by swapping round the dialogue entries for *Samples* and *Appraisers*, we see that the effect of Temp is also significant with $p = 0.016$.

Generalized linear model (ordinal logistic): SPSS analysis leading to Fig 6.40. See also 3.4.7. Scan here to watch the video or find it via www.oxfordtextbooks.co.uk/orc/currell/

6.4.7 **Generalized linear model**

We could not reliably use a GLM/ANOVA for the 'Fingerprint' case study because the response data was *ordinal*, and we used the nonparametric analyses to test the two factors separately. However, it is possible to use the GsdLM (3.4.7), which performs a single hypothesis test that analyses both factors together by transforming the *ordinal* data values using a cumulative logit function (8.3.3).

SPSS

> **SPSS > Analyze > Generalized Linear Models> Generalized Linear Model...**
> **> Type of model: ⊙-Ordinal logistic**
> **>Response: Dependent variable:** *Quality*

> **Predictors: Factors:** *Temp Method*
> **Model:** *Temp Method*

The results in Fig 6.40 show that both factors are significant, which is consistent with the conclusions from Fig 6.39.

Tests of Model Effects

Source	Type III		
	Wald Chi-Square	df	Sig.
Temp	7.136	2	.028
Method	11.202	2	.004

Dependent Variable: Quality
Model: (Threshold), Temp, Method

Fig 6.40 SPSS: Output from the GsdLM

6.4.8 Analysis of covariance, ANCOVA

The typical analysis in an ANOVA has factors which have a limited number of specific levels, e.g. *bout*, *method*, but we now consider the additional effect of a variable with a continuous range of possible values.

ANCOVA 2: SPSS analysis leading to Figs 6.41 and 6.42. See also 3.3.3. Scan here to watch the video or find it via www. oxfordtextbooks. co.uk/orc/ currell/

> **Case study: Ink analysis / 6. ANCOVA analysis 2**
>
> —continued from 3.3.3
>
> The graph in Fig 5.6 gives the percentage transmissions, *%T*, for three different *inks*, A, B, and C, which could be considered as a one factor problem. However, the *%T* value is also affected linearly by the *wavelength* of the measurement, which is then called a *covariate*. The modification of the simple ANOVA to accommodate this covariate factor is called an ANCOVA.

The calculation using this data is carried out in 3.3.3 using Excel and Minitab, in which it is assumed that the *%T* value varies linearly with the *wavelength*, and the first step is to perform a linear regression to measure the relationship (i.e. slope) between the two variables, *%T* and *wavelength*. Once the best-fit relationship is known, a customized 'correction' can be applied to each value of *%T* to compensate for the wavelength effect, and then it is possible to treat the problem as a simple ANOVA analysis.

The analysis using SPSS is as follows:

> **SPSS > Analyze > General Linear Model > Univariate...**
> **Dependent variable:** *%T*
> **Fixed factor(s):** *Ink*
> **Covariate(s):** *Wavelength*
> **> Options...**
>   **Display Means for:** *Ink*

☑-**Compare main effects**
Confidence interval adjustments: ▼ Bonferroni
☑-**Parameter estimate**

The results for the between-subjects effects give $p = 0.000$ for *wavelength,* confirming that it is indeed a significant covariate. The values for the coefficients, B, in the parameter estimates shown in Fig 6.41 give the linear relationship between them (with the inks coded with integer values) as:

$$\%T = 0.875 \times \text{Wavelength} - 569$$

which is consistent with the calculations in 3.3.3 using data for inks A and B.

Parameter Estimates

Dependent Variable: T

Parameter	B	Std. Error	t	Sig.	Lower Bound	Upper Bound
					95% Confidence Interval	
Intercept	-569.422	16.126	-35.311	.000	-603.061	-535.784
Wavelength	.875	.023	38.328	.000	.827	.922
[Ink=A]	6.039	.426	14.189	.000	5.151	6.927
[Ink=B]	.894	.426	2.100	.049	.006	1.782
[Ink=C]	0[a]	.	.	.	.	.

a. This parameter is set to zero because it is redundant.

Fig 6.41 Parameter estimates in ANCOVA results (SPSS)

The between-subjects effects also give $p = 0.000$ for *ink,* which detects a significant difference between at least two inks. The post hoc test for this difference is conducted by using the Bonferroni comparison of the main effect, giving the results in Fig 6.42, which detects a significant difference between *A* and both *B* and *C,* but no significant difference ($p = 0.146$) between *B* and *C.* This is consistent with the graphs as presented in Fig 5.6.

Pairwise Comparisons

Dependent Variable: T

(I) Ink	(J) Ink	Mean Difference (I-J)	Std. Error	Sig.[b]	Lower Bound	Upper Bound
					95% Confidence Interval for Difference[b]	
A	B	5.145*	.426	.000	4.033	6.257
	C	6.039*	.426	.000	4.927	7.151
B	A	-5.145*	.426	.000	-6.257	-4.033
	C	.894	.426	.146	-.218	2.006
C	A	-6.039*	.426	.000	-7.151	-4.927
	B	-.894	.426	.146	-2.006	.218

Based on estimated marginal means
*. The mean difference is significant at the .05 level.
b. Adjustment for multiple comparisons: Bonferroni.

Fig 6.42 Bonferroni comparison of *inks* (SPSS)

7 Related variables

Introduction

This chapter considers the analysis of *related* data samples, in which there are unique links between pairs of values in each sample. The content builds on Section 4.4 to measure the *agreement* between two different assessments that are measuring the *same* quantity, and refers to Section 3.6 for successive measurements of the *same* variable, possibly 'before' and 'after' an intervention, leading to a *paired* analysis for two samples or *repeated measures* for multiple samples. However, a main focus of the chapter is to build on Sections 2.1 and 2.3 to develop the analysis of interrelated samples that measure *different* quantities, often described by an *x–y* graph, e.g. in an absorbance vs concentration calibration graph. It also develops convolution and spectral analysis techniques to address more complex exploratory *x–y* data and time series data.

> Section 7.1 develops the basic parametric and nonparametric methods with which we can analyse the *regression*, *correlation*, and *agreement* between two variables.
>
> Section 7.2 considers the methods that are available to analyse data that is expected to have a specific *nonlinear* relationship.
>
> Section 7.3 introduces some less common techniques for handling *general x–y data* whose behaviour is not described by a simple mathematical function.

7.1 Regression, correlation, and agreement

A correlation analysis measures the extent to which a change in one variable is related to a change in the other. It can test whether the observed relationship could have occurred by chance, but it can also measure the *strength* of the relationship. Linear correlation measures the extent to which the change in one variable is *proportional* to a change in the other, but nonparametric correlation just measures the extent to which one variable increases (or decreases) *in step with* the other. For two variables that are correlated, a *regression* analysis calculates the values that *quantify* the relationship (e.g. slope and intercept) between a response variable and a predictor variable.

Chapter 2 developed the *statistics* associated with linear correlation and regression. The analyses can be performed when the variables are recording *different* scientific quantities, e.g. between the values of absorbance and concentration with different units of measurement. However, we can also look for *agreement* between variables when they have the *same units*, e.g. when comparing the proportions of cell deaths recorded by different assays. Chapter 4 introduced the techniques associated with the tests and measurements of nonparametric correlation, association, and agreement.

7.1.1 **Example data**

The most common form of *related* data is *x–y* data that is expected to produce a *straight line*. The theory and practice for this is introduced in Chapter 2, and you are referred to the case studies:

Best-fit straight line (2. Introduction)

which develops the statistics of linear regression and its use in analysing scientific data.

The assessment of agreement within *categorical* or *binary* data is introduced in 7.1.6 and analysed primarily using *crosstabulation* and *contingency tables* (8.2.4). Relevant case studies are:

Association (8.2.1)

Forensic questionnaire (8.2.1)

We also meet *related* data in *paired* and *repeated measures* analyses in the case studies:

River *pH* (3.6.1)

Ink analysis (3.6.2)

The following two case studies give examples in which the related variables are measuring *different* quantities (i.e. absorbance and concentration) and in which they are 'repeated' measurements of the *same* quantity (i.e. mortality).

Case study: **Spectrophotometer calibration / 1. Calibration (overview)**

In Fig 7.1(a) the data in columns A and B provide an example of interrelated *interval* data.
A spectrophotometer typically analyses the concentration, *Conc*, of a chemical sample by recording the absorbance, *Abs*, of light passing through the sample. It is initially *calibrated* by measuring absorbance values for the known concentrations of prepared standards, as shown in Fig 7.1(b). In 7.1.3 and 7.1.4 we assess the linearity of the calibration line using residual plots and correlation measurements.

7.1.5 / 2. Measuring an unknown solution. The absorbance of an unknown solution is measured as 0.62, and, using the calibration line, we calculate the confidence interval, 44 ± 3, for the concentration of the unknown solution.

2.1.5 / 3. Linearity range. The linearity of calibration is analysed using residuals and correlation coefficients.

2.2.1 / 4. Calibration result. Calculation of the confidence interval of an unknown solution using Excel.

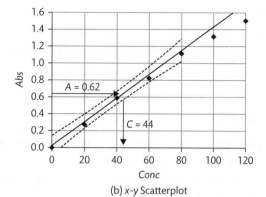

	A	B
1	Conc, C	Abs, A
2	mg/L	
3	0	0.00
4	20	0.27
5	40	0.58
6	60	0.82
7	80	1.12
8	100	1.31
9	120	1.50

(a) Calibration data

(b) x-y Scatterplot

Fig 7.1 Spectrophotometer calibration

Case study: **Toxicity assays / 1. Comparative assays (overview)**

Two types of assay, A and B, have been used to measure the percentage of cell deaths (mortality), Fig 7.2, due to exposure to different concentrations, C mM, (also entered as its logarithm, $\log(C)$) of an antibacterial agent. Each assay was tested twice, and we test for a nonparametric *correlation* between pairs of $A1$, $B1$, and $\log(C)$, etc. and we can look for numerical *agreement* between A and B.

4.1.1 / 2. Correlation. Develops measures of *correlation* between pairs of variables.

4.4.2 / 3. Agreement. Develops measures of *agreement* between pairs of variables.

4.4.3 / 4. Multiple comparisons. Measures agreement and correlation between *more than two* variables.

	A	B	C	D	E	F
1	Conc	Log(C)	A1	A2	B1	B2
2	0	-2.0	2.0	3.5	2.0	4.0
3	0.1	-1.0	25.5	18.5	15.0	16.0
4	1	0.0	37.5	40.0	36.0	50.5
5	5	0.7	57.0	62.0	33.0	54.5
6	10	1.0	69.5	55.0	35.0	65.0
7	20	1.3	78.7	62.5	57.5	75.5

(a) Experimental data

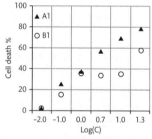

(b) Line graph
(not for interval x-y data)

Fig 7.2 Comparison of assay results

7.1.2 **Analytical options**

We list below *some* of the most common analyses used for related data, together with links for further information. The statistics of linear regression and correlation are developed in depth in Chapter 2, and Chapter 4 develops the association and agreement between variables.

Describing data

- Graphical plots: $x-y$ scatterplots (Note that the 'line' plot in Excel uses a *categorical* x-axis and is not suitable for $x-y$ data), residuals, and Bland–Altman plots (7.1.3).
- Numerical statistics: Slope and intercept of straight line. Correlation coefficients.

Tests / measurements:

- **Nonparametric correlation** to test whether one variable increases *in step with* the other: Spearman's rho and Kendall's tau-b (4.1.2)
- **Linear correlation** to test whether there is a *linear* relationship between the variables: Pearson's correlation coefficient (2.1.3, 4.1.1).
- **Straight line analysis**: See Chapter 2 for linear regression analysis of $x-y$ data and use of Excel, slope and intercept (2.1.1), errors and uncertainty (Section 2.2).
- **Calibration** calculations based on a straight line: Calculations of an unknown x-value (2.2.1), exact x/y intercepts (including standard additions) (2.2.2)
- **Nonlinear** regression: Linearization techniques (Section 2.3), nonlinear regression (2.4.3, Section 7.2).
- **Binary** regression: See 8.3.4.
- **Agreement** and **reproducibility**: Section 4.4 develops relevant methods including Bland–Altman plot for interval data, Kappa for ordinal data, t-test and Wilcoxon test, and for more than two variables, Kendall's coefficient of concordance is a nonparametric measure of correlation and Friedman's test measures rank differences.
- **Association** between categorical variables: Crosstabulation and chi-squared (8.2.4).

7.1.3 Describing the data

The most direct method of presenting the relationship between two numeric values is with an '$x-y$ scatterplot', together with a best-fit trendline and error bars if required. The example in Fig 7.1(b) uses the *Abs/Conc* data, but the trendline shown is based only on the first five points (as explained in 7.1.5 below).

The 'line' graph in Excel is *not* generally appropriate for $x-y$ data, because the x-axis is for *categorical* data. For example, Fig 7.2(b) uses a 'line' graph of *A1*, with each point 'labelled' by the value of $\log(C)$, but it is clear that the x-axis values are not *proportional* to $\log(C)$.

Very often, the important information can be hidden in a simple $x-y$ plot, and Fig 7.3 illustrates two ways in which slight, but important, differences can be highlighted.

Fig 7.3(a) is a plot of *residuals* for the data in Fig 7.1(a), showing the differences between each data point and the best-fit straight line. This plot shows a deviation of the two upper points, away from the direction of the lower portion of the graph, and in the calibration calculations below we decide to use only the lower 'linear' five data points.

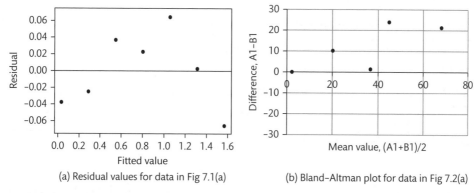

(a) Residual values for data in Fig 7.1(a) (b) Bland–Altman plot for data in Fig 7.2(a)

Fig 7.3 Differentiated values

Fig 7.3(b) is a Bland–Altman plot that gives the *differences* between the values of *A1* and *B1* plotted as a function of their average value, in which it can be seen that *A1* is generally scoring higher than *B1*.

7.1.4 Correlation

The parametric statistics of correlation between two interval variables (bivariate) are developed in 2.1.3 and 4.1.1, with the statistics of nonparametric correlation being developed in 4.1.2. The basic calculations can be accessed in SPSS and Minitab through:

SPSS > Analyze > Correlate > Bivariate...   **Minitab > Stat > Basic Statistics > Correlation...**
Variables: y x **Variables:** y x
☑ **Pearson** ☑ **Kendall's tau-b** ☑ **Spearman's**   ☑ **Display p-values**

SPSS provides the correlation coefficients and *p*-values for Pearson's linear correlation coefficient, r, and also for the nonparametric values of Kendall's tau, τ, and Spearman's rho, ρ. Minitab 16 gives the Pearson's r results directly, and Spearman's rho can be calculated using:

Minitab > Stat > Tables > Cross Tabulation and Chi-Square...
For rows: y **For columns:** x
> Other Stats...: ☑ **Correlation coefficients for ordinal categories**

In Excel it is possible to use the functions CORREL(), PEARSON(), or 'Correlation' in Data Analysis Tools to calculate the linear correlation coefficient. The derivation of the *p*-value in Excel is given in 2.1.3.

Table 7.1 gives results for some of the pairs of variables in Figs 7.1 and 7.2. The calculated correlation coefficients, r, measure the *strength* of the correlation, and the *p*-values give the hypothesis test significances for the *presence* (or not) of correlation.

Table 7.1 Selected correlation statistics for data from Figs 7.1 and 7.2

Data pair	Pearson's r		Spearman's ρ		Kendall's τ–b	
	r	p	ρ	p	τ	p
Abs/Conc	0.997	0.000	1.000	0.000	1.000	0.000
A1/A2	0.954	0.003	0.943	0.005	0.867	0.015
A1/B1	0.918	0.010	0.829	0.042	0.733	0.039

Note that the rounding of values implies that 1.000 is '≥ 0.9995' and 0.000 is '<0.0005'.

For the calibration data, *Abs/Conc*, the *p*-value is not relevant because it is already known that the variables are strongly correlated, but the *strength* of the linear correlation, $r = 0.997$, is an important measure for the quality of the *calibration* line (see 7.1.5). The values for *A1/A2* arise from analyses using the same method and provide a measure of the *reproducibility* of assay *A*, and those for *A1/B1* provide a measure of the *agreement* between the different assays *A* and *B*.

The correlation between two variables may be dependent on their joint correlation with a third variable, and it is possible to take this into account by calculating *partial* correlation coefficients. This analysis is developed in 4.1.4.

7.1.5 Linear regression and calibration

The following case study uses a straight line calibration to calculate the confidence interval of the concentration of an unknown sample.

Calibration uncertainty 2: Minitab and SPSS analysis. See also 2.2.1. Scan here to watch the video or find it via www. oxfordtextbooks. co.uk/orc/ currell/

Case study: Spectrophotometer calibration / 2. Measuring an unknown sample

—continued from 7.1.1, leading to 2.1.5

We use the calibration data that is given in columns A and B of Fig 7.1 to calculate the concentration of an unknown solution. based on the analysis developed in 2.2.1. Three replicate measurements of the unknown solution give an average absorbance of 0.62.

The basic regression calculations in Minitab and SPSS are accessed through:

Minitab > Stat > Regression > Regression > Fit Regression Model...
Responses: *Abs*
Continuous predictors: *Conc*
> Graphs... ☑ Residuals versus fits
→ Output: Fig 7.3(a) and Fig 7.4

SPSS > Analyze > Regression > Linear...
Dependent: *Abs*
Independent(s): *Conc*
→ Output: Gives the same values as in Fig 7.4

```
Abs = - 0.0006 + 0.0140 Conc

Predictor        Coef    SE Coef       T      P
Constant      -0.00060    0.01375   -0.04  0.968
Conc         0.0139850  0.0002808   49.81  0.000
S = 0.0177567   R-Sq = 99.9%   R-Sq(adj) = 99.8%

Analysis of Variance
Source           DF        SS        MS       F      P
Regression        1   0.78232   0.78232 2481.20  0.000
Residual Error    3   0.00095   0.00032
Total             4   0.78327
```

Fig 7.4 Linear regression output values in Minitab or SPSS

It is useful to look at the printout of the residuals, Fig 7.3(a), as this give a good indication of the *linearity* of the calibration line, and in this example we see a straight line portion up to a concentration of about 80 mg/L and then a distinct curvature. In addition, for a good calibration line we would expect values of linear correlation, *r*, greater than 0.999, but for *Abs/Conc* in Table 7.1 the correlation coefficient *r* = 0.997. However, if we only use the values up to *Conc* = 80, we now find good linear correlation with *r* = 0.999—see also the analysis in 2.1.5.

As it appears that the calibration data is curved above *Conc* = 80, we decide to use only the first five calibration values and, repeating the regression calculation, we obtain the results in Fig 7.4, which gives the slope, *m* = 0.0140, and intercept, *c* = −0.0006, of the best-fit straight line. If we were confident that the zero adjustment of the instrument was exactly set so that 0.0 absorbance corresponded to 0.0 concentration, then we could *force* the best-fit line through the origin giving zero intercept, *c*.

The confidence deviations, *Cd*, of slope and intercept can be calculated by multiplying the standard error (SE Coef in Fig 7.4) by the relevant *t*-value (3.18) for *n*−2 degrees of freedom (where *n* = 5 is the number of data pairs in this example). This then gives the confidence interval (Eqn 1.23) for the slope.

$$m = 0.0140 \pm 0.00028 \times 3.18 = 0.0140 \pm 0.0009$$

The relevance of the R^2 and R^2 (adj) values (99.9% and 99.8%) as measures of the 'quality of fit' are explained in 2.1.3 and 4.4.1.

The unknown solution has an absorbance, *Abs* = 0.62, which allows us to calculate the unknown concentration by rearranging Eqn 2.1:

$$Conc = (0.62 - (-0.0006)) / 0.0140 = 44.3$$

We can calculate the confidence interval for this result by using the Excel method developed in 2.2.1. However, the measured absorbance is close to the *middle* of the calibration range and it is possible to use the approximate method. The first step is to calculate the standard error of regression, SE_{REG} from the sum of squares of the residuals from Fig 7.4, using Eqn 2.14:

$$SE_{REG} = \sqrt{\frac{SS_{RESID}}{n-2}} = \sqrt{\frac{0.00095}{5-2}} = 0.0178$$

Then, using Eqn 2.19, we calculate the standard uncertainty, u_x, in the x-intercept (where k is the number of replicates in the measurement of the unknown value):

$$u_x \approx \frac{SE_{REG}}{m} \times \sqrt{\left\{\frac{1}{k}+\frac{1}{n}\right\}} = \frac{0.0178}{0.0140} \times \sqrt{\frac{1}{3}+\frac{1}{5}} = 0.928$$

The confidence deviation is then calculated using Eqn 2.21 with a t-value calculated for $n-2$ degrees of freedom:

$Cd = 0.928 \times 3.18 = 2.95$

This then gives the confidence interval for the unknown concentration (to 1 dp) as:

$Conc = 44.3 \pm 3.0$ (95% CI)

Note that if the unknown absorbance value was towards the ends of the calibration range, or beyond, then the uncertainty would increase and the full uncertainty calculation in Excel (2.2.1) would be required.

7.1.6 Agreement between results

It is not appropriate to talk about agreement between the values of absorbance and concentration in Fig 7.1, because they are two different *quantities*. However, it is relevant to consider the agreement between the assays' results of Fig 7.2 because they are aiming to record the *same values*.

The analysis of the data in Fig 7.2 is developed in detail in 4.4.1, 4.4.2, and 4.4.3, together with the underlying statistics. In overview, for the assessment of the agreement between two variables, we can use:

● tests for *correlation* which should show strong significance and a *low* p-value.

● linear regression and correlation between the variables to assess whether the change in one is *proportional to* or *equal to* the change in the other. A slope of $m = 1.00$ in regression indicates that the *changes* are *equal* to each other.

● a *paired* t-test or Wilcoxon test to test for a zero *overall* difference between values, which should give a *high* p-value.

● Bland–Altman plot to display the *differences* on a scatterplot as a function of the *averages*, which should give near zero differences across all values.

For the two assays *A1* and *B1* in Fig 7.2, Spearman's correlation gives $p = 0.042$ which shows that there is correlation between the assays, but the slope of linear regression, $m = 1.48$, (Table 4.5) shows a difference in the value response, which is confirmed by the Wilcoxon test with $p = 0.043$ indicating a significant overall difference between the assays. These results are also consistent with the Bland–Altman plot in Fig 7.3(b) which shows that the assays are *correlated*, but that they do not give the same *values*, with *A1* giving increasingly higher values than *B1*.

For comparison between *more than two* variables, we see the use of Kendall's Coefficient of Concordance (4.4.3) in combining the multiple correlations between variables to produce an overall statistical value. Friedman's repeated measures analysis also provides a complimentary analysis for the difference in values between multiple variables.

McNemar's test and Cochran's Q (4.4.5) can be used to test for the agreement between two or more samples of binary data.

7.2 Nonlinear relationships

In this section, we consider the analysis of x–y relationships that are expected to follow mathematical relationships other than that of the simple straight line (7.1). The following section (7.3) deals with more *general* experimental x–y data that cannot be modelled directly using these methods, including periodic data.

7.2.1 Example data

Different scientific mechanisms lead to data defined by a wide variety of mathematical models:

- Exponential relationships (2.3.3), e.g. radioactive decay and the growth and decay of bacterial populations.
- Power and reciprocal relationships (2.3.1, 2.3.2), e.g. thermal emission of radiation and the gas laws.
- Complex relationships (2.3.6), e.g. the Michaelis–Menten and Arrhenius equations.

Fig 7.5 illustrates different approaches to nonlinear data which we develop through the case studies:

Exponential decay (7.2.3) which uses radioactive decay as an example for the different methods of analysis that are available to describe any form of exponential growth or decay.

Rowing (7.2.4) which determines the (mathematical) power relationship between the power, W watts, and pace, P seconds per metre, of competitive rowers.

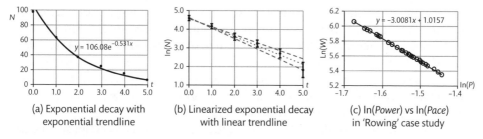

(a) Exponential decay with exponential trendline

(b) Linearized exponential decay with linear trendline

(c) ln(*Power*) vs ln(*Pace*) in 'Rowing' case study

Fig 7.5 Graphical examples of nonlinear relationships

7.2.2 **Analytical options**

Describing data

- Graphical plots: x–y scatterplots using nonlinear trendlines (e.g. Fig 7.5(a)). A 'line' graph can be used to plot *interval y*-data against *categorical x*-data (e.g. 7.2(b)). Convolutes can be used to produce localized 'best-fit' lines to nonlinear data (7.3.5 and 7.3.6).

Tests/Measurements:

- **Transform to a linear relationship**

  The basic techniques of linearization should be reviewed first by reference to Section 2.3, giving the main options for transforming one or both variables such that a linear relationship can be obtained. The mathematics of linear regression is then used to analyse the relationship.

- **Fit to a mathematical model**

  Section 2.4 develops an iterative approach to nonlinear regression using the Excel add-in 'Solver', which allows the use of different data distributions (e.g. Poisson) and optimization criteria (e.g. maximum likelihood estimation). The 'Exponential decay' case study in 7.2.3 demonstrates this analysis performed for nonlinear regression by Minitab and SPSS, and in 3.4.7 it demonstrates the use of the GsdLMmodel.

- **Derive the mathematical model**

  The experimental data may be used to derive the model itself, as a way of understanding the science of the relationship (7.2.4, 7.2.5).

An important point to check when using a nonlinear analysis is that the criteria for the analysis are still met, i.e. the standard process of linear regression assumes that the uncertainties in the data show a normal distribution and that the variances are the same throughout the data set. We use the 'Exponential decay' case study to explore these issues, with the use of *weighting* in 2.2.4 for linearized data, and the iterative process of Solver in 2.4.3 for handling different underlying statistical distributions.

Nonlinear regression (Minitab): Analysis for Fig 7.6(a). See also 2.4.3. Scan here to watch the video or find it via www.oxfordtextbooks.co.uk/orc/currell/

7.2.3 **Iterative nonlinear regression**

The common approach to nonlinear data is to use one of the linearization techniques developed in Section 2.3. These transform the data into a linear relationship in which it is possible to use the statistics of linear regression to *calculate* the constants (slope and intercept) of the best-fit straight line. In this section, we aim to fit a *nonlinear* mathematical model to the data. However it is not always possible to calculate the coefficients of a complex model directly, and it is sometimes necessary to use a process of *iteration*, which works through repetitive cycles of 'trial and error'.

In 2.4.3 we use the Excel add-in 'Solver' to introduce the *iteration* process for an exponential decay, and in this section we demonstrate the use of Minitab and SPSS in using iteration to perform a nonlinear regression, which

1. assumes that the best-fit line can be described by a specific mathematical relationship, e.g. an exponential, $y = Ae^{Bx}$,

2. starts with a reasonable *guess* of the values of the constants, i.e. A and B, that describe that relationship, then

3. measures the *sum of squares of the residuals*, SS_{RES}, between the values predicted by the model and the experimental values, and then

4. adjusts the values of the constants, A and B, to minimize the value of SS_{RES}.

Nonlinear regression (SPSS): Analysis for Fig 7.6(b). See also 2.4.3. Scan here to watch the video or find it via www.oxford textbooks.co.uk/ orc/currell/

There are some limitations with iterative processes, particularly with more complex systems.

For example, when trying to find the *minimum* value in a test statistic, the iteration might get stuck with different parameter values that give a *local* minimum value. Sometimes this problem can be overcome by starting the iteration with different initial values.

The radioactive decay calculation below is used as an example representative of the many forms of exponential growth and decay.

Case study: **Exponential decay / 6. Nonlinear regression using Minitab and SPSS**

—continued from 2.3.4, 2.4.3, and 3.4.7

Nuclear radiation occurs when atomic nuclei decay from one state to another. Each nucleus has the same probability of decay at any time, with the result that the intensity of radiation decreases as the number of original nuclei falls. The rate of decay is defined by the time it takes for half of the atoms to decay—the half-life, $T_{1/2}$ (2.3.3).

Fig 7.5(a) plots the exponential decay of the radioactive count, N, as a function of time, t, using the data from Fig 2.20.

Using the data in Fig 2.20, it is necessary, in both Minitab and SPSS, to define the mathematical relationship, either by entering the equation directly or by using the menu options to select from example relationships.

Minitab > Stat > Regression > Nonlinear Regression...

Response: N

There are *three* options for entering the relationship to be fitted to the data:

> Use Catalog...: Enables the selection of one of a menu of standard relationships.

> Use Calculator...: Enables the use of functions to create the relationship.

or **Edit directly:** We can use this to enter: A*exp(B*t)

It is necessary to enter *starting* values for the constants.

> Parameters...: Enter guesses for starting values of A and B for the iteration,

e.g. A = 50, B= -0.5 (if you have no idea, try just zero values!)

→ Output: Fig 7.6(a)

The process for SPSS is very similar:

SPSS > Analyze > Regression > Nonlinear...

Dependent: N

Model Expression: A*EXP(B*t)–select functions and operators to help.

> Parameters...: It is necessary to enter *starting* values for the constants

e.g. A(50), B(−0.5)(as for Minitab)

→ Output: Fig 7.6(b)

```
Starting Values for Parameters
Parameter  Value
A             50
B           -0.5

Equation
N = 97.8601 * exp(-0.476281 * t)

Parameter Estimates
Parameter  Estimate  SE Estimate
A          97.8601    2.11541
B          -0.4763    0.01916

Summary
Iterations        6
Final SSE   20.6051
DFE               4
MSE         5.15127
```

Iteration history[b]

Iteration number[a]	Residual sum of squares	Parameter A	Parameter B
1.0	3849.550	50.000	−.500
1.1	34.589	97.808	−.451
2.0	34.589	97.808	−.451
2.1	20.609	97.884	−.476
3.0	20.609	97.884	−.476
3.1	20.605	97.861	−.476
4.0	20.605	97.861	−.476
4.1	20.605	97.860	−.476
5.0	20.605	97.860	−.476
5.1	20.605	97.860	−.476

Derivatives are calculated numerically.

a. Major iteration number is displayed to the left of the decimal, and minor iteration number is to the right of the decimal.

b. Run stopped after 10 model evaluations and 5 derivative evaluations because the relative reduction between successive residual sums of squares is at most SSCON = 1.000E−008.

(a) Output from Minitab   (b) Output from SPSS

Fig 7.6 Nonlinear regression using Minitab and SPSS

The result from Minitab using six iterations and SPSS using ten model evaluations give the same derived equation:

$$N' = 97.86 \times e^{-0.476}$$

This agrees with the values calculated by Solver in 2.4.3. Fig 7.6(b) shows the iteration process for SPSS from the starting values of $No = 50$ and $k = −0.5$ to the final values of $No = 97.86$ (3 sf) and $k = −0.476$, when the iteration stops because successive iterations make no significant reduction on the residual sums of squares. There are only a few steps in this particular analysis because the starting value for k was chosen quite close to the true value, but the number of steps would obviously increase the further the iteration has to travel.

7.2.4 Deriving the mathematical model

The following case study now demonstrates the use of nonlinear regression to calculate the unknown *power* of a relationship between variables (2.3.5).

Deriving the model: Excel analysis for Fig 7.7. Scan here to watch the video or find it via www.oxford textbooks.co.uk/orc/currell/

Case study: Rowing performance/power and pace

Ten students, all with some years rowing experience, took part in an investigation into the effect of four different 'warm up' periods on their performance over a subsequent 2.0 km timed trial using rowing machines. The rowing machine for each student was calibrated using an initial trial to adjust the drag factor, in order to compensate for differences between students. Fig 7.7 gives a selection of results showing average power values, W watts, recorded as a function of the time, t seconds, taken to complete the 2.0 km course. The aim of this part of the analysis is to test the validity of the following relationship (assuming that it is independent of any warm up period):

$$W = 2.8 / P^3$$

where P is the pace measured in seconds per metre, and is given by P = t/2000.

We start by assuming that the relationship between power, W, and pace, P, is a general mathematical power equation given by:

$$W = A \times P^B$$

taking natural logs of both sides of the equation (2.3.2) gives:

$$\ln(W) = \ln(A) + B \times \ln(P)$$

and, if we plot ln(W) against ln(P), the slope will be the value of B.

The first step is to calculate in column E the values of *Pace*, using *Pace* = *Time*/2000, and then to transform both *Power* and *Pace* by taking the natural logs of columns C and E to produce columns of ln(W) and ln(P). We use Excel functions to perform the linear regression of ln(W) against ln(P), arriving at the values:

Slope, m = −3.008
Intercept, c = 1.016

The power, B, in the equation is given directly by the slope, m, giving B = −3.0

	A	B	C	D	E	F	G	H	I	J
1	Student	WarmUp	Power, W	Time	Pace, P	ln(W)	ln(P)			
2			watts	seconds	s/m					
3	9	0	259	442	0.221	5.557	-1.510		Slope =	-3.008
4	2	0	272	435	0.2175	5.606	-1.526		Intercept =	1.016
5	8	2	291	425	0.2125	5.673	-1.549			
6	6	0.5	226	463	0.2315	5.421	-1.463		B =	-3.008
7	10	1.25	268	437	0.2185	5.591	-1.521		A =	2.761
8	9	1.25	274	434	0.217	5.613	-1.528			

Fig 7.7 Random selection from 40 results for power and pace in rowing

The constant, A, in the equation is given by $\ln(A) =$ intercept, c, which can be rearranged to give:

$$A = e^c = \exp(1.016) = 2.76$$

The resultant equation:

$$W = 2.76 \times P^{-3} = 2.76 / P^3$$

agrees very closely, within experimental error, with the accepted equation.

General regression (Minitab): Analysis for Fig 7.8. See also 3.4.3. Scan here to watch the video or find it via www.oxfordtextbooks.co.uk/orc/currell/

7.2.5 General regression

We demonstrate the use of general regression analysis, including the Box–Cox transformation (5.4.7), by using the data in Fig 7.8 which shows the *numbers of occurrences* of two types of animal behaviours, *B1* and *B2*, observed *together* on different days under varying environmental conditions. We wish to investigate whether we can use regression to derive an approximate relationship between the number of *B1* behaviours and the number of *B2* behaviours.

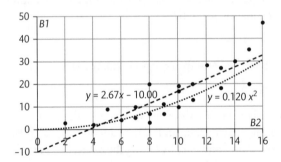

Fig 7.8 Frequency of observations of behaviours *B1* and *B2* occurring together

We start the analysis by performing a *simple linear regression* between the two variables, entering the *related* values of *B1* and *B2* into two columns in Minitab:

Minitab > Stat > Regression > Regression > Fit Regression Model...
Responses: *B1*　　　　　　　　　**Continuous predictors:** *B2*
> Graphs: Residuals for plots: ☑ *Standardized*
　　　　　☑ *Residuals versus fits* to assess the equality of variance

The regression derives the best-fit *straight line* as shown by the straight, dashed trendline in Fig 7.8 with the equation:

$$B1 = -10.0 + 2.67 \times B2$$

The analysis also plots the residuals shown in Fig 7.9(a).

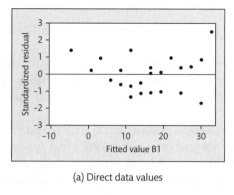 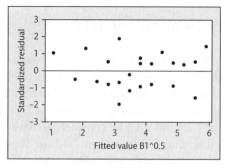

(a) Direct data values   (b) With Box–Cox transformation

Fig 7.9 Residuals against the fitted value for data from Fig 7.8

This interpretation is unsatisfactory for two reasons:

- the straight line with its '−10' intercept on the *B1* axis is not a good representation of the science as the true frequency *B1* cannot be negative, and
- the residuals in Fig 7.9(a) show increasing variance with value.

We repeat the regression, but now request a Box–Cox transformation:

> **Options...** Box–Cox transformation
> ◉ *Optimal λ*

Minitab calculates the 95% confidence interval for the optimum value of lamda, λ, (5.4.7) as being between −0.054 and 0.695, and then chooses to use the rounded value of $\lambda = 0.5$. This gives a square root transformation for *B1*, resulting in a best-fit regression equation:

$$(B1)^{0.5} = 0.362 + 0.346 \times B2$$

However, the value of the constant, 0.362 with a standard error of 0.441, is not significantly different from *zero*, so we can approximate this equation to

$$\sqrt{(B1)} \approx 0.346 \times B2$$

which, by squaring both sides of the equation, gives:

$$B1 \approx 0.120 \times (B2)^2$$

The transformed regression gives a better fit to the data, with

- the square relationship plotted on Fig 7.8 as the dotted curve, and
- the residuals showing a more even variance in Fig 7.9(b) across the range of fitted values.

However, this analysis is only indicative of a *possible* mathematical relationship, and a more direct approach, as in 7.2.4, would be required to test whether a simple power relationship was relevant in this case.

7.3 General x–y data

Sections 7.1 and 7.2 developed the analyses of *x–y* curves that are expected to follow a single mathematical model. In this section, we consider *x–y* relationships that cannot be analysed by using an *overall* mathematical description, but do contain sections or patterns of behaviour that can be analysed by various techniques.

In 7.3.4 and 7.3.5 we consider methods of drawing best-fit curves through the experimental data points, and in 7.3.7 we illustrate the use of autocorrelation and spectral analysis to identify any periodic patterns hidden within the data. We also introduce the techniques of convolution in 7.3.5 and 7.3.6 as localized best-fit methods of analysing varying data.

7.3.1 Example data

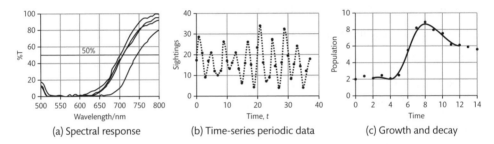

Fig 7.10 Examples of general *x–y* data

Fig 7.10(a) relates to the case study:

> **Ink analysis** (5.1.6) giving the spectral responses of four black inks from different sources. The problem is to identify a forensic characteristic that can be used to differentiate the inks.

Fig 7.10(b) relates to the case study:

> **Porpoise sightings** (7.3.7), an example of 'time-series' data in which there appears to be *periodically* varying components. The dashed line is not a best-fit line, but is a smoothed line 'joining the dots', which is included as a visual aid.

Fig 7.10(c) relates to the case study:

> **Bacterial growth** (5.2.2, 7.3.5), an example of time-dependent data, recording sections of growth and decay. The localized line of best-fit was created using the smoothing convolute developed in 7.3.5.

7.3.2 **Analytical options**

Describing data

- Graphical plots: x–y scatterplots with a best-fit line. 7.3.4 identifies the problems of using polynomials for a best-fit line and 7.3.5 develops the use of smoothing convolutes.

Tests / Measurements:

- **Identify relevant characteristics**: The human brain is one of the best analytical tools for picking out patterns in graphs, giving another reason for displaying the data graphically.
- **Isolate specific characteristics**: The data recorded by the exploratory study may contain superfluous data outside the range required for analysis, and it is sometimes necessary to identify just the specific sections or characteristics that may address the scientific objectives of the investigation (7.3.3).
- **Transform to a linear relationship**: If the whole data set can be described by a specific mathematical relationship (e.g. exponential decay), then it may be possible to use linearization techniques (Section 2.3) or nonlinear analysis (7.2.3).
- **Identify periodic components**: In time series data we often see the influence of periodic factors. These can be analysed using autocorrelation techniques (7.3.7).

7.3.3 **Identifying relevant analytical characteristics**

The data produced by an exploratory investigation often asks more questions than it answers. The first step is to produce a graphical presentation of the data that can highlight aspects worthy of more detailed analysis. In 7.3.4 we consider fitting an overall polynomial curve to smooth out the random variations in the data, but then, in 7.3.5, we use localized polynomial curves to calculate *best-fit values* for the data which can be used to draw the best-fit curve in 7.10(c). In 7.3.6 we use similar convolutes to *differentiate* the curve and plot its slope, highlighting such features as maxima and minima and points of maximum slope.

The graphs given in Fig 7.10 give examples of identifying different analytical features that are capable of statistical analysis:

(a) In distinguishing between the spectral curves of the four inks, there is very little difference in transmission over the main visible spectrum (they all appear black), but there are differences in the wavelengths at which the lines pass the 50% transmission mark at the high wavelength 'cut-off'. The case study (5.1.6) uses different methods to measure and test for these differences.

(b) The graph appears to show strong *periodic patterns*, which we can analyse using autocorrelation and spectral analysis (7.3.7).

(c) The portion of scientific interest is the bacterial growth period in the middle section of the curve. Initially in 5.2.2, the case study differentiates between different responses by comparing the *slopes* of this growth, and then in 7.3.5, we use the best-fit values from the smoothing convolute to estimate the confidence interval of the *difference between maximum and minimum* values.

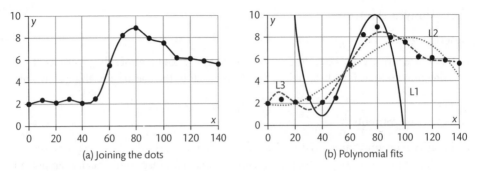

Fig 7.11 Options for fitting a curved line to a set of data points

7.3.4 Describing the data

The data points in Fig 7.11(a) are based on the data in rows 1 and 2 of Fig 7.13, and it is natural to want to 'improve' the visual quality of the graph by adding a best-fit curve. As a starting point, we could just 'join the dots' by using the *smoothed* line in Excel, but this is not generally appropriate because it implies that each point is 100% accurate, when, in fact, we know that the 'dots' should appear randomly on *either side* of a best-fit curve.

We can try fitting a curve defined by a polynomial equation as in Eqn 7.1

$$y = b_0 + b_1 x + b_2 x^2 + b_3 x^3 + \cdots + b_n x^n \qquad (7.1)$$

The maximum value of n in the polynomial in Eqn 7.1 is called the *degree* of the polynomial. A polynomial with degree $n = 2$ is called a quadratic equation, $n = 3$ gives a cubic, $n = 4$ a quartic, and $n = 5$ a quintic, etc. The number of maximum and minimum values that can appear in the graph is equal to $n-1$, with a quadratic equation, $n = 2$, having *one* turning point in the graph (either a maximum or a minimum), and a cubic equation, $n = 3$, having *two* turning points in the graphs (see lines L1 and L2 in Fig 7.11(b)).

A polynomial of degree n can make an *exact* fit through $n + 1$ points, i.e. it can pass through every point. For example, a straight line is a polynomial with degree 1, and will fit exactly through two points. Similarly, a quadratic equation can make a line fit *exactly* through $2 + 1 = 3$ data points. However, trying to fit a polynomial of degree n through *more* than $n + 1$ points will require a 'best-fit' curve, which does not necessarily pass through any point exactly.

The use of polynomials to fit experimental data is illustrated in Fig 7.11(b).

- **L1** (solid line) is a best fit for a *cubic* equation, but only for the *six points* $x = 40$ to 90. Although it produces a reasonable fit for this limited range of data, it clearly cannot fit the rest of the data.

- **L2** (dotted line) uses another best-fit *cubic* equation, but for *all* of the data points. The need to include all the data dramatically reduces the overall goodness of fit.

- **L3** (dashed line) uses a polynomial of degree $n = 6$, which gives a closer overall *statistical* fit, but the oscillation between $x = 0$ and 40 does not have any equivalence in the real data.

The main problems in using increasingly complicated polynomials to fit experimental data points are that

- different sections of the real experimental data may be generated by different scientific processes, which cannot be matched by a single *overall* mathematical relationship, and that

- an artificially complex *mathematical* relationship is unlikely to represent the underlying simple *scientific* processes that are generating the data points.

In 7.3.5 we see how we can use 'convolutes' to achieve a compromise by fitting a polynomial to short ranges of data, but still derive a best-fit curve to the whole data set.

7.3.5 Smoothing convolutes

> ## Case study: **Bacterial growth / 4. Using smoothing convolutes**
> —continued from 5.2.2
>
> The data for this analysis is drawn from Fig 7.13, and is similar to that depicted in Fig 5.5. We wish to develop a localized best-fit curve to smooth experimental variations, from which we can also derive an 'averaged' best-fit value for each data point.
>   We then use the best-fit values to estimate the confidence interval between maximum and minimum values.

Using convolutes: Excel analysis for Figs 7.13 and 7.14. Scan here to watch the video or find it via www. oxfordtextbooks. co.uk/orc/ currell/

Instead of aiming to fit one *single* polynomial equation simultaneously to all of the data points, the process of convolution derives best-fit curves that cover the data over short *local* ranges. A small set of values (convolutes) are combined with a section of data values to produce a single derived 'best-fit' value, and then, *by moving* the convolute throughout the whole data set, it generates a new modified data set. The relationship between the original data set and the derived data set depends on the choice of convolute values, and we will consider two types:

- Smoothing convolutes to provide a best-fit curve to complex data,

- Differentiating convolutes (7.3.6) to calculate slopes and identify maxima, minima, etc.

In Fig 7.13 the y-values in row 2 give luminescence measurements of bacterial populations as a function of time, x, in row 1. We wish to produce a best-fit curve for this data.

As an illustration, we first fit a cubic polynomial in Fig 7.12(a), using a least squares fit with Solver (Section 2.4), to *only the five* data points around $x = 70$. The y-value of this polynomial at $x = 70$ is calculated to be $y = 8.03$, and this is the *best-fit y-value at this point*.

This process is then repeated using the five data points around $x = 80$ (see Fig 7.12(b)), and we get a best-fit y-value at $x = 80$ of $y = 8.72$. In Fig 12(c) the process is repeated again around $x = 90$, giving the best-fit $y = 8.26$.

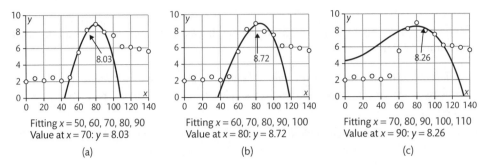

Fitting x = 50, 60, 70, 80, 90
Value at x = 70: y = 8.03

(a)

Fitting x = 60, 70, 80, 90, 100
Value at x = 80: y = 8.72

(b)

Fitting x = 70, 80, 90, 100, 110
Value at x = 90: y = 8.26

(c)

Fig 7.12 Fitting a cubic curve to five points at a time

	A	B	C	D	E	F	G	H	I	J	K	L	M	N	O	P	Q
1	x	0	10	20	30	40	50	60	70	80	90	100	110	120	130	140	
2	y	2.1	2.4	2.1	2.5	2.1	2.5	5.5	8.2	8.9	8.0	7.5	6.2	6.1	5.9	5.6	
3							↓	↓	↓	↓	↓						Sum
4	C1:						-3	12	17	12	-3						35
5																	
6	y'			2.34	2.23	2.08	2.90	5.40	8.02	8.76	8.27	7.23	6.48	5.99			
7	y'-y	(residuals)	0.24	-0.27	-0.02	0.40	-0.10	-0.18	-0.14	0.27	-0.27	0.28	-0.11				

Fig 7.13 Applying the convolute

The above procedure of calculating a best-fit polynomial for *every* point would be too time consuming for practical use, but we can obtain very similar results by using the process of convolution with convolute values published by Savitsky A and Golay M J E ('Smoothing and Differentiation of Data by Simplified Least Squares Procedures'. *Analytical Chemistry* 36(8): 1,627–1,639, 1964).

The derivation in Fig 7.13 uses a set of 'convoluting integers', {−3,12,17,12,−3}, held in G4:K4, together with the sum of their values, 35, in Q4. The calculation in I6 for the five data points around x = 70 multiplies each of the five data values by the corresponding convolute value in the same column, adding the five products, and dividing by the convolute sum:

$$[I6] = \frac{-3 \times 2.5 + 12 \times 5.5 + 17 \times 8.2 + 12 \times 8.9 - 3 \times 8.0}{35} = 8.02$$

This gives a best-estimate y-value at x = 70. We can then 'move' this convolute left and right to repeat the process to calculate the best-fit values for all values of x. In practice, we perform this operation in Excel by simply copying the equation along cells in row 6, provided that we have entered the appropriate '$' signs in the equation to lock the column values for the convolute values:

$$[I6] = (\$G4 * G2 + \$H4 * H2 + \$I4 * I2 + \$J4 * J2 + \$K4 * K2)/\$Q4$$

The convolute calculations give values similar to those calculated using the separate polynomial calculations. They represent the weighted average of the data at each point, taking into account the values of the five local points, and we use them to plot a best-fit curve as in Fig 7.10(c). The curve must stop two values from each end, otherwise the convolute would overlap the end of the raw data values.

We can also use the best-fit data for further calculations, but first we will need to estimate the *uncertainty* in the data by comparing the experimental values with the best-fit values. We do this by calculating the individual residual differences in row 7, and use Eqn 1.20 to estimate the experimental standard deviation as:

$$s = \sqrt{s^2} = \sqrt{\frac{\sum (x_i - x)^2}{(n-1)}} = 0.244$$

The normal calculation for the standard error is given by Eqn 1.21:

$$SE = \frac{s}{\sqrt{n}}$$

but, in this case, the $n = 5$ data points do not have an equal 'averaging' effect, and we must *weight* (2.2.4) their individual variances using their convolute values:

$$\left(\frac{-3}{35}\right)^2 + \left(\frac{12}{35}\right)^2 + \left(\frac{17}{35}\right)^2 + \left(\frac{12}{35}\right)^2 + \left(\frac{-3}{35}\right)^2 = 0.697$$

so that we can estimate the standard error:

$$SE = 0.697 \times 0.244 = 0.17$$

In calculating the difference between the maximum and minimum values, the best esti-mate maximum and minimum values in row 6 are 8.76 and 2.08 respectively, which give a *difference* between maximum and minimum of 6.68. Combining (1.4.4) the standard error in each measurement, gives a standard error for the *difference* of

$$0.17 \times \sqrt{2} = 0.24$$

The original experimental uncertainty calculation is based on 11 values, which gives a *t*-value of 2.23, which then gives a confidence deviation:

$$Cd = 2.23 \times 0.24 = 0.54$$

The confidence interval for the difference between maximum and minimum values can be rounded conservatively to give:

$$CI = 6.7 \pm 0.6$$

Other convolutes are possible. The simplest has a constant value convolute, $\{1,1,1,1,1\}/5$, which just adds up the five data points and divides by five. This is a simple average which is moved through the data–a *moving average*, also called *boxcar* averaging. You should note that the moving average *trendline* in Excel calculates the average of the data points at the *end* of the convolute range, and not, as we have done it, in the middle. This means that there is a 'delay' along the *x*-axis and the averaged trendline is displaced along the *x*-axis, for example by two units for a five element convolute.

Savitsky and Golay also published larger convolutes and higher order polynomials, e.g. for a convolute with seven values:

Cubic, $n = 3$:	−2	3	6	7	6	3	−2	Sum:	21
Quintic, $n = 5$:	5	−30	75	131	75	−30	5	Sum:	231

7.3.6 Differentiating convolutes

In addition to a smoothing function, convolutes can be used to derive the *slope* of the curve. For example, Fig 7.14(a) gives the results of a potentiometric titration, recording potential, E, versus the amount of added titrant, V, with the end point of the titration being the point of *steepest* slope.

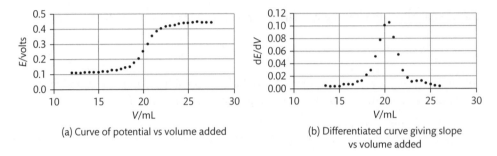

(a) Curve of potential vs volume added   (b) Differentiated curve giving slope vs volume added

Fig 7.14 Potentiometric titration

The convolute $\{-22, +67, +58, 0, -58, -67, +22\}/252$ fits a cubic curve to the data and provides the slope of the data at the central point with the assumption that there is unit distance between each point along the x-axis. Fig 7.14(b) gives the result of this convolution, showing that the point of maximum slope is for an added volume of about $V = 20.3$ mL.

In the calculations for Fig 7.14(b) we have also divided each value by the separation between data points, 0.5 mL, so that the curve becomes the *first* differential, dE/dV, of the original E vs V curve. It is also possible to use the same convolution again to obtain the *second* differential which will then show the top of the peak in Fig 7.14(b) as a zero slope crossing point on the axis, precisely identifying the end point of the titration.

7.3.7 Spectral analysis

Spectral analysis: SPSS and Minitab analyses for Figs 7.16 and 7.18. Scan here to watch the video or find it via www.oxfordtextbooks.co.uk/orc/currell/

Variables that are recorded over a period of time often show some form of periodic behaviour (e.g. diurnal, annual), and statistical analysis can help in identifying specific periods hidden within the data.

We start with the concept of *autocorrelation*, and, in order to understand this technique, it is useful to use the simple function F(0) in Fig 7.15 which has a varying value with a well-defined repetition period of $T = 4$ time units.

The process of autocorrelation measures the correlation of a function with a *time-shifted* version of itself. If we take the original function, F(0), and shift it by two units, called the *lag* time, we obtain the function F(2), and similarly for F(3), F(4), and F(8). The autocorrelation function, ACF, is then a plot of the correlation values against the lag time.

For a lag time = 2, autocorrelation measures the *correlation* between F(0) and F(2), getting a value $r = -1.0$ because F(2) behaves in exactly the *opposite* way to F(0), i.e. the peaks in F(0) correspond to the troughs in F(2). Autocorrelation repeats this process for different lag times. Function F(3) has a lag time of three units, and the peaks of F(3) fall in between the peaks and troughs of F(0) resulting in zero correlation, $r = 0.0$, but function F(4) with a lag time of four units now falls exactly *in step* with F(0) resulting in perfect correlation, with $r = 1.0$. The function F(8) with a lag time of eight units is again in perfect correlation with $r = 1.0$.

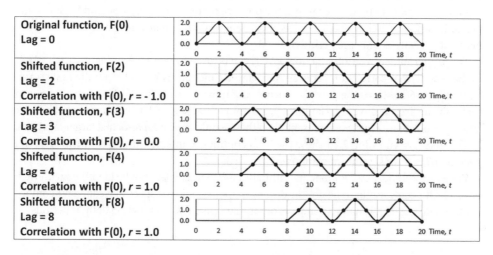

Fig 7.15 Autocorrelation

The autocorrelation function, ACF, for the function F(0), entered as *Data*, can be obtained in Minitab or SPSS via:

Minitab > Stat > Time Series >
    Autocorrelation...
        or
    Partial Autocorrelation...
Series: *Data*
◉ **Number of lags:** e.g. 10
→ Output: Similar to Fig 7.16

SPSS > Analyse > Forecasting >
    Autocorrelations...
Variables: *Data*
        ☑ *Autocorrelations*
        ☑ *Partial autocorrelations*
> Options
        Maximum number of lags: e.g. 10
→ Output: Fig 7.16

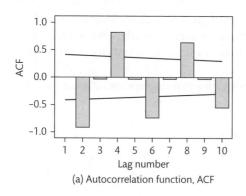

(a) Autocorrelation function, ACF

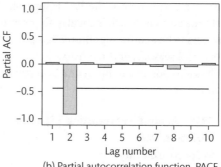

(b) Partial autocorrelation function, PACF

Fig 7.16 ACF and PACF using SPSS

We can see in Fig 7.16(a) that the autocorrelation function, ACF, confirms the interpretation in Fig 7.15 with *positive* correlation for lag times of four and eight, and *negative* correlation for lag times of two, six, and ten. We would like to use the ACF as an analytical tool to identify hidden periodic components in the data. However, the basic ACF can lead to possible *ambiguity* in that, for example, the positive correlation at lag = 8 could be due to the periodic variation in the data of $T = 4$ as above, but it could also be due to a different component with exactly double the period of $T = 8$.

The *partial* autocorrelation function, PACF, can be used to avoid the ambiguity in the ACF, in that any further correlation between the delayed functions is 'cut-off' after the first correlation. Fig 7.16(b) shows the partial autocorrelation for the function F(0), and it can be seen that only the first negative correlation appears uniquely at *Lag* = 42, which means that if other larger period components were present, then these would appear *separately* at larger lag times.

The lines drawn in Fig 7.16 show the 95% confidence interval limits, confirming that a period of $T = 4$ is a significant component in the data.

We see how frequency components within time-dependent data can be identified with ACF or PACF. It is now useful to be able to display these periodic (or spectral frequency) components within a spectral analysis that plots out the components as a function of their periods.

Case study: Porpoise sightings / Spectral analysis

Fig 7.17 records the number of sightings of porpoises in equal periods, recorded in each of four seasons over 14 years. We wish to analyse the variations to identify any significant patterns of variation.

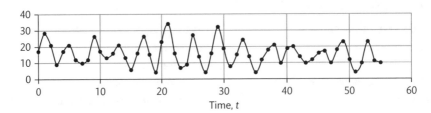

Fig 7.17 Frequency of porpoise sightings per quarter year

We can derive the partial autocorrelation function, PACF, given in Fig 7.18(a). The first negative correlation in the PACF suggests a significant frequency with a period of $2 \times 2 = 4$ time periods, which is equal to four seasons, i.e. an annual variation. However, there also appears to be another significant component hidden within the data with a period of about $2 \times 6 = 12$ seasons, or three years.

It is also possible to produce a 'spectral density' plot that displays the different frequency components as peaks against a horizontal period or frequency axis.

SPSS > Analyse > Forecasting > Spectral Analysis...
Variables: *Data*
☑ **Spectral density**
◉ **By period**
→ Output: Fig 7.18 (b)

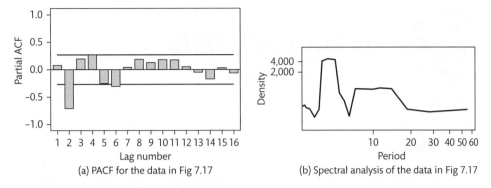

Fig 7.18 Partial autocorrelation and spectral analysis

The spectral analysis plot in Fig 7.18(b) shows the major component with a period of about four quarters, and also shows a less defined component with an approximate period of about ten quarters or two and a half years. The existence of frequency components in the data might suggest further investigation into whether the additional variations could be linked to any other factor over the 14-year period.

Spectral analysis can be a useful exploratory tool for identifying hidden periodic patterns within the data, and we have only given a simplified introduction to its use. The general analysis of 'time-series' data rapidly becomes more complex, and seeking the help of specific statistical advice is recommended.

Frequency data

Introduction

This chapter develops techniques which analyse the frequencies and probabilities of observed events. It first introduces methods for describing and managing the data either as individual observations or as tabulated frequencies, and then builds on the contingency table statistics developed in Sections 3.7 and 4.2. Finally, the content addresses the modelling of binary systems and analysing the probabilities of a system being in a particular state.

Section 8.1 presents analyses relevant to *single samples* of categorical data, e.g. data descriptions, 'goodness of fit' test.

Section 8.2 develops the statistics of the *contingency table* and the use of *cross-tabulation* to generate the table from individual observations.

Section 8.3 develops the analysis of binary systems using *logistic regression* and the description of probabilities using *ROC* plots.

8.1 Single variable

In this section we consider single samples of data for which we can perform a *frequency* analysis. We may have simple *categorical* data in which the individual observations already fall into specific categories. Alternatively, the raw data may consist of a *distribution* of many *interval* values, and the first step is to group the values into specific categories (binning), and then count the numbers (frequencies) of records that fall into each category in a table (tabulation). Once this is done it is possible to plot the distribution of frequencies on a frequency bar or column graph (Fig 8.2) or histogram (Fig 8.5). The actual processes of binning and tabulation are described in 8.1.7 and 8.1.8. The chi-squared statistics for testing the distribution of observations across categories is developed in 3.7.2.

8.1.1 Example data

Case study: **Chi-squared / 1. Genotypes (overview)**

This case study introduces examples of the chi-squared family of analyses. In Fig 8.1, column B records 200 observations in which the response variable is recorded as one of four genotypes, defined by the *text* categories, GTypeT: AB, Ab, aB, or ab. Column C gives the same values, but the categories are *numerically* coded, GTypeN: 1, 2, 3, and 4 respectively. It is sometimes necessary to recode text labels

	A	B	C	D	E	F	G	H	I	J	K	L	M	N
1	Type:	Nominal	Nominal		Category	Freq	Ratios		Interval			Bin		Freq
2	Variable:	*GTypeT*	*GTypeN*		*GTypeT*		(expected)		*BAlc*		Cutpoints	Range	BinNo	*BAlc*
3		AB	1		ab	8	1		78.8		75	< 75	1	1
4		AB	1		aB	39	3		82.7		76	75 - 76	2	3
5		Ab	2		Ab	28	3		79.8		77	76 - 77	3	6
6		ab	4		AB	125	9		78.9		78	77 - 78	4	22
7		AB	1		Total	200	16		77.4		79	78 - 79	5	30
8		aB	3						76.6		80	79 - 80	6	26
9		aB	3						78.8		81	80 - 81	7	27
10		AB	1						77.3		82	81 - 82	8	22
11		AB	1						86.3		83	82 - 83	9	13
12		AB	1						81		84	83 - 84	10	7
13		aB	3						79.4		85	84 - 85	11	1
14		ab	4						78.5		86	85 - 86	12	1
15		AB	1						76.6			> 86	13	1
16		AB	1						80.8				Total	160
17		aB	3						81					
162		AB	1						77.2					
163		AB	1											
202		Ab	2											

Fig 8.1 Typical data sets using a *frequency* analysis (Rows 18 to 161 and 164 to 201 are 'hidden' in the worksheet)

as numbers, or vice versa, to suit the input requirements of different software analyses. Column F gives the *tabulated* frequencies for the same data, in which the occurrences of each category are counted and recorded alongside the category label in column E. In this section, we use this data to demonstrate the use of frequency bar (or column) graphs (8.1.3), the one-way 'goodness of fit' chi-squared test (8.1.5), and tabulation (8.1.7).

> 3.7.2 / 2. One-way 'goodness of fit' test: Develops the underlying statistics of the one-way analysis.
>
> 3.9.3 / 3. Monte Carlo analysis: Uses a re-sampling technique to perform a 'goodness of fit' test when the expected frequencies are too low for the standard chi-squared analysis.

Case study: **Blood alcohol / 9. One sample analysis**

—introduced in 1. Introduction

In Fig 8.1, column I records 160 replicate measurements of blood alcohol level, *BAlc*, (mg of alcohol per 100ml of blood) from a source population with a mean of 80 and a standard deviation of 2.0. Column L defines possible value ranges (*categories* or *bins*), and column N records the tabulated numbers (or frequencies) of measurements that fall into each range.

The main difference between the genotype and blood alcohol data in Fig 8.1 is that the categories of *GTypeT* are defined by the *science involved*, and are few in number, but the categories (bins) used for *BAlc* are defined purely as a way of *describing* the statistical distribution of values. Data similar to *GTypeT* or *GTypeN* would be either *nominal* (in this example) or *ordinal* (e.g. assessment levels) but the data similar to *BAlc* would typically be *interval*.

Another form of 'frequency data' occurs where the measured response variable is *itself* a frequency. For example in radioactive decay (where the recorded activity can be measured as the 'number of counts per second'), or in the growth of a bacterial population (measured as the number of 'colony forming units', *cfu*). We see in 6.1.1 that we can treat higher frequency values as continuous *interval* data, and, in particular, we analyse radioactive decay as interval data in 2.3.4.

8.1.2 Analytical options

We list below *some* of the most common 'frequency' analyses, together with links for further information.

Describing data

- Graphical plots: Boxplot (1.1.2), bar chart, histogram (8.1.3), stem and leaf plot (Fig 6.3).
- Numerical statistics: Mean, median, standard deviation, confidence interval, etc. (6.1.3).

Frequency data (Minitab): Descriptives and graphs. Scan here to watch the video or find it via www.oxford textbooks. co.uk/orc/ currell/

Tests / measurements:

- Chi-squared 'goodness of fit test' tests whether the **distribution** between defined categories is different from a *test distribution* of frequencies (8.1.5, 3.7.2).
- Tabulation and binning of data are the processes of **counting** and **grouping** frequencies (8.1.7/8).
- The **normality** of a distribution of data values can be tested (8.1.6) using Anderson–Darling, Kolmogorov–Smirnov, and Ryan–Joiner (similar to Shapiro–Wilk) tests.
- The Kolmogorov–Smirnov test can be used to test whether a **distribution** of values is different from a *standard distribution* (normal, Poisson, binomial, or uniform) (8.1.6).
- Testing the **proportion** of data in specific categories (6.1.7 and Section 3.8).

Frequency data (SPSS): Descriptives and graphs. Scan here to watch the video or find it via www.oxford textbooks. co.uk/orc/ currell/

8.1.3 Describing categorical data

For categorical data, given as a list of *individual* categories, e.g. *GtypeT* in column A, we can use Minitab or SPSS to plot bar graphs directly and print out the tabulated frequencies for each category.

Minitab

Minitab > Graph > Bar Chart...
Bars represent: For a column of values select: ***Counts of unique values***
              For a single sample of values select: ***Simple***
Categorical variable: *GtypeT*
→ Output Fig 8.2(a)

The tabulated counts can be read from the bar chart or printed using:

Minitab > Stat > Tables > Tally Individual Variables... or **Descriptive Statistics...**
→ Output: Plot as in Fig 8.3(a)

SPSS

SPSS > Analyze > Descriptive statistics > Frequencies...
Variable(s): Identify categorical data, e.g. *GtypeT*
> Charts: ⊙-**Bar charts** (for categorical *GtypeT* data)
→ Output: Bar chart as in Fig 8.2(b) and tabulated frequencies in Fig 8.3(b).

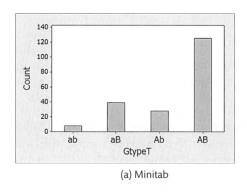

(a) Minitab

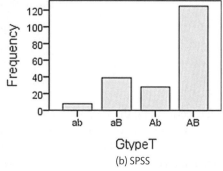

(b) SPSS

Fig 8.2 Frequency bar charts for data from Fig 8.1

Tally for Discrete Variables: GtypeT

GtypeT	Count
ab	8
aB	39
Ab	28
AB	125
N=	200

(a) Tally chart (Minitab)

GtypeT

		Frequency	Percent	Valid Percent	Cumulative Percent
Valid	ab	8	4.0	4.0	4.0
	aB	39	19.5	19.5	23.5
	Ab	28	14.0	14.0	37.5
	AB	125	62.5	62.5	100.0
	Total	200	100.0	100.0	

(b) Frequencies and percentages (SPSS)

Fig 8.3 Tabulated frequencies

The tabulated results in Fig 8.3 are useful when the data is presented as individual records (as in column B). If the data has already been tabulated as *frequencies* (as in column E), the same bar graphs can be obtained using:

Minitab

Minitab > Graph > Bar Chart...
Bars represent: For a column of values select: ***Values from a table***
　　　　　　　　For a single sample of values select: ***Simple***
Graph variables: *Freq*
Categorical variable: *GtypeT*
→ Output Fig 8.2(a)

SPSS

In SPSS it is necessary to *weight* categories, *ab*, *aB*, *Ab*, and *AB*, with the respective frequencies, by entering the values as in Fig 8.4 and using:

SPSS > Data > Weight Cases...
⊙-Weight cases by:
   Frequency variable: *Freq*

	GtypeT	Freq
1	ab	8.00
2	aB	39.00
3	Ab	28.00
4	AB	125.00

Fig 8.4 Data entry for 'weighting' data in SPSS

After performing the weighting operation, each row in the data will be treated as being repeated the number of times given by the *Freq* value in that row. For example, when the variable, *GtypeT*, is used it will be treated as 8 values of *ab*, 39 of *aB*, 28 of *Ab*, and 125 of *AB*, allowing the same set of instructions for a bar chart as above.

8.1.4 Editing histograms

The process of obtaining the histogram (1.3.1) of a sample of *interval* data (as in column I) involves *binning* and *tabulation*. Binning is a process of splitting up the data range into categories or *bins* (column L), and then tabulation is the process of counting how many data values fall into each bin. The *cutpoints* (column K) are the divisions that separate each of the bins. Binning and tabulation are considered as separate processes in 8.1.7 and 8.1.8 but we now see that they also become an integral part of editing a histogram when describing interval data.

Minitab

Minitab > Graph > Histogram...
For a single sample of values select: ***Simple***
Graph variables: *BAlc*
→ Output: Produces a graph that can be edited, Fig 8.5(a)
To change the 'binning':
> Right click on the *x*-axis, and select **Edit X Scale...**
> Select **Binning** tab:
Interval type: Check ⊙-Cutpoint
Interval definition: Check ⊙-Midpoint/Cutpoint positions and
   Enter values with spaces: 74 75 76 77 78 79 80 81 82 83 84 85 86

SPSS

SPSS > Analyze > Descriptive Statistics > Frequencies...

Variable(s): Identify data, e.g. *BAlc*

> Charts: ⊙**-Histograms** (for scale *BAlc* data)

☑**-Show normal curve on histogram**

→ Output: Produces a graph that can then be edited, Fig 8.5(b)

To change the 'binning':

> Double click on the graph to open the editing window

> Right click on the histogram, and **Select ▶ This Histogram Bar**

> Select **Binning** tab:

x-axis: Check ⊙**-Custom, ⊙-Interval width**: Enter 1.0

For ☑**-Custom value for anchor** Enter 80 or 80.5 to define position of cutpoints

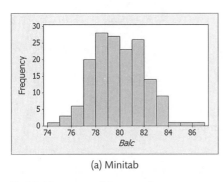

(a) Minitab

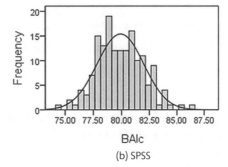

(b) SPSS

Fig 8.5 Histograms of the *same* data

Chi-squared goodness of fit (Minitab): Analysis for Fig 8.6. See also 3.7.2. Scan here to watch the video or find it via www.oxford textbooks. co.uk/orc/ currell/

The reason for the difference in the histogram shapes in Fig 8.5 (a) and (b) is a difference in the cutpoints and bin widths chosen to define the bins, but these can be edited as described in the derivations above.

8.1.5 Chi-squared 'goodness of fit' test

The chi-squared 'goodness of fit' test, which compares the observed and expected frequencies within a single set of categories, is developed in 3.7.2. We now use Minitab and SPSS to analyse the data in columns E and F, which is the same as in Fig 3.46

Minitab

Minitab > Stat > Tables > Chi-Squared Goodness-of-Fit Test (One Variable)...

Enter the data either as one of:

⊙**-Observed counts:** Column containing 8, 39, 28, 125 *or*

⊙**-Categorical data:** *GtypeT*

Under **Test** enter the frequency values for comparison as one of: ⊙*-Equal proportions* or

⊙**-Specific proportions:** 0.0625, 0.1875, 0.1875, 0.5625

Chi-squared goodness of fit (SPSS): Analysis for Fig 8.7. See also 3.7.2. Scan here to watch the video or find it via www.oxford textbooks. co.uk/orc/ currell/

(the proportions can be calculated: $1/(1+3+3+9) = 0.0625$, etc. or

◉-**Proportions defined by historical counts:** 1, 3, 3, 9

For either of the last two above options, select either: **Input Column** if the comparison values are in a column or **Input Constants** to enter values directly as shown in this example.

→ Output in Fig 8.6

Chi-Square Goodness-of-Fit Test for Categorical Variable: GtypeT

Category	Observed	Historical Counts	Test Proportion	Expected	Contribution to Chi-Sq
ab	8	1	0.0625	12.5	1.62000
aB	39	3	0.1875	37.5	0.06000
Ab	28	3	0.1875	37.5	2.40667
AB	125	9	0.5625	112.5	1.38889

N	N*	DF	Chi-Sq	P-Value
200	0	3	5.47556	0.140

Fig 8.6 Output from chi-squared 'goodness of fit' test (Minitab)

SPSS

SPSS > Analyze > Nonparametric Tests > One Sample...
Objective: ◉-**Customize analysis**
Fields: Select data, e.g. *GtypeT*
Settings: ◉-**Customize tests**
☑-**Compare observed probabilities to hypothesized (chi-squared)**
   Options: Select either
      ◉-All categories have equal proportions **or**
      ◉-Customize expected probability. Enter relative frequencies as in Fig 8.7(a).
→ Output in Fig 8.7(b)

Expected probabilities:

Category	Relative Frequency
ab	1
aB	3
Ab	3
AB	9

Hypothesis Test Summary

	Null Hypothesis	Test	Sig.	Decision
1	The categories of GtypeT occur with the specified probabilities.	One-Sample Chi-Square Test	.140	Retain the null hypothesis.

Asymptotic significances are displayed. The significance level is .05.

(a) Entering relative frequencies   (b) Summary output

Fig 8.7 Chi-squared 'goodness of fit' test (SPSS)

Alternatively in SPSS it is possible to use an earlier version of the test retained within the 'Legacy Dialogs' as described in 3.9.3.

The standard outputs from chi-squared tests typically present the observed frequencies, together with the calculated expected frequencies, the calculated chi-squared value, and *p*-value. The output from SPSS in Fig 8.7(b) is a summary with just the *p*-value, but the full information can be obtained by double-clicking on the summary output in the Output window. The reported statistics in Figs 8.6 and 8.7 agree with those calculated using Excel in Fig 3.46, and since $p = 0.140$ is more than the default significance of 0.05 we conclude that the observed distribution is not significantly different from the expected ratios and may have occurred by random chance.

8.1.6 Testing distributions

The use of Minitab and SPSS to test specifically for the *normality* of a set of experimental data is developed in Section 5.4. In addition, we can use the Kolmogorov–Smirnov test in SPSS for differences from a range of *standard* distributions:

Testing distributions (Minitab): Analysis for distributions and Fig 8.8(b). Scan here to watch the video or find it via www.oxford textbooks. co.uk/orc/ currell/

SPSS

> **SPSS > Analyze > Nonparametric Tests > One Sample...**
> **Objective: ⊙-Customize analysis**
> **Test Fields:** Select data (only scale values), e.g. *BAlc*
> **Settings: ⊙-Customize tests**
>   ☑-**Kolmogorov–Smirnov**
>   **> Options: Select** Normal, Uniform, Poisson, Exponential
> → Output in Fig 8.8(a)

It is possible to *define* a specific test distribution for comparison by entering values for the mean and standard deviation of the test distribution. However, it is common to use the default which allows the software to calculate the distribution that is *closest* to the data, and then perform the test for a significant difference from that calculated distribution.

Testing distributions (SPSS): Analysis for distributions and Fig 8.8(a). Scan here to watch the video or find it via www.oxford textbooks. co.uk/orc/ currell/

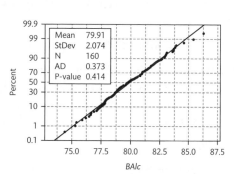

Hypothesis Test Summary

	Null Hypothesis	Test	Sig.	Decision
1	The distribution of BAlc is normal with mean 79.91 and standard deviation 2.07.	One-Sample Kolmogorov-Smirnov Test	.646	Retain the null hypothesis.
2	The distribution of BAlc is uniform with minimum 74.40 and maximum 86.30.	One-Sample Kolmogorov-Smirnov Test	.000	Reject the null hypothesis.
3	The distribution of BAlc is Poisson with mean 0.00.	One-Sample Kolmogorov-Smirnov Test	.	Unable to compute.
4	The distribution of BAlc is exponential with mean 79.91.	One-Sample Kolmogorov-Smirnov Test	.000	Reject the null hypothesis.

Asymptotic significances are displayed. The significance level is .05.

(a) SPSS distribution tests

(b) Minitab probability plot and test for normality

Fig 8.8 Distribution tests for *BAlc*

The output in Fig 8.8(a) shows that the data does *not* show a significant difference from a normal distribution, but that it is significantly different from a uniform or exponential distribution. It is unable to compare the *continuous* values of the *BAlc* data with a Poisson distribution because a Poisson distribution must have *integer* values.

It is also possible to select the Kolmogorov–Smirnov test directly from:

SPSS > Analyze > Nonparametric Tests > Legacy Dialogs...

SPSS also provides normality plots (5.4.5) and Kolmogorov–Smirnov and Shapiro–Wilk tests using:

SPSS > Analyze > Descriptive statistics > Explore...
Dependent List: *BAlc*
> Plots: ☑-Normality plots with tests
→ Output similar to Fig 8.8(b)

Minitab provides a specific test for normality:

Minitab > Stat > Basic Statistics > Normality Test
Variable: *BAlc*
Select a specific normality test, e.g. **⦿-Anderson–Darling**
→ Output in Fig 8.8(b)

Fig 8.8(b) shows a normality plot (5.4.5) in which the data is closely aligned with the neutral diagonal line which indicates that the data is close to a normal distribution. This is confirmed by the $p = 0.414$ result of the Anderson–Darling (AD) test in the data box, from which we conclude that there is no evidence that the distribution is *not* normal.

Minitab also provides a specific test for a Poisson distribution:

Minitab > Stat > Basic Statistics > Goodness-of-Fit Test for Poisson...

which uses a one-way chi-squared 'goodness of fit' test.

8.1.7 Tabulation of data

If the data has a limited number of discrete categories (e.g. *GTypeT* in Fig 8.1), we can *count* the occurrences within each category, and list these frequencies in a table. This is the process of *tabulation*. One of the oldest ways of recording a count is to use a *tally chart* where occurrences are recorded as strikes which build up in blocks of five, for example:

/̶N̶/ // = 7.

Minitab uses this terminology when counting the occurrences of specific values in data.

Minitab

Minitab > Stat > Tables > Tally Individual Variables...
***Variables:** GTypeT*
☑-Select required statistics and/or storage of output values in data table
→ Output: Returns the frequency values given in column F, and the count values can also be stored in new columns under Tally1 and Tally 2

SPSS can also produce tabulated results using

>   **SPSS > Analyze > Descriptive statistics > Frequencies...**

8.1.8 Binning

Binning is a process in which numeric values are identified as falling into specific *value ranges*, called *bins*. The *total* range of possible values is divided into a limited number of individual bins, defined by 'cut-points' which are the values between adjacent bins—see columns K and L in Fig 8.1.

Binning can be done interactively during the editing of a histogram (8.1.4) for both Minitab and SPSS.

SPSS also has a specific binning option:

>   **SPSS > Transform > Visual binning...**
>   **Variables to bin**: *BAlc*
>   **Binned variable:** Enter a name for a new variable to hold bin values, e.g. *BinNo*
>   **> Make Cutpoints:** Define the bin locations and sizes, e.g.
>   ***First Cutpoint Location:*** 75.0   ***Number of cutpoints:*** 12   ***Width:*** 1.0
>   → Output: Saves a new variable, *BinNo*, in the next available column such that each entry records the bin number (1 to 13 in this example) for each value in *BAlc*.

Having identified the bin number for each datum, it is possible to tabulate the values, and find the frequency within each bin, using the methods in 8.1.3.

Excel

We can bin frequency data in Excel, in Fig 8.1, by first using the functions, MAX() and MIN(), to find the maximum and minimum values, 74.4 and 86.3, in the *BAlc* data. We can then define (for example) 12 cutpoints, 75 to 86, in cells K3:K14, which will create 13 bins. The range for each bin is shown by text in cells L3:L15.

The binning process uses an 'array' function and the process must be carefully followed:

- Highlight the target cells N3:N15 to hold the output frequency values.

- Using the function FREQUENCY(), identify the *data range* I3:I162 and the *bin range* K3:K14.

- Holding down the Shift and Ctrl key *together*, press Enter.

Excel can also produce a histogram with the above data using data analysis tools.

8.2 Contingency tables

Section 8.1 developed the frequency statistics for categories defined by the *levels* of just one factor or variable, e.g. genotypes, blood alcohol levels. In this section, the categories are defined by the levels of *two* factors, which provide the rows and columns to create the cells of a 2-D contingency table. The statistics of the contingency table have been developed in 3.7.4 and then Sections 4.2, 4.3, and 4.4 developed tests/measures for *association* and *agreement* between these two factors.

8.2.1 **Example data**

	A	B	C	D	E	F	G	H	I	J	K	L
1	Improvement:	None	Some	Good	Excellent			*Sex*	*SexN*	*Improv*	*ImprovN*	*Freq*
2		*N* / 1	*S* / 2	*G* / 3	*E* / 3	Totals:		F	1	N	1	13
3	Female, F	13	18	159	72	262		F	1	S	2	18
4	Male, M	26	28	166	68	288		F	1	G	3	159
5	Totals:	39	46	325	140	550		F	1	E	4	72
6								M	2	N	1	26
7	Improvement:	*N+S*	*G+E*	Totals:				M	2	S	2	28
8	Female, F	31	231	262				M	2	G	3	166
9	Male, M	54	234	288				M	2	E	4	68
10	Totals:	85	465	550								

Fig 8.9 Case study: Association / 1. Bacteriophage treatment

Case study: **Association / 1. Bacteriophage treatment (overview)**

The contingency table, B3:E4, in Fig 8.9 shows the numbers of male and female patients who have received bacteriophage treatment and shown no, *N*, some, *S*, good, *G*, or excellent, *E*, improvement. The data in columns H to L contain the same information as in the table, with the sex of the patient identified both as a nominal (text) variable *Sex* and a numeric variable *SexN*. The improvement is given both by the nominal variable *Improv* and by the numeric variable *ImprovN*. We have included a numeric code for each variable because some analyses require the data presented in specific formats. In this section we use Minitab and SPSS to test for an association between the level of improvement and the sex of the patient.

> 3.7.4 / 2. Contingency table: Develops the underlying test statistics by testing for an association between the sex of children and their preferred study subjects.

	A	B	C	D	E	F	G	H	I	J	K	L
1	*Grp*	*Q1*	*Q2*	*Q3*	*Q4*		*G1+G2:*	*Q1\Q2*	N	Y		*Q6*
2	G2	N	N	Y	Y			N	24	19		0
3	G1	Y	Y	Y	Y			Y	7	16		-2
4	G2	N	N	N	N							-1
5	G2	Y	N	Y	Y		*G1:*	*Q1\Q2*	N	Y		1
6	G2	Y	N	N	N			N	11	16		-2
7	G1	N	N	N	N			Y	1	8		2
8	G1	N	Y	Y	N							2
9	G1	N	Y	N	N		*G2:*	*Q1\Q2*	N	Y		2
10	G1	N	N	Y	N			N	13	3		1
11	G1	N	Y	N	N			Y	6	8		1
12	G1	N	Y	N	Y							3
67	G2	Y	N	Y	Y							0

Fig 8.10 Case study: Forensic questionnaire / 2. Crosstabs and contingency table (Rows 13 to 66 are 'hidden' in the worksheet)

Case study: **Forensic questionnaire / 2. Crosstabs and contingency table**

—continued from 9.2.1, leading to 4.4.5 and 6.2.1

Fig 8.10 is derived from forensic questionnaire results in Fig 9.12, and gives the *binary* responses to four questions, Q1, Q2, Q3, and Q4, in a questionnaire on the interpretation of forensic evidence for two groups of people defined by G1 and G2. The contingency table in I2 : J3 records the number of specific Q1 / Q2 pairs in the whole data set, but the tables in I6 : J7 and I10 : J11 are *layered* to include only groups G1 and G2 respectively. Q6 gives a response on a −3 to +3 Likert scale, which is analysed in 8.2.7.

8.2.2 **Analytical options**

Describing data

- **Graphical plots**: Clustered and 3-D bar charts (8.2.3).
- **Numerical statistics**: Present the frequencies using the contingency table values.
- **Layering tables**: Differentiates between members of different groups (8.2.8).

Tests / Measurements:

- **Cross-tabulation** (8.2.4) is the process of creating a *contingency table* from lists of paired observations.
- **Contingency table** statistics (3.7.4, 8.2.4) provide tests for the *existence* of an association between factors, in addition to measures for *strength* of this association and *agreement* between values.
- Identify a **progressive** association between ordinal factors (8.2.5).
- **Data consolidation** (8.2.6) can be used to vary the detailed objective of the analysis.
- Due to **low count values** it may be necessary to combine factor levels together or to use a Monte Carlo analysis (8.2.7).
- 2×2 **binary** tables are effectively two *proportions* and are considered in 6.1.7 and 6.2.9.

The basic tests for the *existence* of an association in contingency tables are developed in 3.7.4 and Section 4.2:

- Pearson's chi-square, χ^2, and the likelihood ratio, G, test for an association between *nominal* factors, but they do not test for any sense of progression from one category to the next.
- Yates continuity correction is used when the degrees of freedom in the calculation is equal to 1.0.
- Fisher's exact test (4.2.3) provides the exact binomial test for association in a 2×2 table, which can also be viewed as a test for a difference in *proportion* between rows or columns.
- The linear by linear (Mantel–Haenszel) association (4.2.4) tests for a progressive *linear* association between the values of two *interval* variables, measured using Pearson's correlation coefficient.

Measures for the *strength* of association between nominal factors include:

- Phi and Cramer's V (4.3.3) take into account the number of data values involved.
- Goodman and Kruskal's Lambda (4.3.4), Gamma and Somer's D (4.3.5), Kendall's tau-b (4.3.5), Spearman's, ρ, and Pearson's r measure the strength of association of ordinal and interval variables.
- Eta, nominal-by-interval association, (4.3.6) measures the strength of association between nominal and interval factors.

Measures for *agreement* between nominal factors include:

- Kendall's coefficient of concordance (4.4.3) and Cohen's and Fleiss's kappa (4.4.4) measure the amount of *agreement* between related variables.
- McNemar's and Cochran's Q tests (4.4.5) measure the agreement between *binary* data values.

8.2.3 Describing the data

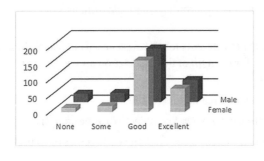

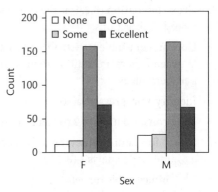

(a) 3-D bar chart (Excel)   (b) Clustered column graph (SPSS)

Fig 8.11 Plotting categorical data

The 3-D bar chart in Excel is a very convenient way of visualizing the frequency values in each category.

Excel > Insert > Column chart ▼ Select from the 3-D options
> Select data source:
Chart data range: B3:E4 (from Fig 8.9)
Edit each data series by entering series names: Male / Female
Edit horizontal labels by entering B1:E1 as the axis label range
→ Output: Fig 8.11(a)

Clustered column graphs also show the grouping of data within the categories:

Minitab

Minitab > Graph > Bar Chart...
Bars represent ▼ Values from a table
One column of values: *Cluster*
Graph variables: *Freq*
Categorical variables for grouping: *Sex Improv*
→ Output: Similar to Fig 8.11(b)

For **SPSS** it is necessary to *weight* each row by the *Freq* values, using the method given below in 8.2.4, and then:

SPSS > Graphs > Legacy Dialogs > Bar...
Clustered and then
> Define
  Category Axis: *Sex*
  Define Clusters by: *Improv*
→ Output: Fig 8.11(b)

Crosstabs and contingency tables (Minitab): Analyses for Figs 8.12/13/14/15. See also 3.7.4. Scan here to watch the video or find it via www.oxford textbooks.co.uk/orc/currell/

8.2.4 Contingency tables and cross-tabulation

We may already have the data in terms of frequency values within a contingency table. If so, it is possible to use Minitab for a basic chi-squared analysis by entering a table of *frequency values*, as in B3:E4 in Fig 8.9, *directly* into the worksheet, and using the selection 'Summarized data in a two-way table' in either of menu selections:

Minitab > Stat > Tables > Chi-Square Test for Association... or
Minitab > Stat > Tables > Cross Tabulation and Chi-Square...
and then: **Columns containing table**: Enter the columns for the table

The full range of analyses in SPSS is only available using *crosstabs* which require every observation to be listed separately. For example, for the data in Fig 8.9 with columns for *Sex*, *SexN*, *Improv*, *ImprovN*, it would require:

Crosstabs and contingency tables (SPSS): Analyses for Figs 8.12/13/14/15/16. See also 3.7.4. Scan here to watch the video or find it via www.oxfordtext books.co.uk/orc/currell/

13 rows with values 'F, 1, N, 1',
18 rows with   'F, 1, S, 2',
159 rows with   'F, 1, G, 3', etc.

The need for individual observations can be avoided in SPSS by using the process of 'weighting' to enter each combination only once, but identify the frequency with which each row is to be 'duplicated'. Each different combination of values, corresponding to one cell in the table, is entered in a separate row as in columns H to K and the frequencies for each particular combination cell are entered as *Freq* values in a separate column, L.

After running the instruction to weight cases:

SPSS > Data > Weight Cases...
⊙ **Weight cases by**
   Frequency variable: *Freq*

each row would be treated as though it was repeated the number of times given by the *Freq* value.

We now use cross-tabulation, or *crosstabs*, which is the process of counting the numbers of observation *pairs* that fall into each cell of a contingency table, and allows related data values, such as *Q1* and *Q2*, to be analysed using contingency table statistics.
In SPSS:

SPSS > Analyze > Descriptive statistics > Crosstabs ...

and Minitab

Minitab > Stat > Tables > Cross Tabulation and Chi-Square...

and then both:

Rows: Select the factor to define rows of the table. e.g. *Sex* or *SexN* or *Q1*

Columns: Select the factor to define columns of the table. e.g. *Improv* or *ImprovN* or *Q2*

Layer: If required it is possible to group the data into 'layers' using a further factor. e.g. *Grp*

The reason for entering a *numeric* value is to enable the factor to be treated as an ordinal value with a sense of *progression* from one category to the next (8.2.5). Having 'set up' the contingency table, the menu options then allow the choice of a variety of analysis options summarized above in 8.2.2.

SPSS:

> **Statistics:** Select specific analyses as required:

☑-**Chi-square**   ☑-**Correlation**   ☑-**Phi and Cramer's V**
☑-**Lambda**   ☑-**Gamma**   ☑-**Somer's d**
☑-**Kendall's tau-b**   ☑-**Kappa**   ☑-**McNemar**

> **Cells:** Check for required output, e.g.

☑-**Observed counts**   ☑ **Expected counts**

Minitab:

> **Chi-Square...** Check for required output, e.g.

☑-**Chi-square analysis**   ☑-**Expected cell counts**

> **Other Stats...** Select specific analyses as required:

☑-**Fisher's exact test for 2×2 tables**   ☑-**Cramer's V-square statistic**
☑-**Goodman and Kruskal lambda and tau**   ☑-**Measures of concordance for ordinal categories**

The results from these analyses are given in Figs 8.12/13/14/15.

```
Expected counts are printed below observed counts
Chi-Square contributions are printed below expected counts

      None, N  Some, S  Good, G  Excellent, E  Total
  1      13      18       159        72          262
       18.58    21.91   154.82     66.69
        1.675    0.699    0.113      0.423

  2      26      28       166        68          288
       20.42    24.09   170.18     73.31
        1.524    0.636    0.103      0.384

Total    39      46       325       140          550

Chi-Sq = 5.556, DF = 3, P-Value = 0.135
```

(a) 'Contingency table' case study (Minitab)

Q1 * Q2 Crosstabulation

			Q2		
			N	Y	Total
Q1	N	Count	24	19	43
		Expected Count	20.2	22.8	43.0
	Y	Count	7	16	23
		Expected Count	10.8	12.2	23.0
Total		Count	31	35	66
		Expected Count	31.0	35.0	66.0

(b) 'Forensic' case study (SPSS)

Fig 8.12 Contingency table calculations

The contingency table output from Minitab in Fig 8.12(a) prints the observed value, the expected value, and contribution to the chi-squared value for each cell (e.g. 13, 18.58, and 1.675 respectively in the top left cell). The sum of the individual contributions give a total chi-squared value, $\chi^2 = 5.556$ (Eqn 3.19) which, for degrees of freedom, $df = 3$ (Eqn 3.20), gives a p-value $= 0.135$.

Chi-Square Tests

	Value	df	Asymp. Sig. (2-sided)	Exact Sig. (2-sided)	Exact Sig. (1-sided)
Pearson Chi-Square	3.875[a]	1	.049		
Continuity Correction[b]	2.923	1	.087		
Likelihood Ratio	3.958	1	.047		
Fisher's Exact Test				.070	.043
McNemar Test				.029[c]	
N of Valid Cases	66				

a. 0 cells (0.0%) have expected count less than 5. The minimum expected count is 10.80.
b. Computed only for a 2×2 table
c. Binomial distribution used.

Fig 8.13 Hypothesis tests for significant association between Q1 and Q2 including McNemar's test (SPSS)

For the 'Forensic questionnaire' case study, SPSS gives the contingency table for the 'Forensic' case study as part of its output in Fig 8.12(b), and, in Fig 8.13, the calculations for the basic 'chi-squared' tests from Section 3.7, using the observed and expected counts:

Pearson's chi-square, Eqn 3.19 :

$$\chi^2 = \frac{(24-20.2)^2}{20.2} + \frac{(19-22.8)^2}{22.8} + \frac{(7-10.8)^2}{10.8} + \frac{(16-12.2)^2}{12.2} = 3.9$$

Yates continuity correction, Eqn 3.21:

$$\chi^2 = \frac{(3.8-0.5)^2}{20.2} + \frac{(3.8-0.5)^2}{22.8} + \frac{(3.8-0.5)^2}{10.8} + \frac{(3.8-0.5)^2}{12.2} = 2.9$$

Likelihood ratio, Eqn 3.22 :

$$\chi^2 = 2 \times \left\{ 24 \ln\left(\frac{24}{20.2}\right) + 19 \ln\left(\frac{19}{22.8}\right) + 7 \ln\left(\frac{7}{10.8}\right) + 16 \ln\left(\frac{16}{12.2}\right) \right\} = 4.0$$

The p-value for Fisher's exact test is derived in 4.2.3 and a different (not binomial) calculation for McNemar's test in 4.4.5 gives a similar value of $p = 0.031$.

In terms of deciding on the significance of an association, the results in Fig 8.13 appear to be contradictory. The basic chi-squared test and the likelihood ratio both suggest significance, however their reliability is suspect for 2×2 tables with just one degree of freedom. The Yates continuity correction and the Fisher's exact test both give more cautious results, suggesting that there is not enough real evidence for a significant association.

The interesting result is for the McNemar test which is actually testing for a difference in the way the respondents change their minds between questions (4.4.5). The result concludes that significantly more people change their mind, Q1 to Q2, from N to Y (19) than from Y to N (7).

Figs 8.14 and 8.15 give the additional symmetric and directional measures of association calculated in Sections 4.2 and 4.3. Minitab produces the same values as SPSS except that it gives Cramer's V $squared = 0.2423^2 = 0.587$.

Symmetric Measures

		Value	Asymp. Std. Error[a]	Approx. T[b]	Approx. Sig.
Nominal by Nominal	Phi	.242			.049
	Cramer's V	.242			.049
Ordinal by Ordinal	Kendall's tau-b	.242	.117	2.049	.040
Measure of Agreement	Kappa	.226	.111	1.968	.049
N of Valid Cases		66			

a. Not assuming the null hypothesis.

b. Using the asymptotic standard error assuming the null hypothesis.

Fig 8.14 Symmetrical measures of association between Q1 and Q2 (SPSS)

Directional Measures

			Value	Asymp. Std. Error[a]	Approx. T[b]	Approx. Sig.
Nominal by Nominal	Lambda	Symmetric	.093	.116	.766	.444
		Q1 Dependent	.000	.000	[c]	[c]
		Q2 Dependent	.161	.194	.766	.444
	Goodman and Kruskal tau	Q1 Dependent	.059	.057		.051[d]
		Q2 Dependent	.059	.056		.051[d]
Ordinal by Ordinal	Somers' d	Symmetric	.242	.117	2.049	.040
		Q1 Dependent	.231	.113	2.049	.040
		Q2 Dependent	.254	.122	2.049	.040

a. Not assuming the null hypothesis.

b. Using the asymptotic standard error assuming the null hypothesis.

c. Cannot be computed because the asymptotic standard error equals zero.

d. Based on chi-square approximation

Fig 8.15 Directional measures of association between Q1 and Q2 (SPSS)

8.2.5 Progression within the table

	A	B	C	D	E	F	G	H	I	J	K	L
1		1	2	3	4	Totals:		1	2	3	4	Totals:
2	1	13	18	159	72	262		13	159	18	72	262
3	2	26	28	166	68	288		26	166	28	68	288
4	Totals:	39	46	325	140	550		39	325	46	140	550

Fig 8.16 Contingency tables with categories swapped

The table B2:E3 in Fig 8.16 reproduces the data from Fig 8.9 with *ordinal* values describing the *progression* between levels of improvement, *ImprovN*. In Fig 8.12(a) the result, $p = 0.135$, suggests that there is no significant improvement factor, but we now take into account the fact that there is a sense of progression in *ImprovN*, and carry out an analysis using SPSS as in 8.2.4, requesting the statistics of:

☑-**Chi-square**   ☑-**Correlation**   ☑-**Kendall's tau-b**   ☑-**Eta**

and find that:

Pearson's chi-square		$p = 0.135$
Linear by linear association, Eta		$p = 0.027$
Interval by interval	Pearson's correlation	$p = 0.027$
Ordinal by ordinal	Spearman's correlation	$p = 0.049$
	Kendall's tau-b	$p = 0.047$

The analysis gives the same p-value for the chi-square value (for an association between *nominal* levels), but the linear by linear association test gives $p = 0.027$, which suggests that there is indeed a progressive association between the *ordinal* levels. The same p-value occurs for the Pearson's measurement of linear correlation, and the nonparametric measurements of correlation also record significant associations. The direction of correlation can be assessed by the correlation coefficients, which are negative, indicating that the improvement obtained by females is significantly greater than that by males.

As a confirmation of the effect of the ordinal measures, the second table H2:K3 in Fig 8.16 has the same values as in B2:E3 but the columns 2 and 3 have been reversed. Applying the same analysis to this table we obtain the *same* significance value for chi-square, $p = 0.135$, in respect of a possible difference between rows in the *distribution* of values across columns. However, the tests for a *progression* in values across the columns now show that there is no longer a significant relationship between *ImprovN* and *SexN*, with the linear by linear association, $p = 0.270$, and Spearman's rho and Kendall's tau-b, $p = 0.279$.

This analysis shows that even when the standard chi-squared test fails to find a significant difference, it is possible to test specifically for a progressive (or correlation) relationship between the two variables.

8.2.6 **Data consolidation**

It is important here to be quite clear about the *scientific* conclusion that can be drawn from the *statistical* results of the contingency table in Fig 8.12(a). In this case, the *non-significant* result of $p = 0.135$, actually answers the question 'Is there a difference in the *distribution* of responses by male and female patients over the *four* categories?' and reports that the observed distribution could have been observed by chance.

However, if our research is *exploratory*, we would be entitled to ask whether different *questions* might have given different conclusions. For example, we can ask the question 'Is there a difference between male and females in whether the improvement can be rated good or not?', in which case, *N* and *S* would combine to a 'No' category and *G* and *E* would combine to a 'Yes' category and giving the 2×2 table B8:C9 in Fig 8.9. Fisher's exact test for two proportions (3.8.3) then gives a significant result of $p = 0.025$ for a difference between males and females who show *good* improvement.

The process of *combining* factor levels increases the frequencies in the remaining cells of the table thereby increasing the *power* of the analysis, but at the same time reduces the range of questions that can be asked of that data. However, it is clearly not appropriate to claim a significant effect just because you have sliced up the data in different ways until you get p less than 0.05. This is acceptable as an exploratory tool, but if you believe that you have found a possible significance, the next step would be to redesign and rerun the data collection focussing on the new hypothesis for confirmation.

Data consolidation is also used below in dealing with the problem of too *few* observations in too *many* factor categories.

8.2.7 Low expected frequencies

⊿	A	B	C	D	E	F	G	H
1	Score:	-3	-2	-1	0	+1	+2	+3
2	G1	1	1	2	7	10	10	5
3	G2	3	3	5	9	8	1	1
4								
5	Score:			-3/2/1	0	+1	+2/3	
6	G1			4	7	10	15	
7	G2			11	9	8	2	

Fig 8.17 Questionnaire scores on a Likert scale of −3 to +3 for question Q6

If we take crosstabs between the group, *Grp*, and the response, *Q6*, from Fig 8.10, we obtain the contingency table A1:H3 in Fig 8.17, in which rows 2 and 3 give the number of responses from two groups *G1* and *G2* to question *Q6* on a Likert scale of −3 to +3. If we perform a chi-squared analysis on the table to test whether the distribution of values in row 2 is different from the distribution in row 3, we get $p = 0.038$, which suggests that there is a difference. However we also get a warning that there are eight cells with expected counts of less than five, which means that we cannot be confident that the p-value is reliable.

As the Likert scale is for an *ordinal* value, we could use data consolidation (8.2.6) by combining the −3, −2, and −1 results and the +2 and +3 results as shown in rows 6 and 7. The chi-squared analysis then gives $p = 0.004$, confirming a significant difference between the groups. Although this gives a significant result for the new range of answer categories, it is necessary to be cautious and avoid 're-grouping the answers' until the p-value becomes less than 0.05! However, it is a useful result for exploratory data as it can help design more effective and focussed questions in any future questionnaire.

The situation is different if the categories are *nominal* values (e.g. observations of different animal species), in that we may not be able to combine them into meaningful groups. In this case we can use the Monte Carlo analysis (3.9.3) which employs a resampling technique to estimate a *p*-value, even for low expected frequencies. Using the Monte Carlo option in SPSS:

SPSS > Analyze > Descriptive statistics > Crosstabs ...
Row(s): *Grp*   **Columns(s):** *Q6*
> Exact: ☑-**Monte Carlo**
> Statistics: ☑-**Chi-squared**
→ Output: Fig 8.18

Chi-Square Tests

| | Value | df | Asymp. Sig. (2-sided) | Monte Carlo Sig. (2-sided) | | |
| | | | | Sig. | 99% Confidence Interval | |
					Lower Bound	Upper Bound
Pearson Chi-Square	13.353[a]	6	.038	.030[b]	.026	.035
Likelihood Ratio	14.807	6	.022	.042[b]	.037	.047
Fisher's Exact Test	13.191			.028[b]	.024	.032
N of Valid Cases	66					

a. 8 cells (57.1%) have expected count less than 5. The minimum expected count is 1.82.
b. Based on 10000 sampled tables with starting seed 508741944.

Fig 8.18 Monte Carlo analysis for chi-squared statistics

Fig 8.18 gives the calculated value for Pearson's chi-squared, $p = 0.038$, but it also gives a 99% confidence interval range for the *p*-value calculated by the Monte Carlo method, between 0.026 and 0.035, which confirms the significant association.

8.2.8 Layered contingency tables

It is possible that the data used to generate a contingency table can be divided into sub-groups according to another variable in the data. For example, in Fig 8.10 the people answering questions *Q1* and *Q2* can be identified as belonging to either group, *G1* or *G2*, and we may wish to investigate whether there is a difference in the association of *Q1* and *Q2* between the two groups. We can do this by producing contingency tables for each sub-group by entering *Grp* into the **Layer** option in the analysis.

The results of layering by *Grp* for *Q1/Q2* is given in Table 8.1.

Table 8.1 Results from layered contingency tables

Group	Pearson's chi-squared	Fisher's exact test	McNemar's test
Combined *G1* and *G2*	$p = 0.049$	$p = 0.070$	$p = 0.029$
G1	$p = 0.102$	$p = 0.219$	$p = 0.000$
G2	$p = 0.029$	$p = 0.057$	$p = 0.508$

Using Fisher's test for the 2×2 contingency table there is no *p*-value less than 0.05, although it is possible to see a difference in the results given for the two groups. However, the interesting statistics are the results of McNemar's test which identifies differences in the numbers of respondents *changing* their answers in different directions. It is significant ($p < 0.0005$) that the majority of changes in group *G1* are 16 from *N* (for *Q1*) to *Y* (for *Q2*) with only one in the opposite direction, whereas there is no significant difference ($p = 0.508$) for *G2*, which suggests that the intervention occurring between *Q1* and *Q2* is having an effect on how those in *G1* answer the question, but not on those in *G2*.

8.3 Binary output data

Binary data has just two possible categories that could be defined in a variety of ways: True/False, Heads/Tails, Yes/No, 0/1, etc. Many important scientific outcomes depend on the *probability* of finding an individual in one of the two possible states. For example, this could be in the diagnosis of whether a single patient is suffering from a particular disease or not, and then, for a whole population of individuals, this translates into an expected *proportion* of people with this disease. This section develops statistical models that predict the underlying binary probabilities.

8.3.1 Example data

The analysis of binary *proportions* using Minitab and SPSS is introduced in 6.1.7 and 6.2.9, with reference to the underlying statistics developed in Section 3.8, and the use of 2×2 contingency tables for binary data developed in Section 8.2.

Two examples of data with *binary* output values dependent on *interval* input variables are given in Figs 8.19 and 8.20.

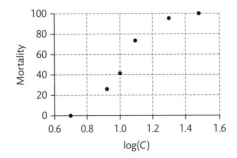

Fig 8.19 Percentage *mortality* of nematodes as a function of the log of a drug concentration

	C1	C2-T	C3	C4	C5	C6	C7	C8
	RecNo	Group	v1	v2	v5	v6	v9	EPRO1
1	1	A	74.7	41.4	58.1	68.2	20.8	0.785
2	2	B	89.5	30.1	39.7	51.1	23.8	0.139
3	3	A	61.7	22.9	48.5	67.9	49.2	0.974
4	4	A	77.2	35.7	57.3	64.4	50.2	0.931
5	5	B	70.5	14.8	41.8	46.5	9.3	0.069
6	6	A	72.6	47.0	64.5	70.5	48.9	0.965

Fig 8.20 Variables, *v1, v2, v5, v6, v9,* and *EPRO1* (in Minitab 16 or FITS1 in Minitab 17) as possible *predictors* of group *A* or *B*.

In Fig 8.19 the binary state, alive or dead, of *individual* nematodes determines the percentage of dead nematodes (mortality) in an overall *population*. The 'LC50' case study (8.3.3) develops nonlinear regression techniques to analyse this data.

Fig 8.20 shows the first 6 of 100 subjects from the 'Screening test' case study (8.3.4) which develops binary regression and ROC curves to model the prediction of the group *A* or *B* of an individual subject as a function of the interval variables, *v1*, *v2*, *v5*, *v6*, and *v9*.

8.3.2 **Analytical options**

For binary variables, we analyse the *probability* of individuals being in a particular state. If we are considering a sample of *n* individuals, or items, each of which may be in one of two states, then this probability determines the overall *proportion* of individuals/items that are in a particular state.

Describing data

- Techniques for describing categorical data are given in 8.1.3 and 8.2.3. We introduce here deviance curves (Fig 8.25) and ROC curves (Fig 8.27).

Tests / Measurements:

- Fisher's exact binomial tests to compare **proportions** with a target value or with another proportion (Section 3.8, 4.2.3, 6.1.7, 6.2.9).
- **Probit and logit** transformations of a proportion (or percentage) to a linear function (8.3.3).
- **Binary regression** to develop a model to predict binary probability (8.3.4).
- Use of **ROC curves** for the relationship between sensitivity and specificity (8.3.5).

8.3.3 **Logit and probit linearization**

The following case study considers the *probabilities* of individuals being in a specific state, and the consequential *proportion* of a population in that state.

> #### Case study: **LC50 / Logit and probit**
>
> In preparation for a chemotaxis measurement (6.3.1), it is necessary to find the LC50 value for the concentration of a chemotherapeutic drug which kills 50% of nematodes within a specified time. The percentage dying (mortality) for a range of relative concentrations, *C*, of a standard solution is recorded in Fig 8.19, and we wish to use a process of 'regression' to obtain a best-fit relationship that allows us to estimate the concentration at which 50% of the nematodes die.

Logit and probit: Excel analysis for Fig 8.22. Scan here to watch the video or find it via www.oxford textbooks. co.uk/orc/ currell/

Mortality is the overall *proportion*, *P*, (or percentage, *P%*) resulting from the binary condition that each nematode in the population is either alive or dead. The graph in Fig 8.19 records mortality plotted against the *log* of the concentration because it is often convenient to use log to base 10 for the *x*-axis as the *unit* values then give the concentration axis displayed in 'powers of 10'.

A typical calculation derives the LC50 value which is the concentration at which 50% ($P = 0.5$) of the population has died, and to do this we need to derive a best-fit line through the data points and calculate the concentration at which the proportion equals 50%. The problem is that we do not have an easy way of fitting a best-fit equation to such a curve, as the relationship between proportion, P, and the log of the concentration, $\log(C)$, is clearly nonlinear. In fact the relationship cannot be linear because $\log(C)$ can have theoretical values from $-\infty$ to $+\infty$ whilst the values of P are limited by 0 and 1.

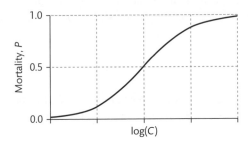

Fig 8.21 Sigmoid curve

The data in Fig 8.19(a) can be modelled by the mathematical *sigmoid* curve in Fig 8.21, which is given by the logistic function

$$P = \frac{e^x}{e^x + 1} = \frac{1}{1 + e^{-x}}$$

where P is the proportion ($= P\%/100$) and x is a value proportional to $\log(C)$.

There are two transformations that can be used to linearize the data in Fig 8.19(a):

Logit(P) is the actual inverse of the above logistic function:

$$\text{logit}(P) = \ln\left(\frac{P}{1-P}\right) \tag{8.1}$$

which gives a value of 0 when $P = 0.5$ (50%).

Probit(P) is an alternative transformation that was developed for this type of problem, with values derived from a set of probit tables. However the values can also be calculated by using the inverse of the normal distribution in Excel:

$$\text{Probit}(P) = 5 + \text{NORM.S.INV}(P) \tag{8.2}$$

which gives a value of 5 for $P = 0.5$ (50%). The '5' was included to avoid negative values (as occur with Logit) which were thought to cause confusion.

The logit transformation has a more widespread use in software analysis, but the probit analysis has been developed for quick dose–response calculations of LC50/LD50 using published tables. We can compare their use with the following analysis given in Fig 8.22.

In the experiment, initial measurements were made by diluting a standard drug solution giving concentrations of 0, 1%, 5%, 10%, and 20%, but, when it was observed that the LC50

	A	B	C	D	E	F	G	H
1	Conc, C%	log(C%)	Alive	Dead	Prop, P	Corr, P	Logit(P)	Probit(P)
2	0		95	1	0.010			
3	1	0.00	88	2	0.022			
4	5	0.70	82	0	0.000			
5	8.3	0.92	67	23	0.256	0.248	-1.11	4.32
6	10	1.00	59	42	0.416	0.410	-0.37	4.77
7	12.5	1.10	29	81	0.736	0.734	1.01	5.62
8	20	1.30	0	92	1.000			

Fig 8.22 Numbers of nematodes killed by different concentrations of a drug

value falls between the 5% and 20% values, further measurements were made at 8.3% (1/12) and 12.5% (1/8).

The proportion dead (mortality) at each concentration is calculated in column, E, e.g.

Mortality proportion: $[E2] = D2 / (C2 + D2)$

The proportion, P_0, for zero concentration (in E2) is used as a 'control' for the nematodes that die independently of the treatment. The values in column F then give the corrected proportion using Abbott's formula which includes the 'control' data:

$$\text{Corrected, } P = 100 \times \frac{P - P_0}{100 - P_0}. \tag{8.3}$$

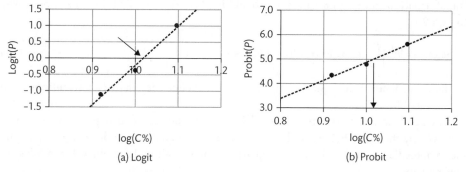

(a) Logit   (b) Probit

Fig 8.23 Logit and probit transformations

The two transformations, Eqn 8.1 and Eqn 8.2, give the results in columns G and H respectively, which now both give best-fit *straight* lines in Fig 8.23. The intercepts at both logit(P) = 0, and probit(P) = 5, give the log of the LC50 concentration as 1.02 from which LC50 = $10^{1.02}$ = 10.5%.

It is important to note that, for the type of problem in this case study, proportion values close to 0 (0%) or 1.0 (100%) will be dependent on the states of a very few individuals, alive or dead, which can lead to large *relative uncertainties* at these limits. For this reason we have only used proportions between 25% and 75%.

Binary regression: Minitab analysis for Fig 8.24. Scan here to watch the video or find it via www.oxford textbooks. co.uk/orc/ currell/

8.3.4 Binary regression

In 8.3.3 we used a data transformation to help us predict, given a specific drug concentration, whether the probability of nematodes being alive or dead was greater or less than 50%. We now develop this into a more sophisticated regression model that enables us to predict actual *probability* values.

Case study: Screening test / 3. Binary regression

—continued from 9.1.4, leading to 8.3.5

In 9.1.4 we use principal component analysis to derive the elements of the principal components, PC1, PC2, etc. that can describe the separation between the two subject groups, A and B. These groups are given in Fig 8.20, together with the values of the variables, *v2, v5, v6*, and *v9*, which are identified as the elements of the main principal component, PC1. We wish to develop a mathematical model in which we can use these values to predict the state of a given *individual* subject.

The first step in modelling binary outcomes, is to develop a regression equation that relates the binary *probability* to the values of one or more predictor variables. Section 2.3 introduces the use of transformations to *linearize* a regression equation, and we meet the concept of *multiple regression* in 9.1.6 in which several variables are used to predict the value of a single outcome. Combining these techniques we start by using the logit transformation from Eqn 8.1 for the probability, P, and expressing this in a similar way to Eqn 9.1:

$$\text{logit}(P) = \ln\left(\frac{P}{1-P}\right) = B_0 + B_1 x_1 + B_2 x_2 + B_3 x_3 + \cdots. \tag{8.4}$$

The coefficients of the equation, $B_0, B_1, B_2, B_3\ldots$ define the relative contributions from each x variable.

When combining probabilities it is useful to use the concept of odds, the values of which are given by the ratio P/(1-P), and we can express the equation in odds simply by taking the exponential of both sides of the equation. The overall odds are then the *product* of power factors:

$$\text{Odds} = \frac{P}{1-P} = e^{B_0 + B_1 x_1 + B_2 x_2 \cdots} = e^{B_0} \times \left(e^{B_1}\right)^{x_1} \times \left(e^{B_2}\right)^{x_2} \cdots \tag{8.5}$$

The process of regression calculates the values of the coefficients in the equation that produce the best fit to the experimental data. A binary *regression* uses an iterative maximum likelihood procedure (2.4.2) to get the best-fit of a model to the data and calculate the coefficients in Eqn 8.4.

In the following analysis we choose *not* to use variable *v1*, because this is not identified (9.1.4) as one of the cluster of variables that best predict the subject group probability. We compare our results here with *v1* in Fig 8.27.

Minitab

Minitab > Stat > Regression > Binary Logistic Regression >
Fit Binary Logistic Regression Model ...
▼ *Response in binary response / frequency format*
Response: *Group*
Response event: *A* (makes *A* the value of *Group* for probability = 1.0)
Continuous predictors: *v2 v5 v6 v9*
> Options...: Choose link function, e.g. **Logit**
> Storage...:
   ☑-**Fits (event probabilities)** - stores the probabilities for each subject in a new variable
FITS1 (EPRO1 in Minitab 16)
   ☑-**Delta deviance** - stored as new variable DDEV1
→ Output: Fig 8.24

```
Link Function: Logit

Response Information
Variable  Value  Count
Group     A         50    (Event)
          B         50
          Total    100

Logistic Regression Table
                                                 Odds      95% CI
Predictor        Coef    SE Coef      Z      P  Ratio  Lower  Upper
Constant     -12.7234    2.90043  -4.39  0.000
v2         -0.0573760  0.0431040  -1.33  0.183   0.94   0.87   1.03
v5          0.0368401  0.0643224   0.57  0.567   1.04   0.91   1.18
v6          0.190982   0.0782711   2.44  0.015   1.21   1.04   1.41
v9          0.0591785  0.0352363   1.68  0.093   1.06   0.99   1.14

Log-Likelihood = -33.027
Test that all slopes are zero: G = 72.576, DF = 4, P-Value = 0.000
```

Fig 8.24 Extract of binary regression results (Minitab)

This first part of the results in Fig 8.24 confirms that the *link function* is given by the logit function. It then gives the observed numbers in each of the two categories *A* and *B* and shows that *A* has been defined as the 'event' related to a probability of 1.0.

The results then give the coefficients $B_0 = -12.72$, $B_1 = -0.0574$ etc. for the logit(P) equation:

$$\text{logit}(P) = \ln\left(\frac{P}{1-P}\right) = -12.72 - 0.0574 \times v_2 + 0.0368 \times v_5 + 0.1910 \times v_6 + 0.0592 \times v_9$$

If we now consider the first subject, *RecNo* = 1 in Fig 8.20, we can use the logit(P) equation to predict its group. We enter the variable values of *v2* = 41.2, *v5* = 58.1, etc. into the above equation and we get:

$$\text{logit}(P) = 1.303$$

It is then possible to calculate the probability, P, that this particular subject is in group A by evaluating

$$P = \frac{e^{\text{logit}(P)}}{1 + e^{\text{logit}(P)}} = 0.786$$

Hence our model gives a 78.6% probability that subject 1 is in group A, which matches the fact that subject 1 is known to be group A.

Within the analysis, we requested that the 'event probabilities' be stored, and in a new column under the default title *FITS1* (or *EPRO1* in Minitab 16), we see that, for subject 1, the result 0.785 in Fig 8.20 agrees, within rounding errors, with our calculation above.

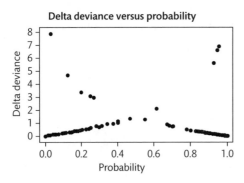

Fig 8.25 Deviance when using the variables *v2, v5, v6,* and *v9* as predictors for group *A* or *B* (Minitab)

The other data stored was delta deviance, which we have plotted in an *x–y* graph against the event probability in Fig 8.25. (This graph can be requested directly in Minitab 16.)

Deviance is a measure of the goodness of fit of the model:

$$D = -2 \ln\left(\frac{\textit{Likelihood of fitted model}}{\textit{Likelihood of perfect fit}} \right) \tag{8.6}$$

An overall value for D provides a similar role to R^2 for the goodness of fit for analyses using the least squares method. Values of D can be calculated for individual points and Fig 8.25 shows the variation of D with the individual subject probabilities. If the probability of a subject being in group A is close to 1 and it is in group A (correct), then this will give very low deviance as shown by the data values in the *bottom right* of the plot, but, if a subject with a probability close to 1 is actually in group B, then this is an incorrect prediction and the point will have a high deviance as shown by the few points leading to the *top right* of the diagram. Points from group A form the line from the top left to bottom right and those from group B the line from bottom left to top right. The few points where the prediction fails to put the subjects in the correct groups are those in the top branches of the two lines.

ROC curves: SPSS analysis for Fig 8.27. Scan here to watch the video or find it via www.oxford textbooks. co.uk/orc/ currell/

8.3.5 **Binary probabilities and ROC plots**

In 8.3.4, we developed the mathematics of calculating the *probabilities* of binary values, and it is clear from Fig 8.25 that *decisions* based on those probabilities can produce a percentage of wrong choices. We now introduce a graphical presentation of the process of making choices based on two overlapping probabilities, and the name of the plot, receiver operating characteristics (ROC) derives from its early use to describe the properties of signal detectors in recording either a *true* or *false* signal.

We will illustrate the calculations by considering the medical diagnostic case where the relevant test results, *v*, of a large survey of *healthy* adults (no disease, $\bar{D}$) are shown to have a mean value of 20 and a standard deviation of 4.0, whereas the test results of another large survey of adults with a *disease* condition, *D*, have a mean value of 30 and a standard deviation of 5.0. These probability distributions are shown in Fig 8.26(a). This is an idealized example because real diseased states would often show a much larger standard deviation than the healthy state and probably with a pronounced positive skewness.

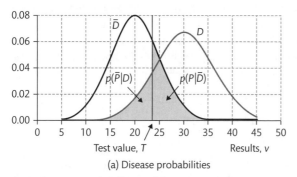
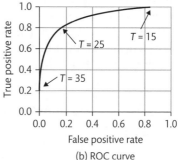

(a) Disease probabilities

(b) ROC curve

Fig 8.26 Disease probabilities and the associated ROC curve

The aim is to use this data to develop a decision criterion, by which it is possible to put a person into a specific category depending on their personal test result, *v*. We need to decide on a test value, *T*, such that we classify the result according to the simple choice:

- If $v > T$ we classify the result as positive (*D*) and decide that the person *does* have the disease.
- If $v < T$ we classify the result as negative ($\bar{D}$) and decide that the person does *not* have the disease.

It is clear that both distributions have 'tails' that overlap the test value, and, in each case the shaded region represents an *incorrect* classification. This leads to two key performance criteria for a binary test:

- **Sensitivity** of the test is the probability of recording a *positive* result, *P*, if the person *does* have the disease, $p(P|D)$.
  Sensitivity is the probability, between 0 and 1.0, of a *true positive*.
- **Specificity** of the test is the probability of recording a *negative* result, $\bar{P}$, if the person does *not* have the disease, $p(\bar{P}|\bar{D})$.
  Specificity is the probability, between 0 and 1.0, of a *true negative*.

It is useful to note that the probability of a *false* positive is given by $p(P|\bar{D}) = 1.0 - \text{Specificity}$.

We can trace the characteristics of the test in a ROC curve shown in Fig 8.25(b), where we plot the *true positive rate* (sensitivity) against the *false positive* rate (1.0 − specificity) for different chosen values of *T*.

If the value of *T* is set low (e.g. *T* = 15) then most measured values of *v* will result in positive decisions, making the probabilities of *true* positives and *false* positives both close to 1.0. As the value of *T* is increased the probability of a *false* positive initially drops more quickly than that of a *true* positive, drawing out the line in Fig 8.26(b). At the other extreme, a large value of *T* (e.g. *T* = 35) will result in most decisions being negative, with low probabilities for both *true* positives and *false* positives.

The optimum choice of test value will aim to maximize the probability of a true positive without unduly increasing the rate of false positives. In the ROC plot, this optimum situation will be a position on the graph towards the top left of the plot, with an intermediate value of *T*, e.g. *T* ≈ 25. The choice of the optimum value for the decision criterion, *T*, will depend on the relative *consequences* of recording false positives and negatives.

The overall quality of a test is greatest if the curve fits closely into the top left-hand corner of the plot. In this case the area under the curve would approach 1.0. If the test has no diagnostic quality, the characteristic line would be a straight diagonal with a cumulative area of 0.5.

Case study: **Screening test / 4. Binary classification**

—continued from 8.3.4

Referring to the data in Fig 8.20, we now compare the abilities of the variable, *v1*, and the probability, *EPRO1*, derived in 8.3.4 as a combination of *v2*, *v5*, *v6*, and *v9*, to predict the group, *A* or *B*, of a specific subject.

In practice, the characteristics of a diagnostic test are usually developed, not through theoretical modelling as above, but through the analysis of experimental results. Fig 8.27 compares the effectiveness the two variables *v1* and *EPRO1* in diagnosing whether each subject is in group *A* or *B*.

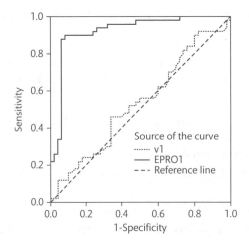

Fig 8.27 ROC curves for *v1* and *EPRO1*

Using SPSS

> **SPSS > Analyze > ROC curve...**
> Test variable: *v1 EPRO1*
> **> State variable:** *Group*
>   **Value of State Variable:** Identify the 'true' level of the state being tested, e.g. *A*
> ☑-**ROC curve**
>   ☑-**With diagonal reference line**
> → Output: Fig 8.27

We find in Fig 9.6 that the variable *v1* does not appear in the principal components, PC1 or PC2, and thus has little predictive capability for identifying the group *A* or *B*. This is confirmed in Fig 8.27 in that it lies close to the *neutral* reference line with an area (to the bottom right) of 0.531. However, the calculated probabilities, *EPRO1*, for predicting the group *A* or *B*, produce a line which passes close to the top left-hand corner of the ROC plot with an area of 0.922. This confirms that the probability variable, *EPRO1*, is a good predictor for identifying the group of an individual subject.

The area is not the only important factor, as the relative importance of sensitivity and specificity can depend on the *consequences* of false positives or false negatives, making the *shape* of the curve important.

Multiple variables

Introduction

This chapter considers situations in which the investigation produces several measured variables, which can be confusing for students who have only been introduced to the uni-variate ANOVA, the two sample *t*-test, and linear regression between two variables. The first section develops techniques of data reduction through principal component analysis, and data modelling through cluster analysis and multiple regression. The second section takes the example of questionnaire data, which could also represent multiple experimental measurements, and identifies a wide range of possible analytical techniques, together with specific references to analyses developed previously. This final section could be used as a review of the analyses and techniques that have been introduced throughout the book.

Section 9.1 develops the statistical analyses in which a data set of *multiple input variables* can be rationalized to create a *model* that predicts the state of a scientific system.

Section 9.2 uses a data set with multiple experimental variables as an example to consider the different analytical *questions* that could be asked, particularly in reference to questionnaire data.

9.1 Modelling multiple variables

As the ability to measure and record multiple variables with greater ease and accuracy continues to increase, we are being faced with very large data sets from which we wish to draw out hidden information. A general type of approach is to investigate whether different *groups of variables* provide related information and then whether it is possible to represent the actions of many variables with a *small number* of composite variables. This is a process of *data reduction*, and we investigate this using both graphical and numerical representations.

9.1.1 Example data

Referring to the screening data in Fig 9.1, other investigations producing a similar structure could include:

- interpreting questionnaire responses by looking for common patterns in the opinions of respondents on nine issues. In this case the data values would probably be on an ordinal scale.

- identifying common survival traits by comparing the occurrences of nine species in different environments. In this case the data values may be counts or frequencies.

- developing a method to identify and classify bacteria using hyperspectral imaging at nine possible wavelengths.

Case study: **Screening test / 1. Clustering of variables and subjects (overview)**

The data in Fig 9.1 shows the measurements of nine variables relating to each of 100 subjects. These were recorded by gas chromatography as part of an investigation to develop a screening sensor to detect human abnormalities by 'sniffing' a range of aromatic compounds released from saliva. Fifty 'normal' subjects are identified in set A and 50 subjects with the abnormal condition in set B. We start in 9.1.3 by using cluster analysis to identify groups of variables and subjects within the experimental data.

9.1.4 / 2. Principal component analysis: Uses PCA to derive the combination of variables that best discriminates between the two main groups of subjects.

8.3.4 / 3. Binary regression: Uses multiple regression to predict the probability (odds) of each individual being in a specific state.

8.3.5 / 4. Binary classification: Develops the ROC curve as a graphical indication of the sensitivity and specificity of a binary choice prediction.

	A	B	C	D	E	F	G	H	I	J	K	L
1	RecNo	Group	v1	v2	v3	v4	v5	v6	v7	v8	v9	Cluster
2	1	A	74.7	41.4	66.6	30.2	58.1	68.2	69.7	17	20.8	
3	2	B	89.5	30.1	49	21.7	39.7	51.1	47.7	12.5	23.8	
4	3	A	61.7	22.9	46.6	22	48.5	67.9	58	21.7	49.2	
5	4	A	77.2	35.7	72.8	35.1	57.3	64.4	81.7	18.8	50.2	
6	5	B	70.5	14.8	85.3	43.5	41.8	46.5	96.3	13.5	9.3	
7	6	A	72.6	47	59.8	30.9	64.5	70.5	69.1	15.6	48.9	
8	7	A	73.7	44.2	54.7	16.1	54.7	64.9	57.4	20.5	46.4	
101	100	B	89	45.5	67	31.7	49.6	62.2	80.2	20.6	25.5	

Fig 9.1 100 subjects in groups *A* or *B* each recording nine variables, *v1* to *v9* (Rows 9 to 100 are 'hidden' in the worksheet)

We also treat factors and their interactions as multiple variables in the case study:

Boxing performance

9.1.6 / 2. Multiple regression: Uses stepwise regression and general regression to identify the significant factors in a best-fit model.

9.1.2 **Analytical options**

Data reduction. The overall aim of many investigations is to develop a mathematical model that describes how an observed response variable can be predicted with just a *subset* of a large number of possible input variables. For example, we might know that a range of symptoms are related to the existence of a disease, but it would be valuable to develop a model which identifies a specific combination of a few symptoms as being strongly predictive of the existence of the disease. There are several steps in the process:

1. Quantifying the relationships between possible input and output variables.
2. Identify *relationships* (e.g. correlations) between variables that are possible predictor variables.
3. Develop a practical model which maximizes the accuracy in *predicting* the output with a *minimum* of input variables.

Cluster analysis (9.1.3) is a technique that enables us to *visualize* how input variables might be grouped together, and in so doing help in simplifying our understanding of the system. It is also possible to use clustering techniques to identify similarities and differences between the *subjects* being measured, and to act as a diagnostic test between different subject states, e.g. between well and ill patients.

Principal component analysis, PCA, (9.1.4) is a technique that is used to both simplify the number of variables that are used to predict the output and also to optimize their relative inputs to the model.

Factor analysis (9.1.5) is a development of PCA that performs additional calculations ('rotations') to derive mathematical 'factors' that are new combinations of the original variables. The advantage of these new factors is that they can provide a more effective way of summarizing the whole data set.

Multiple regression (9.1.6) expresses the response variable in relation to multiple predictor variables, and the analyses in SPSS and Minitab will add and/or reject possible inputs depending on whether they have a significant effect on the response variable.

9.1.3 **Cluster analysis**

Cluster analysis: Minitab and SPSS analyses for Fig 9.2. Scan here to watch the video or find it via www.oxfordtext books.co.uk/ orc/currell/

It is very difficult to see any patterns within a table of *numbers*, but, if these numbers can be converted into suitable *graphics*, then the human brain is very good at picking out visual patterns. Clustering is a useful method of identifying and displaying groups within the data, occurring in two main forms:

- Hierarchical clustering looks for groupings either between different experimental *variables* or between different *records* (also called *subjects*, *cases* or *observations*) in the data set.
- *K*-means clustering looks for *groupings* of *records* in an 'area' defined by the calculated components or factors.

We use the data in Fig 9.1 to introduce hierarchical clustering, with the immediate objective of identifying which *combinations* of the nine compounds respond in similar ways. Using analytical software, the *hierarchical cluster analysis* analyses the relationships between variables, and produces the dendrogram (from the Greek, *dendron*, for tree) in Fig 9.2 which shows the *similarity* between different variables.

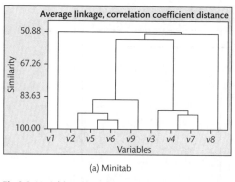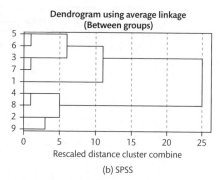

(a) Minitab   (b) SPSS

Fig 9.2 Variables cluster dendrogram

Minitab

Minitab > Stat > Multivariate > Cluster Variables...
Variables: *v1-v9*
Linkage method: Try different methods to improve discrimination of variables.
☑ **Show dendrogram**
→ Output: Fig 9.2(a)

SPSS

SPSS > Analyze > Classify > Hierarchical Cluster...
Variables: *v1-v9*
Cluster: Choose to cluster either ⊙-**Cases** (subjects, see below) or ⊙-**Variables**
Plots...: ☑-**Dendrogram**
> Method...: Try different methods to improve discrimination of variables
→ Output: Fig 9.2(b)

In Fig 9.2(a) all the variables are separated along the 100% baseline, showing that there are *no* variables that are 100% similar. At about 85% similarity, the variables *v2, v5, v6,* and *v9* form one similar group, and *v3, v4,* and *v7* form a second group, but variables *v1* and v8 show little similarity with any other variables.

The dendrogram suggests that variables, *v2, v5, v6,* and *v9,* are similar measures for one main *component* and *v3, v4,* and *v7* are similar measures for a *second* main component, with *v1* and v8 unrelated variables. This interpretation is developed further in the next section.

The two dendrograms in Fig 9.2, although in different orientations, are derived from the same data, using Minitab and SPSS respectively. The reason for the different grouping of variables is due to different default choices in the way in which the statistics links and groups the data values. Fig 9.2(a) uses the Minitab default 'correlation' calculation and Fig 9.2(b) uses the SPSS default 'squared Euclidean distance' calculation. In practice, it is often useful to try different methods to see if they are more effective at grouping the variables. If the correlation calculation is used in SPSS, it produces the same grouping as for Minitab.

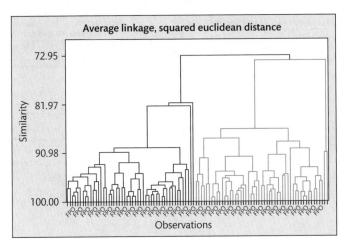

Fig 9.3 Dendrogram separation of subjects into clusters

We can also investigate if there is any clustering of the *subjects* (*records* or *observations*) in the data. Using Minitab we find a good separation into two clusters is achieved in Fig 9.3 by using 'average linkage' with 'squared Euclidean distance'.

Minitab

> **Minitab > Stat > Multivariate > Cluster Observations...**
> **Variables:** *v1-v9*
> **Linkage method:** *Average*
> **Distance measure:** *Squared Euclidean*
> **Specify final partition by:** ⊙ **Number of clusters:** Choose 2 in this example
> ☑ **Show dendrogram**
> **> Storage...:** Enter new column to receive the cluster allocation for each subject
> → Output: Fig 9.3

Principal component analysis: Minitab analysis for Figs 9.6 and 9.7. Scan here to watch the video or find it via www.oxfordtext books.co.uk/ orc/currell/

Fig 9.3 shows the individual subjects along the horizontal axis, but with 100 subjects it is not possible to identify separate labels in this diagram. Nevertheless, we can see from the plot that the subjects split into two clear 'clusters' separated at about 73% similarity.

In SPSS, selecting ⊙ **Cases** under 'Hierarchical Clusters' gives the same diagram but it is turned through 90 degrees and stretched vertically to separate the subjects.

Using the option **> Storage**, Minitab allows us to record the 'cluster' allocation, '1' or '2', for each subject in a new column. We find that 43 group A subjects were correctly allocated to cluster '1' and 44 group B subjects to cluster '2', leaving only seven group A and six group B subjects misclassified.

9.1.4 **Principal component analysis**

The aim of principal component analysis (PCA) is to describe as much *variability* in the data as possible by using as *few variables* as possible. We introduce the *concept* of PCA with reference to a very simple data rotation example, using three points, A, B, and C, on the x–y diagram in Fig 9.4(a) which are described by the related variables $v1$ and v2 with values (0.5, 1.0), (0.8, 0.6), and (1.1, 0.2) respectively.

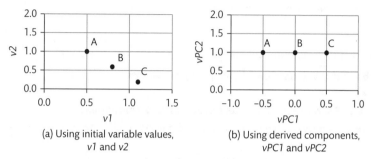

(a) Using initial variable values, $v1$ and $v2$

(b) Using derived components, $vPC1$ and $vPC2$

Fig 9.4 Three data points, A, B, and C, defined by values of the variables, $v1$ and $v2$

	A	B	C	D	E	F	G	H	I	J	K	L
1		v1	v2		Variable	PC1	PC2	giving:			vPC1	vPC2
2	A	0.5	1		v1	0.6	0.8	vPC1 = 0.6 × v1 + (-0.8)× v2		A	-0.5	1
3	B	0.8	0.6		v2	-0.8	0.6	vPC2 = 0.8 × v1 + 0.6 × v2		B	0	1
4	C	1.1	0.2							C	0.5	1
5												
6	Initial variables				Transforming variables from v1/v2 to vPC1/vPC2					Principal components		

Fig 9.5 Deriving principal components

We can choose, in Fig 9.5, to calculate two *new* variables, $vPC1$ and $vPC2$, called *principal components*, derived from the original variables, $v1$ and $v2$, by using the equations in H2 and H3 respectively. These equations are based on the *coefficients* in the table in shaded cells E1:G3. The calculated values of $vPC1$/$vPC2$, equivalent to $v1$/$v2$ for each of the three data points, are then given in cells K2:L4.

For example, the value of $vPC1$ for the first data point is given by:

$$vPC1 = 0.6 \times 0.5 - 0.8 \times 1.0 = 0.3 - 0.8 = -0.5$$

and calculated in Excel with

$$[K2] = F\$2 * \$B2 + F\$3 * \$C2.$$

If we now use $vPC1$ and $vPC2$ to plot the three data points, we produce Fig 9.4(b). We can see that, in this very simple example, the *variability* in *both* $v1$ and $v2$ in the original data is now described just by the *variability* in a *single* new variable, $vPC1$, with the other variable, $vPC2$, remaining constant. The use of the new principal components, as combinations of the previous variables, has identified and *simplified* the description of the key *pattern* within the data.

The mathematical operation in the example was a simple rotation of coordinates in *two* dimensions. However, the concept is the same for more than two variables, in that the 'rotation' occurs in a multi-dimensional mathematical space.

> **Case study: Screening test / 2. Principal component analysis**
> —continued from 9.1.1, leading to 8.3.4
>
> Referring to the data in Fig 9.1, we now endeavour to find up to nine *principal components* that can simplify the description of the variability in the data set.

We apply principal component analysis (PCA) to the data in Fig 9.1:

Minitab > Stat > Multivariate > Principal Components...
Variables: *v1-v9*
> Graphs...: ☑ **Scree plot** and ☑ **Score plot for first 2 components**
> Storage...: Identify (perhaps four) empty columns to receive the PC coefficients as given in Fig 9.6.

Variable	PC1	PC2	PC3	PC4
v1	0.047	0.066	-0.725	0.654
v2	0.466	0.171	0.040	0.128
v3	0.241	-0.509	-0.094	0.016
v4	0.201	-0.541	-0.042	-0.045
v5	0.484	0.183	-0.029	0.004
v6	0.482	0.199	0.017	-0.059
v7	0.178	-0.561	-0.006	-0.006
v8	0.025	-0.065	0.673	0.721
v9	0.429	0.153	0.086	-0.176

Fig 9.6 Coefficients of the first four principal components

The output gives results in Fig 9.6, similar in format to cells E1:G3 in Fig 9.5, except with up to nine new principal components (although we have only stored the first four, PC1, PC2, PC3, and PC4).

The more significant coefficients in Fig 9.6 are shown in bold print for each of the four principal components, and we see that

- PC1 is mainly a combination of *v2, v5, v6,* and *v9,* and

- PC2 is mainly a combination of *v3, v4,* and *v7.*

which is consistent with the two main groups of variables that we detected by cluster analysis in Fig 9.2. PC3 and PC4 are mainly combinations of the remaining variables, *v1* and *v8*.

The **scree plot** in Fig 9.7(a) shows the contribution (recorded as 'eigenvalues') that each component makes to the *variability* in the overall data, and for this data set we can see that most of the variance can be described by the first two components, PC1 and PC2. This is an example of effective data reduction in which the scree plot drops quickly within the first two or three components before levelling off under an eigenvalue of '1.0'.

The other important plot produced by principal components analysis is the **score plot** shown in Fig 9.7(b) which gives the position of each record (subject) in the data set, plotted using its values of *vPC1* and *vPC2* as the *x–y* coordinates, centralized around zero. It is possible to edit the score plot such that different groups are identified by different markers, giving a good visual picture of differentiation within the data. Although not very clear in Fig 9.7(b), the round markers are from group A and the square markers are from group B, which allows us to see that *most* of group B are to the left of the median value and *most* of group A are to the right, with only a small number misplaced in opposite groups. The first principal component *vPC1* proves to be successful in predicting the group of each subject.

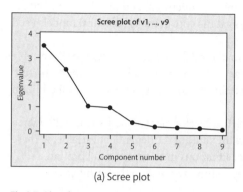

(a) Scree plot

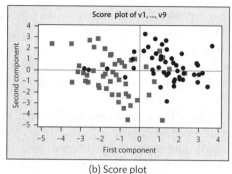

(b) Score plot

Fig 9.7 Plots for principal component analysis

The set of values in each principal component can be calculated in the same way as in Fig 9.5. For example, to calculate the values *vPC1* for the first component, we can use

Minitab > Calc > Calculator...

as shown in Fig 9.8 to store the values in a new column which we could call *vPC1*. For each row of data, the expression multiplies the PC1 coefficient in row 1 by the *v1* value, PC1(1)*v1, and adds the multiplication of the PC1 coefficient in row 2 by the *v2* value, PC1(2)*v2, and so on.

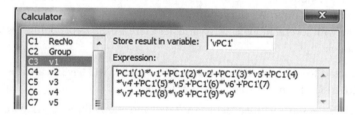

Fig 9.8 Calculation of *vPC1* values using the coefficients of PC1

9.1.5 Factor analysis

We introduced the concept of principal component analysis by giving the simple example of rotating axes in Fig 9.4. The technique of factor analysis performs a more complex mathematical 'rotation' which ensures that the variability in the original data is shared out *equally* between the new factors.

Factor analysis in data analysis software usually offers different types of 'rotation', but the 'varimax' rotation is often a suitable choice as a default starting point. However, in the particular example above, we would find that the effect of the data 'rotation' in factor analysis makes relatively little difference in differentiating between groups.

Multiple regression (Minitab): Analysis for Fig 9.10(a). Scan here to watch the video or find it via www.oxfordtext books.co.uk/ orc/currell/

9.1.6 Multiple regression

The general concept of regression in *statistics* mirrors the desire in many *scientific* investigations to describe how one (or more) variables are dependent on, and may be predicted by, other variables. For this reason, we see regression at the core of many statistical models used to describe our scientific world.

Multiple regression develops an equation that models the output (response) variable based on *several* input (predictor) variables. The software analyses have the ability to 'test' the significance of a given input variable and either include or reject it as a useful addition to the overall model.

For **basic linear regression** see Chapter 2, with the development of the statistics with which an independent variable, x, is assumed to *predict* the value of the dependent variable, y. These statistics find application in a very wide range of situations in science, e.g. calibration.

We see (2.1.1) that the behaviour of y as a simple function of x can be written as

$$y = mx + c \text{ or } y = b_0 + bx$$

and that the process of linear regression calculates the best-fit values for the coefficients of slope, m or b, and intercept, c or b_0. If a system responds to more than one factor variable x, we can express y using a linear combination of x_A, x_B etc.:

$$y = b_0 + b_A x_A + b_B x_B + b_{AB} x_A x_B + b_C x_C \ldots \tag{9.1}$$

Multiple regression (SPSS): Analysis for Fig 9.10(b). Scan here to watch the video or find it via www.oxfordtext books.co.uk/ orc/currell/

where the $x_A x_B$ term represents an *interaction* (3.3.2) between the variables x_A and x_B. Multiple regression is then the process of calculating the best-fit values for b_0, b_A, b_{AB}, etc.

However, it is not always necessary or beneficial to include all possible factor terms in the model. If it is known that a specific factor is not a significant term, then it is appropriate to exclude it from the model. The process of multiple regression tests whether, or not, the inclusion of specific factors improves the goodness of the fit.

For a system with several input factors/variables there are different *approaches* to identifying a final model:

- *Force all factors* to be included, in which case each factor will have an associated p-value relating to its significance in that particular model.

- *Forward selection* starts with *no* factors included and allows the analysis to *include* factors, one at a time, for which their p-values are less than a defined significance level, α, (typically 0.1). Several steps may be required, with the procedure stopping when *no new* factor would have $p < \alpha$.

- *Backward elimination* starts with *all* the factors and allows the analysis to *reject* any for which their p-values are greater than a defined significance level, α, (typically 0.1). Note that as one factor is rejected, the remaining p-values may change and further steps may be required. The procedure stops when *all* factors have $p < \alpha$.

- *Stepwise* regression allows the analysis to include *or* reject factors, forwards and backwards, as required to move towards the best-fit.

The operation of multiple regression is illustrated here for Minitab and SPSS, using the 'Boxing' case study.

Case study: Boxing performance / 2. Multiple regression

—continued from 6.4.1

Fig 9.9 reproduces the data from Fig 6.33, showing the numbers of punches recorded over six rounds (two bouts) by each of six amateur boxers for two levels of hydration, E and D, with the objective of testing for the effect of hydration on performance.

	A	B	C	D	E	F	G	H	I
1	RecNo	Punches	Subject	Hydrat	HydratN	Round	Bout	H*R	H*B
2	11	131	e	E	1	2	1	2	1
3	3	129	c	E	1	1	1	1	1
4	62	117	b	D	2	5	2	10	4
5	42	127	f	D	2	1	1	2	2
6	24	135	f	E	1	4	2	4	2
7	16	150	d	E	1	3	1	3	1
8	55	147	a	D	2	4	2	8	4
9	51	132	c	D	2	3	1	6	2
10	40	132	d	D	2	1	1	2	2
11	59	122	e	D	2	4	2	8	4
12	31	159	a	E	1	6	2	6	2

Fig 9.9 'Boxing performance' case study data

In Fig 9.9, the data set *HydratN* gives the levels of hydration, E and D, coded with the scale values 1 and 2. We wish to include possible *interactions* (3.3.2) within the model, and, for the purposes of this regression model, these have been coded as *additional factors*:

HxR— hydration and *round*, coded by multiplying the values of *HydratN* and *Round*.

HxB— hydration and *bout*, coded by multiplying the values of *HydratN* and *Bout*.

Software analyses can be performed using:

Minitab

Mintab > Stat > Regression > Regression > Fit Regression Model...
Response: *Punches*   **Continuous predictors:** *HydratN, Round, Bout, HxR, HxB*
Method: ▼ **Choose method from:** *Stepwise, Forward selection, Backward elimination*
    Potential terms: *HydratN, Round, Bout, HxR, HxB*
    ☑ **Display the table of model selection details**
    ▼ *Include details of each step*

SPSS

SPSS > Analyze > Regression > Linear...
Dependent: *Punches*   **Independent(s):** *HydratN, Round, Bout, HxR, HxB*
Method: Select method from the drop down menu

Step	1	2	3
Constant	89.42	82.33	82.33
HydratN	41	45	47
T-Value	2.33	2.75	2.88
P-Value	0.023	0.008	0.005
Round	7		
T-Value	0.71		
P-Value	0.482		
Bout	38	59	59
T-Value	1.11	3.63	3.64
P-Value	0.272	0.001	0.001
HxR	-6.0	-1.7	
T-Value	-0.94	-0.85	
P-Value	0.351	0.396	
HxB	-26	-38	-43
T-Value	-1.18	-3.22	-4.22
P-Value	0.241	0.002	0.000
S	22.0	21.9	21.8
R-Sq	33.19	32.68	31.95
R-Sq(adj)	28.13	28.66	28.95
Mallows Cp	6.0	4.5	3.2

Coefficients[a]

Model		Unstandardized Coefficients B	Unstandardized Coefficients Std. Error	Standardized Coefficients Beta	t	Sig.
1	(Constant)	89.417	27.769		3.220	.002
	HydratN	40.931	17.563	.795	2.331	.023
	Round	7.083	10.029	.470	.706	.482
	Bout	37.972	34.256	.738	1.108	.272
	HxR	-5.958	6.343	-.745	-.939	.351
	HxB	-25.625	21.665	-1.085	-1.183	.241
2	(Constant)	82.333	25.798		3.191	.002
	HydratN	45.181	16.438	.878	2.749	.008
	Bout	59.222	16.316	1.151	3.630	.001
	HxR	-1.708	1.998	-.214	-.855	.396
	HxB	-38.375	11.934	-1.625	-3.216	.002
3	(Constant)	82.333	25.747		3.198	.002
	HydratN	46.889	16.284	.911	2.880	.005
	Bout	59.222	16.284	1.151	3.637	.001
	HxB	-43.500	10.299	-1.842	-4.224	.000

a. Dependent Variable: Punches

(a) Minitab 16   (b) SPSS

Fig 9.10 Extracts from multiple regression output using backward elimination (The output for Minitab 17 provides the same data in a different layout)

Using the *backward elimination* method, multiple regression gives the results in Fig 9.10, with (a) Minitab proceeding in three *steps* from left to right, and (b) SPSS proceeding from top to bottom through three *models*.

Starting with all possible factors, *HydratN*, *Round*, *Bout*, *HxR*, and *HxB*, the *Round* factor is the first to be removed, followed by *HxR* in the next step. The model concludes that hydration, bout, and the interaction between them are the appropriate predictors of boxing performance. A knowledge of the round does not add significant information to the model. If we review the quality of fit, we find that with all factors, the R^2 values (4.4.1) are $R^2 = 33.19$ and $R^2(\text{adj}) = 28.13$, and the removal of two 'factors' actually reduces the *total* fit of the model and R^2 drops to 31.95. However, with *fewer* factors, the $R^2(\text{adj})$ *increases* to 28.95, giving a more *efficient* model for describing the data. We can also see in Fig 9.11 that a general regression calculation follows the same analytical logic.

The results in Fig 9.10 include the coefficients, b, (B in SPSS) related to each factor plus the constant, b_0, and, from the above results, the equation to describe the number of punches becomes

$$n = 82.33 + 46.9 \times HydratN + 59.2 \times Bout - 43.5 \times (H \times B)$$

where $H \times B$ is the interaction term of *HydratN* and *Bout*.

The *forward selection* method arrives at the same model for the data. However, it is possible using *stepwise selection* with different *starting* factors in Minitab for the regression process to stop with a different, and less optimal, model. For example, starting with *Round*, *Bout* and *HxR*, the *Bout* factor provides little additional input and is rejected and then *HydratN* is included. However, the final model in this sequence is a slightly less efficient fit to the data as $R^2(\text{adj}) = 28.76$ as compared to 28.95 for the backward elimination model.

We could also use the general regression option in Minitab, entering the factors and inter-action as in the model below:

Minitab > Stat > Regression > Regression > Fit Regression Model...
Response: *Punches*   **Continuous predictors:** *HydratN, Round, Bout*
> Model...
        Highlight *HydratN* and *Bout* in **Terms in the model**
        Cross predictors and terms in the model: Add, giving
        Terms in the model: *HydratN, Round, Bout, HydratN*Bout*
        (Delete any other cross-terms that may appear)
$\rightarrow$ Output: Fig 9.11

Source	DF	Seq SS	Adj SS	Adj MS	F	P
Regression	4	15402.5	15402.5	3850.63	7.9898	0.000025
HydratN	1	6068.3	3957.4	3957.42	8.2113	0.005556
Round	1	818.1	165.0	165.02	0.3424	0.560410
Bout	1	0.9	5648.4	5648.38	11.7199	0.001059
HydratN*Bout	1	8515.1	8515.1	8515.13	17.6682	0.000080
Error	67	32290.4	32290.4	481.95		
Lack-of-Fit	7	1060.5	1060.5	151.50	0.2911	0.954930
Pure Error	60	31229.8	31229.8	520.50		
Total	71	47692.9				

Fig 9.11 Output from general regression using Minitab

The output in Fig 9.11 agrees with the results in Fig 9.10 by showing that *HydratN, Bout,* and the interaction *HydratN*Bout* are all significant factors ($p < 0.05$).

These results also show the effect of *sequencing* the analysis as discussed in 3.4.5. When *first introduced* into the model, *Round* was identified as contributing $SS_{SEQ} = 818.1$ and *Bout* only $SS_{SEQ} = 0.9$ to the variability in the data, which suggests a greater significance for *Round*. However, when these values were adjusted, taking into account all other factors, the analysis reduces the contribution from *Round* to $SS_{ADJ} = 165.0$ and identifies the major contribution of *Bout* with $SS_{ADJ} = 5648.4$.

9.2 Multiple questions

Section 9.1 considered how multiple predictor variables could be used to model the state of a system. In this section, we consider a variety of different variables recorded in relation to a 'system', both before and after a possible intervention, and the different types of 'questions' that can be asked in the analysis. This naturally applies to the analysis of questionnaires, but similar data sets could be produced by measuring multiple laboratory and field variables and factors. We have already met elsewhere most of the analyses that can be applied to this data set, but we will develop here a general overview with links to the relevant techniques.

9.2.1 Example data

Fig 9.12 gives the questionnaire responses for the Forensic Questionnaire case study, but it can also be viewed as an example of a data set that provides information about any scientific

	A	B	C	D	E	F	G	H	I	J
1	Grouping		Background		Responses					
2	Grp	Year	T1	Q1	T2	Q2	Q3	Q4	Q5	Q6
3	G2	3	88	N	87	N	Y	Y	1	0
4	G1	3	57	Y	56	Y	Y	Y	0	-2
5	G2	3	68	N	67	N	N	N	-2	-1
6	G2	2	36	Y	33	N	Y	Y	0	1
7	G2	3	81	Y	78	N	N	N	0	-2
8	G1	2	42	N	31	N	N	N	0	2
9	G1	3	60	N	63	Y	Y	N	2	2
10	G1	3	71	N	77	Y	N	N	0	2

Fig 9.12 Multiple variables data set

Case study: **Forensic questionnaire / 1. Multiple variables (overview)**

Fig 9.12 shows the questionnaire responses for eight out of 66 randomly selected students who took part in a forensic science investigation. The students were either in year 2 or 3 of their course and were further identified as belonging to one of two cohort groups G1 or G2 relating to different programmes being followed.

Initially they were given a short test of multiple choice questions to define their background understanding of a specific topic, with the percentage result, T1. They were also given the evidence presented in an imaginary trial and asked the binary question, Q1, whether they considered that the defendant was guilty. There was then an 'intervention' in the form of specially prepared video tutorials on the chosen topic, and they were then asked to repeat the test, giving result, T2, and asked whether they now believed, Q2, the defendant to be guilty. They were also asked two other binary questions, Q3 and Q4, and two questions with Likert scale responses, Q5 on a scale of -2 to $+2$ and Q6 on a scale of -3 to $+3$.

Section 9.2 Gives an overview of the analysis of the data in Fig 9.12.

8.2.1 / 2. Crosstabs and contingency table. Develops the statistics and technique for analysing related categorical data.

4.4.5 / 3. McNemar's test and Cochran's Q. Develops the statistics for measuring the agreement between related binary variables.

6.2.1 / 4. Ordinal and binary responses. Uses nonparametric analyses for testing for differences between ordinal and binary variables.

system for which each row represents sets of measurements on different 'subjects'. In a very broad classification, it can be useful to group the variables into three main groups:

- Categorical variables, e.g. *Grp* and *Year* that identify **specific groups** within the rows of subjects.

- **Background** variables, e.g. *T1* and *Q1*, that provide further information about each measurement or subject.

- **Response questions** following the **effect of an intervention**.

The variables in Fig 9.12 are summarized in Table 9.1.

The variability for the *T1* and *T2* test data is not initially known. Although the scores of a moderately homogenous group of students sitting a well-structured examination are often normally distributed, this is not the situation here and we can have no expectation that these variables will show a normal distribution.

Table 9.1 Variable characteristics (5.1.2)

Variable	Action	Type	Levels	Variability
Grp	Input	Nominal	*G1 / G2*	Fixed
Year	Input	Ordinal	*2 / 3*	Fixed
T1	Input	Interval	*Continuous* 0–100	Not normal ?
Q1	Input	Binary	*Y / N*	–
T2	Output	Interval	*Continuous* 0–100	Not normal ?
Q2 to Q4	Output	Binary	*Y / N*	–
Q5	Output	Ordinal	Likert scale: -2 to +2	–
Q6	Output	Ordinal	Likert scale: -3 to +3	–

The label of a variable as an 'input' or 'output' can depend on the analysis being performed (1.2.1). If we measure the correlation between *T1* and *T2*, then neither are inputs, but a best-fit straight line could be drawn to predict the value of *T2* (output) based on the value of *T1* (input). In this example, the variables *T1* and *Q1* are recorded as evidence of the respondents initial state and, in general, we can see them as *inputs* into an analysis of the effect of the interaction from which the other variables, *T2, Q2–Q6*, are *outputs*.

9.2.2 Describing the data

The first step is to get a good overall understanding of the data, both *numerically* and *graphically*. It is useful to summarize numerical values to check for any obvious differences, e.g. mean, median, standard deviation for *numeric* data, and frequencies and proportions for *categorical* and *binary* data. Graphs and data plots also provide a valuable visualization within which it is often possible to see patterns that are hidden in lists of numbers.

Various methods are given for *individual* variables in 6.1.3, with, for example, the bimodal distribution of the histogram in Fig 9.13(a) suggesting two distinct groups in the *T1* data. Often of greater interest however is the *simultaneous* display of data from *two or more* variables or groups, developed in 6.2.3, 6.3.3, and 6.4.3. For example, Fig 9.13(b) gives separate boxplots with quite different median values for the two *Year* groups within the *T1* data, which suggests that these two cohorts are responsible for the overall *bimodal* distribution in *T1*. We can also see a datum value of 20 visible as an outlier in both Fig 9.13 (a) and (b).

For categorical data we can plot the frequencies of occurrences within different categories and Fig 9.13(c) gives the numbers of respondents from Fig 8.17 recording values for *Q6*, separated into the two groups, *G1* and *G2*. This provides an immediate suggestion of a possible difference in distribution between the groups which is tested in 8.2.7.

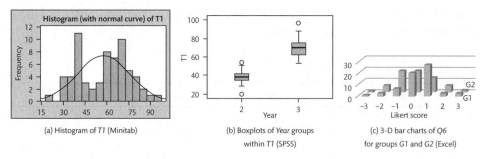

(a) Histogram of *T1* (Minitab)

(b) Boxplots of *Year* groups
within *T1* (SPSS)

(c) 3-D bar charts of *Q6*
for groups *G1* and *G2* (Excel)

Fig 9.13 Describing *T1* and *Q6* values

9.2.3 Testing for normality and homoscedasticity

One of the earliest questions that students learn to ask is 'Is my data normally distributed?' The best approach to this is to consider the *scientific* factors (5.4.3) that may, or may not, cause the random variations to follow a normal distribution. It is then possible, provided that the samples are not too small, to carry out a normality test for an individual variable (8.1.6) and also test for the equality in variance (homoscedasticity) between two sets (6.2.4). However, if we consider *T1*, it is quite clear from Fig 9.13(a) that the overall distribution of values is *not* normal due to the two separate year groups.

The next step is to consider the distribution of residuals after *fitting* an analytical model, e.g. after deriving the best-fit straight line or using an ANOVA calculation (Sections 5.4, 6.3.4, 6.4.5). If we perform a one-way ANOVA analysis on *T1* and produce a model which includes the effect of the two year groups, we can save the residuals as a new data set. The distribution of these residuals is given in Fig 9.14(a), showing that they can now be treated as a normal distribution, with $p = 0.444$ from the Anderson–Darling normality test (Minitab, 5.4.5).

Similarly we can test for a difference in variance between data sets *within* the main statistical analysis. For example, SPSS gives $p = 0.081$ for Levene's test, Fig 9.14(b), when performing either an independent sample *t*-test (6.2.5) or a one-way ANOVA (6.3.5) for a difference between year groups in *T1*, confirming that there is no significant difference in variance and that the choice of parametric test was valid.

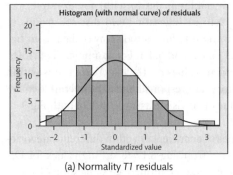

(a) Normality *T1* residuals

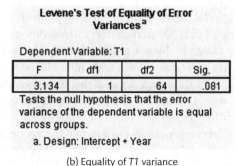

Levene's Test of Equality of Error Variances[a]

Dependent Variable: T1

F	df1	df2	Sig.
3.134	1	64	.081

Tests the null hypothesis that the error variance of the dependent variable is equal across groups.

a. Design: Intercept + Year

(b) Equality of *T1* variance

Fig 9.14 Testing for normality and homoscedasticity

9.2.4 **Analysing an individual variable**

We may wish to compare a measured variable with specific expectations:

- Is the mean of *T1*, or the median of *Q6*, different from an expected value? One sample *t*-test, 6.1.4 or Wilcoxon test, 6.1.5.

- Is the proportion of *Y* answers in the binary questions different from expected proportions? One proportion test, 6.1.7.

- Is the distribution of values in *interval* data, e.g. *T1*, different from an expected distribution? Kolmogorov–Smirnov test, 6.1.6.

- Is the distribution of *categorical* data, e.g. Q5 or Q6, different from an expected distribution? Chi-squared 'goodness of fit test', 8.1.5.

- Is the order of values random? Runs test, 6.1.6.

9.2.5 **Dependence of specific factors**

The next level of analysis for interval variables is often to ask:

- Are the values affected by a grouping defined by a separate categorical variable, e.g. do the groups *G1*, *G2* in *Grp* affect the value of *T1*?

For the effect of just *two levels* (e.g. *G1*, *G2*) we may be testing for a difference in means (*t*-test, 6.2.5), medians (Mann–Whitney test, 6.2.6) or variance (*F*-test, Levene's test, 6.2.4). For example, a *t*-test for the different *Grp* mean values of *T1* gives the non-significant result, $p = 0.790$. More than two levels requires an ANOVA, normally conducted through the GLM (6.3.5), or the nonparametric Kruskal–Wallis test (6.3.7).

A more complicated question involves the effects of more than two categorical variables:

- Are the values affected by groupings and interactions defined by two or more categorical variables, e.g. How do *Grp* and *Year* affect the value of *T1*?

In this case the analysis becomes a multifactorial ANOVA (6.4.4), with which it can be possible to test *separately* for the effects of each factor and a possible *interaction* (3.3.2) between them. For example, the GLM/ANOVA results for *T1* give:

$p = 0.000$ (i.e. < 0.0005) for *Year*, showing the very significant difference between the years that we have already seen in Fig 9.13(b).

$p = 0.790$ for *Grp*, showing that there is no significant difference between groups, *G1* and *G2*, agreeing with the results of the *t*-test above.

$p = 0.263$ for *Year*Grp*, showing that there is no significant interaction (3.3.2) between the groups *Year* and *Grp*.

These results can be viewed graphically in the interaction plot (6.4.3) for *T1* in Fig 9.15, which plots the mean values of the four possible combinations of *Year* and *Grp*. There is a clear difference in *T1* values between years 2 and 3, but there is no vertical difference between the lines for *G1* and *G2*, and the fact that the two lines are nearly parallel suggests that there is no significant interaction between the factors.

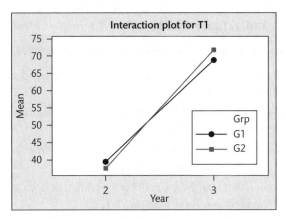

Fig 9.15 Interaction plot for *T1* against *Year* and *Grp*

For ordinal/binary variables, we may wish to ask:

- Are the values affected by a grouping defined by a separate categorical variable, e.g. do the groups *G1* and *G2* in *Grp* affect the value of *Q5*?

In 6.2.6 we use the nonparametric Mann–Whitney test, which records $p = 0.031$, from which we accept that there is a difference between the *median* values of *Q5* for the two groups *G1* and *G2*. This is consistent with Fig 9.13(c) with the *G1* group recording the higher median value.

9.2.6 Comparing variables as unrelated data

The distinction between *related* and *unrelated* variables is developed in Section 6.2. As presented in Fig 9.12, all the variables are related because of the horizontal links through the subject/record rows, and, if we were to treat them as unrelated, we would be ignoring important analytical information.

If the *T2/Q2* results were obtained from a *different sample* of respondents than the *T1/Q1* results, then we would have to ask the *unrelated* questions:

- Is there is an overall difference between two variables, e.g. *T1* and *T2* or *Q1* and *Q2*?

We have to perform the nonparametric Mann–Whitney test between *T1* and *T2* because we know that *T1* is not normally distributed, and this gives $p = 0.90$ which shows no significant difference between the *median* values. To compare *Q1* and *Q2* we test for a difference in the *proportion* of *Y* values between *Q1* and *Q2*, and, in 6.2.9, we obtain the value $p = 0.053$. Neither of these differences are significant when treated as *unrelated* variables, but we see below that there are subtle differences when analysed as *related* variables.

9.2.7 Modelling interrelated variables

We may wish to model a mathematical relationship between two *interrelated* interval variables. This is very common in experimental science, usually described by the familiar *x–y* graph and often leading to the slope and intercept of a best-fit straight line

(Section 2.1), a defined nonlinear relationship (Sections 2.3 and 7.2) or more general patterns of behaviour (Section 7.3). Typically these examples are relating the variation of two *different* quantities, e.g. absorbance and concentration in spectrophotometry, radio-activity decay and time, etc.

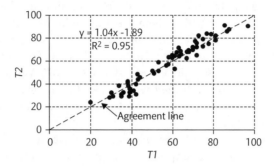

Fig 9.16 Linear regression of *T2* against *T1*

If we plot the interrelationship of two variables measuring the *same* quantity we can display the amount of *agreement* between them. Fig 9.16 plots the values of *T2* as a function of *T1*, with each point showing the responses of one individual. For *perfect* agreement between *T2* and *T1* every point would lie on the agreement line drawn with a slope of 1.0, an intercept of 0.0, and with an R^2 measure of agreement (4.4.1) equal to 1.00.

The actual relationship has the equation:

$$T2 = 1.04 \times T1 - 1.89$$

with $R^2 = 0.95$, which shows fairly good overall agreement. However, we can see that the year 2 students (with *T1* about 30 to 50) tend to give lower values for *T2*, appearing to the lower right of the agreement line, and most year 3 students (with *T1* about 60 to 80) recording an increase to the upper left of the line. This confirms the result of the repeated measures analysis below.

The data in Fig 9.12 does not include any extensive *x-y* relationships, but these can often appear in experimental data results, and we consider a range of possibilities in Section 7.3, including:

- *convolution* (7.3.5) to create a 'best-fit' smoothed line through a whole data set, by using local 'averaging' of data at each point, and convolutes to produce a smoothed 'differential' (7.3.6) of the curve, giving the slope of the curve at each point, and

- *autocorrelation* (7.3.7) and *spectral analysis* (7.3.8) techniques to analyse periodic frequency components hidden within the data.

9.2.8 Comparing related variables

A common example of *repeated* measurements is a 'before' and 'after' situation:

- Does the tutorial intervention cause a difference in the test results *T1* and *T2* of subjects, and is any difference dependent on a categorical grouping of the subjects, e.g. years 2 and 3?

For just two variables the question becomes a *paired* test for means (paired *t*-test 6.2.7) or medians (paired Wilcoxon test, 6.2.8), but for three or more variables we use *repeated measures* (6.3.8, 3.6.2).

In 6.3.8 a repeated measures analysis of *T1* and *T2* with the *Year* as a factor confirms the different *Years* as being highly significant ($p = 0.000$), and, although there is no overall difference between *T1* and *T2* ($p = 0.927$), there is an interaction between the differences, *T2-T1*, and the year group ($p = 0.016$) which shows that the two groups react differently to the intervention of additional videos (see Fig 9.16 above). The lack of *overall* difference between *T1* and *T2* is confirmed using a simple paired *t*-test with $p = 0.670$. We can use the *parametric* analysis for the *paired* test, because the *difference values, T2-T1*, can be considered to be sufficiently close to a normal distribution with $p = 0.076$ for the Anderson–Darling test, even though *T1* itself is not normal.

Alternatively we may have a number of different assessments which might be expected to give the same result, e.g. different assay methods might be used to estimate the same bacterial mortality, or several 'experts' give their opinions on the quality of different wines. We have tested above for agreement between the interval values of *T1* and *T2* by using repeated measures analysis and linear regression, and we now develop the general analysis of agreement between ordinal variables in Section 4.4, involving correlation, Kendall's coefficient of concordance, Kappa, McNemar's test, and Cohen's Q.

In relation to the data in Fig 9.12, we can test for agreement between the *binary* answers.

- Do the binary answers to questions *Q1* to *Q4* agree?

In 4.4.5 we use McNemar's test and Cochran's Q to test for agreement between *Q1, Q2, Q3,* and *Q4*, finding a significant difference just between *Q1* and *Q2* with $p = 0.031$.

9.2.9 Ordinal responses

Choosing the number of categories to use for ordinal responses (e.g. the range Likert style for *Q6*) can be difficult. If there are *too few*, then almost all of the respondents may give the same answer, e.g. on a scale of 1 to 4, everyone might reply '3', making it impossible to do detailed analysis. However, if you have *too many* levels, then the *frequency* of responses in individual categories can become too small for useful analysis. Ideally a pilot study would reveal the ranges of answers that could be expected in a final questionnaire, allowing you to design the questions more sensitively, but this is not always possible for a final year project.

The frequency of responses to levels in *Q6* are given in Fig 9.13(c) in which it can be seen that responses, −3, −2, and +3 all give low numbers. The analytical comparison of the answers from the two groups *G1* and *G2* to *Q6* is given in 8.2.7 which introduces two methods for dealing with low expected frequencies: consolidation of categories and using a Monte Carlo analysis.

9.2.10 Multiple variables

Section 9.1 introduces the analysis of multiple variables to answer the questions:

- Are there any relationships between the variables?
- Is it possible to develop a model that predicts outcomes using multiple input variables?

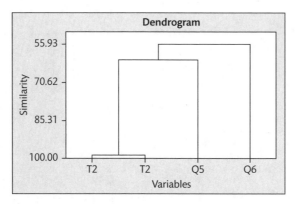

Fig 9.17 Clustering variables dendrogram (Minitab)

Fig 9.17 uses the technique of data clustering (9.1.3) to investigate relationships between variables. It shows the strong similarity in the values of *T1* and *T2*, but a lack of similarity between *Q5* and *Q6*. The binary variables are not included here because they have been entered into the data as text variables 'Y' or 'N', and would need to be coded numerically, e.g. 0 and 1, for inclusion in this analysis.

If we now code the group *Grp* numerically as *GrpN* with values '1' and '2', and the binary answers as *B1, B2, B3,* and *B4* respectively with values '0' and '1', we could develop a mathematical model using multiple regression to predict *GrpN* values. Fig 9.18(a) shows the result of a stepwise regression (9.1.6), which identifies only *Q6* and *B2* as highly significant predictors for *GrpN* ($p = 0.002$ each) with *B1* included because $p = 0.058$ is less than the default rejection criterion of 0.1. However, since *GrpN* is a binary value we can use *binary* regression (8.3.4), with, for example, *Q6, B2, B1,* and *T1* as predictor variables. The results again identify *Q6* and *B2* as highly significant ($p = 0.005$), with *B1* giving $p = 0.089$ and *T1* 'not significant' with $p = 0.734$. The ability to predict the correct group can be displayed with the deviance vs probability graph in Fig 9.18(b), which identifies the two groups as two curves (8.3.4), with correct predictions towards the lower ends of both curves and misclassifications in the upper branches.

```
Response is GrpN on 8 predictors, with N = 66

Step            1       2       3
Constant      1.516   1.677   1.616

Q6           -0.131  -0.132  -0.114
T-Value       -3.61   -3.82   -3.25
P-Value       0.001   0.000   0.002

B2                    -0.30   -0.35
T-Value               -2.80   -3.25
P-Value               0.007   0.002

B1                             0.23
T-Value                        1.93
P-Value                        0.058

S             0.461   0.438   0.429
R-Sq          16.94   26.11   30.32
R-Sq(adj)     15.65   23.77   26.94
Mallows Cp     8.0     2.2     0.7
```

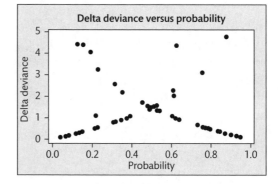

(a) Stepwise regression   (b) Deviance vs probability curve for binary regression

Fig 9.18 Regression modelling of *GrpN* (Minitab)

APPENDIX I
Videos available in the Online Resource Centre

Number	Title	Description	Section
1	**Combining uncertainties**	Excel analysis for Fig 1.13	1.4.4
2	**Propagation of errors**	Excel analysis for Fig 1.14	1.4.4
3	**DIY dice**	Excel analysis for Fig 1.15	1.4.6
4	**Sample statistics and confidence intervals**	Excel analysis for Fig 1.16	1.5.1
5	**Samples and population**	Excel analysis for Fig 1.17	1.5.3
6	**Statistics of linear regression**	Excel analysis for Fig 2.2	2.1.1
7	**Calibration uncertainty 1**	Excel analysis for Fig 2.7. See also 7.1.5	2.2.1
8	**Exact x/y intercepts**	Excel analysis for Figs 2.9 (b) and 2.10 (b)	2.2.2
9	**Weighting data**	Excel analysis for Fig 2.12	2.2.4
10	**Using Solver**	Excel analysis for Fig 2.18	2.4.1
11	**Nonlinear regression**	Excel analysis for Fig 2.20. See also 7.2.3	2.4.3
12	**Difference in slopes**	Excel analysis for Fig 3.1	3.1.1
13	**Two sample *t*-test and *F*-test**	Excel analysis for Fig 3.3	3.1.3
14	**Analysis of variance**	Excel analysis for Fig 3.4	3.2.2
15	**One Way ANOVA**	Minitab and SPSS analyses leading to Fig 3.8	3.2.3
16	**Multi-factorial ANOVA (Minitab)**	Minitab analyses for the data in Figs 3.11 and 3.15	3.3.1
17	**Multi-factorial ANOVA (SPSS)**	SPSS analyses for the data in Figs 3.11 and 3.15	3.3.1
18	**ANCOVA 1**	Excel and Minitab analysis for Fig 3.20. See also 6.4.8	3.3.3
19	**General linear model**	Excel analysis for Fig 3.24	3.4.2
20	**General regression (Minitab)**	Minitab analysis for Fig 3.26 data. See also 7.2.5	3.4.3
21	**Generalized linear model (Poisson loglinear)**	SPSS analysis leading to Fig 3.35. See also 6.4.7	3.4.7
22	**Repeated measures 1**	SPSS analysis leading to Figs 3.41, 3.42 and 3.43. See also 6.3.8	3.6.2
23	**Chi-squared 'goodness of fit'**	Excel analysis for Fig 3.46. See also 8.1.5	3.7.2
24	**Contingency table test for association**	Excel analysis for Fig 3.47. See also 8.2.4	3.7.4

25	**One proportion**	Excel and Minitab analysis for Fig 3.48. See also 6.1.7	3.8.2
26	**Two proportions**	Minitab analysis leading to Fig 3.51. See also 6.2.9	3.8.3
27	**Resampling t-test and Mann-Whitney test**	Excel analysis for Fig 3.52	3.9.2
28	**Resampling chi-squared**	Excel analysis for Fig 3.53	3.9.3
29	**Parametric and nonparametric correlation**	Excel analysis for Fig 4.2	4.1.1
30	**Bivariate and partial correlation**	Analysis for Fig 4.7	4.1.4
31	**Fisher's exact test**	Excel analysis for Fig 4.12	4.2.3
32	**MKT analysis**	Excel analysis for Fig 5.9	5.2.5
33	**Transforming data**	Minitab and SPSS transformations in Fig 5.10	5.3.2
34	**Analysing residuals (Minitab)**	Analysis for Fig 5.14 data. See also 6.4.5	5.4.6
35	**Analysing residuals (SPSS)**	Analysis for Fig 5.14 data. See also 6.4.5	5.4.6
36	**Transforming for normality**	SPSS analysis leading to Fig 5.18 and Table 5.3. See also 5.3.2	5.4.7
37	**Describing sample data (Excel)**	Descriptives and graphs	6.1.3
38	**Describing sample data (Minitab)**	Descriptives and graphs	6.1.3
39	**Describing sample data (SPSS)**	Descriptives and graphs	6.1.3
40	**One sample tests (Minitab)**	Analysis for t-test and Wilcoxon test	6.1.4
41	**One sample tests (SPSS)**	Analysis for t-test and Wilcoxon test	6.1.4
42	**Nonparametric tests (SPSS)**	Nonparametric tests. See also 6.4.6	6.1.6
43	**One proportion (SPSS)**	Analysis for Fig 6.1 data. See also 3.8.2	6.1.7
44	**Two sample tests (Minitab)**	Analysis of variance, means and medians	6.2.5
45	**Two sample tests (SPSS)**	Analysis of variance, means and medians	6.2.5
46	**Paired tests (Minitab)**	Analysis for paired t-test and Wilcoxon test	6.2.7
47	**Paired tests (SPSS)**	Analysis for paired t-test and Wilcoxon test	6.2.7
48	**Two proportions (SPSS)**	Analysis for Fig 6.22. See also 3.8.3	6.2.9
49	**GLM/ANOVA (Minitab)**	Analysis leading to Figs 6.27(a), 6.28 and 6.30	6.3.5
50	**GLM/ANOVA (SPSS)**	Analysis leading to Figs 6.27(b), 6.29 and 6.31	6.3.5
51	**Repeated measures 2**	SPSS analysis leading to 6.32. See also 3.6.2	6.3.8
52	**Factor plots (Minitab)**	Analysis leading to Figs 6.35(a), 6.36(b)	6.4.3
53	**Factor plots (SPSS)**	Analysis leading to Figs 6.35(b), 6.36(a)	6.4.3
54	**Multifactorial GLM/ANOVA (Minitab)**	Analysis leading to Fig 6.37	6.4.4

55	**Multifactorial GLM/ANOVA (SPSS)**	Analysis leading to Fig 6.37	6.4.4
56	**Normality and homoscedasticity (Minitab)**	Analysis of Boxing, case study. See also 5.4.6	6.4.5
57	**Normality and homoscedasticity (SPSS)**	Analysis of Boxing, case study. See also 5.4.6 and 6.3.4	6.4.5
58	**Nonparametric ANOVA (Minitab)**	Analysis leading to Fig 6.39(b)	6.4.6
59	**Nonparametric ANOVA (SPSS)**	Analysis leading to Fig 6.39(a). See also 6.1.6	6.4.6
60	**Generalized linear model (ordinal logistic)**	SPSS analysis leading to Fig 6.40. See also 3.4.7	6.4.7
61	**ANCOVA 2**	SPSS analysis leading to Figs 6.41 and 6.42. See also 3.3.3	6.4.8
62	**Calibration uncertainty 2**	Minitab and SPSS analysis. See also 2.2.1	7.1.5
63	**Nonlinear regression (Minitab)**	Analysis for Fig 7.6(a). See also 2.4.3	7.2.3
64	**Nonlinear regression (SPSS)**	Analysis for Fig 7.6(b). See also 2.4.3	7.2.3
65	**Deriving the model**	Excel analysis for Fig 7.7	7.2.4
66	**General regression (Minitab)**	Analysis for Fig 7.8. See also 3.4.3	7.2.5
67	**Using convolutes**	Excel analysis for Figs 7.13 and 7.14	7.3.5
68	**Spectral analysis**	SPSS and Minitab analyses for Figs 7.16 and 7.18	7.3.7
69	**Frequency data (Minitab)**	Descriptives and graphs	8.1.3
70	**Frequency data (SPSS)**	Descriptives and graphs	8.1.3
71	**Chi-squared goodness of fit (Minitab)**	Analysis for Fig 8.6. See also 3.7.2	8.1.5
72	**Chi-squared goodness of fit (SPSS)**	Analysis for Fig 8.7. See also 3.7.2	8.1.5
73	**Testing distributions (Minitab)**	Analysis for distributions and Fig 8.8(b)	8.1.6
74	**Testing distributions (SPSS)**	Analysis for distributions and Fig 8.8(a)	8.1.6
75	**Crosstabs and contingency tables (Minitab)**	Analyses for Figs 8.12/13/14/15. See also 3.7.4	8.2.4
76	**Crosstabs and contingency tables (SPSS)**	Analyses for Figs 8.12/13/14/15/16. See also 3.7.4	8.2.4
77	**Logit and probit**	Excel analysis for Fig 8.22	8.3.3
78	**Binary regression**	Minitab analysis for Fig 8.24	8.3.4
79	**ROC curves**	SPSS analysis for Fig 8.27	8.3.5
80	**Cluster analysis**	Minitab and SPSS analyses for Fig 9.2	9.1.3
81	**Principal component analysis**	Minitab analysis for Figs 9.6 and 9.7	9.1.4
82	**Multiple regression (Minitab)**	Analysis for Fig 9.10(a)	9.1.6
83	**Multiple regression (SPSS)**	Analysis for Fig 9.10(b)	9.1.6

Case studies used throughout this book

The case studies are listed in alphabetical order and then stage order within each case study. Note that, in some cases, the stage order does not follow linearly within the book.

Case study	Stage	Section
Association	1. Bacteriophage treatment (overview)	8.2.1
Association	2. Contingency table	3.7.4
Bacterial growth	1. Exploratory phase (overview)	5.2.2
Bacterial growth	2. Difference in slopes using t-test	3.1.1
Bacterial growth	3. Difference in slopes as an interaction	3.4.3
Bacterial growth	4. Using smoothing convolutes	7.3.5
Best-fit straight line	1. Overview	2.0.0
Best-fit straight line	2. Slope and intercept	2.1.1
Best-fit straight line	3. ANOVA table	2.1.2
Best-fit straight line	4. Correlation	2.1.3
Best-fit straight line	5. Uncertainty in regression	2.1.4
Best-fit straight line	6. Confidence interval	2.2.1
Best-fit straight line	7. Standard additions	2.2.2
Best-fit straight line	8. Least squares fit using Solver	2.4.1
Best-fit straight line	9. Maximum likelihood using Solver	2.4.2
Blood alcohol	1. Overview	1.0.0
Blood alcohol	2. Simple boxplot	1.1.2
Blood alcohol	3. Boxplots and interval plots	1.1.3
Blood alcohol	4. Data distribution	1.3.1
Blood alcohol	5. Sample statistics	1.5.1
Blood alcohol	6. Samples and populations	1.5.3
Blood alcohol	7. Hypothesis test	1.6.2
Blood alcohol	8. One sample t-test	3.1.2
Blood alcohol	9. One sample analysis	8.1.1
Boxing performance	1. Multifactorial analysis (overview)	6.4.1
Boxing performance	2. Multiple regression	9.1.6
Catalyst	1. One way ANOVA (overview)	3.2.3
Catalyst	2. Two way ANOVA	3.3.1
Catalyst	3. Interactions	3.3.2

Index

See Appendix I for Video index
See Appendix II for Case Studies index
Excel functions are given here in upper case